# Extreme Weather and Society

**Series Editors**

*Extreme Weather and Society* examines people, place and extreme weather, from an emerging trans-disciplinary field of study. The series explores how abrupt and trending changes in weather alter physical environments and force community responses challenged by cultural practices/interpretations, politics, policy, responsibilities of education and communication, community health and safety, and environmental sustainability for future generations. The series highlights extreme weather alterations to different physical and social environments to better explore how people react and respond to extreme weather. The hallmark of this series is the innovative combined inclusion of social and physical science expertise.
*Extreme Weather and Society* contains single and multi-authored books as well as edited volumes. Series Editors are currently accepting proposals, forms for which can be obtained from the publisher, Ron Doering (ron.doering@springer.com).

More information about this series at http://www.springer.com/series/15334

Shirley Laska
Editor

# Louisiana's Response to Extreme Weather

## A Coastal State's Adaptation Challenges and Successes

Springer Open

*Editor*
Shirley Laska
University of New Orleans Emerita
and Lowlander Center
New Orleans, LA, USA

ISSN 2367-3397 ISSN 2367-3400 (electronic)
Extreme Weather and Society
ISBN 978-3-030-27204-3 ISBN 978-3-030-27205-0 (eBook)
https://doi.org/10.1007/978-3-030-27205-0



This Springer imprint is published by the registered company Springer Nature Switzerland AG
The registered company address is: Gewerbestrasse 11, 6330 Cham, Switzerland

# Foreword

Projections of future extreme weather; environmental history; social diversity, inequality, and vulnerability; environmental justice; impacts of historic disasters; actual and potential impacts of policies designed to mitigate disaster losses and adapt to changing coastal conditions; distinctive local and regional cultural traditions; cultures and livelihoods at risk; and recommendations for future risk reduction and adaptation policies and actions—all these topics and more are addressed in this comprehensive volume, which promises to be an indispensable resource for all those seeking to respond to hazard-related environmental stressors, both in Louisiana and beyond.

These pages contain both broad and in-depth analyses of the "wicked problem" posed by hazards and climate change. As several authors note, in contrast with problems that have specific solutions, even difficult ones, wicked problems are highly complex and novel, difficult to define and frame, and subject to multiple and often divergent interpretations of the nature of the problem and how to address it. When policies and programs are created to deal with wicked problems, those solutions often lead to more problems. Adding to this challenge, climate change has been labeled a "super-wicked problem" (Levin et al. 2012) for four reasons. First, there is pressure to act; time is running out to address the problem. Second, those who are trying to reduce the risks posed by climate change, from individuals to corporations to nation-states, are also contributing to it. This means that adequate responses to climate change will require radical, fundamental changes in behavior and social organization. Third, climate change poses a major global collective action challenge, but there is no overarching authority that can compel or manage action. Fourth, both current and proposed solutions focus on near-term timeframes, but addressing the problem requires long-term thinking and strategies that must be sustained over time—which is almost unthinkable under contemporary governance regimes and in light of conflicts over the significance (and even existence) of the problem. The authors in this volume rightly characterize climate change adaptation as a multigenerational challenge, but policies can shift whenever political power changes hand.

The chapters in this volume explore various facets of this super-wicked problem. Among the things we learn in these pages is that the entire state of Louisiana is at risk from land loss, sea level rise, and intensified storms and flooding. Trying to address those vulnerabilities through federal post-disaster mitigation funding, one disaster at a time, or on a project-by-project basis, is not an effective approach, but current policies, such as FEMA's emphasis on post-disaster mitigation projects (as opposed to pre-disaster ones), shape state and community mitigation options. As the discussions here argue, in the current policy environment, the emphasis should be on forward-facing, system-focused post-disaster strategies that are based on the best available science, but that are also just and sensitive to the needs, values, and preferences of the members of affected communities—what the editor refers to as *exceptional recovery for essential resilience.* At the same time, steps must be taken to move toward policies that call for large-scale programs that are not tied to specific events.

Attempts to address the conundrums of climate change and associated extremes are fraught with difficulty. With respect to coastal land loss, an environmental risk that Louisiana faces more than any other state in the nation, the book documents how the concept of restoration may be embraced in the abstract, but on-the-ground projects to address land loss are often contested. Ambitious restoration projects like the state's Coastal Master Plan represent a major step forward, but they overemphasize technocratic and engineering-based approaches while downplaying the importance of local preferences, and they can be out of step with efforts aimed at strengthening community resilience. Programs that are broad in geographic scope are appropriate for the scale of the problem of land loss, but local voices, particularly those of marginalized groups, may be muted by such efforts. Similarly, managed retreat from rapidly disappearing coastal areas makes complete sense conceptually, but the chapter authors show how complex this process is in practice. For example, what constitutes a fair and just relocation process for tribal peoples with a prior history of forced relocation? How can the distinctive cultures of coastal natural resource communities be maintained when environmental refugees are required to abandon the places where those cultures flourished? What happens to place-based livelihood strategies and family traditions when communities are uprooted? Relocation *from* endangered coastal regions means relocation *to* somewhere else—typically somewhere else in Louisiana. However, what if those new locations are lacking in adequate services, personnel, and administrative capacity and are economically depressed—as many rural communities in Louisiana are—and what if the residents of receiving communities are less than welcoming? What if those receiving communities are also vulnerable to disasters like the so-called unexpected flooding of 2016?

This volume addresses in rich detail the aftermath and recovery following Hurricane Katrina. The importance of a political economy analysis of hazards and disasters is a thread that runs through these discussions. Katrina's devastation was caused in no small measure by the activities of the oil and gas industry and the interest that promoted projects such as the Mississippi River-Gulf Outlet (MR-GO). Occurring in a historical context of spatial racism, that devastation disproportionately fell upon the poor, racialized, and minoritized residents of New Orleans. We learn in these chapters how those residents were further marginalized during the recovery process. Deprived of public housing, ignored in many recovery plans, priced out of a rising rental market in part because of the scarcity of living-wage jobs, and unable to cope with the loss of kin and friendship networks and neighborhood solidary, many never returned. And we learn how, guided by neoliberal logics of recovery such as land speculation, privatization, and the reduction of public services, the "new" New Orleans is becoming whiter, less affordable, and more gentrified, but spatially remains strongly racialized. These and other research findings provide a strong critique of putatively color-blind policies that are blind to their effects on people of color.

Discussions in this volume also show how, in the context of major disasters and accelerating land loss, Louisiana has become a laboratory for new efforts to alleviate those problems, such as the LA SAFE (Louisiana's Strategic Adaptations for Future Environments) program and the Gentilly Resilience District in New Orleans. Both were funded primarily through the National Disaster Resilience Competition, a collaboration of the Department of Housing and Urban Development and the Rockefeller Foundation. In these chapters, questions arise regarding whether such projects, developed with limited funding on a one-time basis, are replicable and scalable. The broader feasibility and efficacy of these and other experiments can only be assessed over time.

As some chapter authors note, issues of federalism inevitably arise in any discussion of hazards and disasters. The contradictions of federalism are yet another example of the wickedness of the problems discussed here. Within the US federal system, decisions in areas that are important for hazard mitigation, such as land use, are largely outside the purview of federal and state jurisdictions. Federal government requirements like the Disaster Mitigation Act of 2000 cannot ensure that the mitigation plans that are required by the law are adequate—or that they are implemented. Federal post-disaster assistance programs require financial participation on the part of states and local jurisdictions, with the latter showing considerable variation in their ability to meet matching requirements. Because specialized skills are needed to even apply for some federal forms of assistance, that assistance favors well-resourced, high-capacity large cities, as opposed to smaller communities and rural areas.

At several points in the book, the authors argue for the importance of state action in disaster risk reduction and climate change adaptation. The importance of states is inarguable, particularly in the current policy environment, in which federal leadership has rolled back measures such as the flood risk management standard and federal climate change action planning and has withdrawn from the Paris Climate Agreement. States are considered closer to local communities than a distant federal bureaucracy and more sensitive to the distinctive concerns of different populations and interest groups within their jurisdiction. At the same time, in a neoliberal political economy, many states lack the capacity to ensure the provision of even basic services, much less those associated with hazard mitigation and climate change adaptation. Some states are not politically inclined to develop such capacity.

Moreover, as the Louisiana case illustrates only too well, state governments are susceptible to capture by powerful economic interests, to the detriment of many of their residents. The same is the case for local communities, whose decisions, absent significant organized opposition, tend to conform to growth machine imperatives.

Researchers have a tendency to identify problems and leave their solutions to others. This is not the approach taken by the authors in this volume. Virtually, every chapter proposes strategies for ameliorating problems like those associated with population relocation and including marginalized groups in adaptation decision-making and ends with policy recommendations. Quite frequently, the recommendations focus on the importance of framing and tackling problems holistically, blending a knowledge of the technical and scientific aspects of particular challenges with an understanding of their human dimensions. Louisiana itself is a testament to the inadequacy of technological fixes, from levees that provided a false sense of protection and then failed repeatedly to projects like MR-GO which caused erosion and contributed to Katrina's storm surge.

At many points throughout this volume, the authors refer to Louisiana as a canary in the coal mine, a harbinger of things to come, and a microcosm in which to analyze national and global disaster and climate change challenges. As these authors show, although undeniably unique in so many respects, Louisiana illustrates starkly the kinds of issues with which many other communities, states, and nations are currently grappling and will continue to be forced to contend with as climate change progresses and intense hydrometeorological events occur with greater frequency. Both the issues and dilemmas identified and the recommended actions proposed here are relevant for communities across the nation and around the world.

University of Colorado Emerita
Boulder, Colorado

Kathleen Tierney

## Reference

Levin, K., Cashore, B., Bernstein, S., & Auld, G. (2012). Overcoming the trajectory of super-wicked problems: Constraining our future selves to ameliorate global climate change. *Policy Sciences*, *45*(2), 123–152.

# Preface

When a particular location is selected to be the topic of an edited volume, the editor has to select contributors who have experience with that location and have done research there on the topic of the book within the discipline(s) sought for it. Simple? Well, no. One has to *interactively* identify appropriate subtopics—in this case for human/social response to extreme weather risk and adaptation to it—with what scientists of that discipline *have* studied within the locale, i.e., Louisiana. However, the assignment from the "Extreme Weather and Society" series editors Bill Sprigg, an atmospheric scientist, and Sheila Steinberg, an environmental/social scientist, added to the requirements: to include multiple disciplines, not just my discipline of sociology, because they believe in and are approaching the series to enhance "an emerging *trans-disciplinary* field of study of extreme weather."

They expected contributors from multiple disciplines to be included in one of their series volumes and that those selected should offer a logical weave of findings and policy/practice implications for extreme weather/climate change adaptation. Given my career of transdisciplinary research collaboration as well as numerous assignments to multidisciplinary advisory boards within Louisiana and in other states and serving on the National Academy of Sciences multidisciplinary teams for environmental/societal risk topics, I was already a "practitioner" of the approach and thus enthusiastically sprung to the challenge.

This book honors the series editors' goals with 10 disciplines represented in the 22 contributors, including 1 by a biophysical scientist and 4 chapters having multiple authors from more than one discipline.[1] I believe that the disciplines of the contributors selected are all very relevant to the serious societal need to swiftly and appropriately address the adaptation to extreme weather including climate change.

In order to ensure that such a transdisciplinary edited volume would be approachable to more than specialists of an individual chapter's author(s)' discipline, 5 of the 11 chapter manuscripts were reviewed by relevant expert reviewers from different disciplines than their authors, 4 chapters had coauthors from different disciplines,

[1] Anthropology, architecture, geography, history, marine science, planning, public administration, sociology, urban studies and GIS.

and the other 2 were reviewed by the book's editor who does not have their specialty but is a specialist of the their topics. The reviewers were selected based on the interaction between the editor and first author of each chapter with those reviewers selected having very solid credentials. All of the authors were very positive about having their manuscripts be subjected for review because they were committed to the series editors' goal of transdisciplinary access and utility. Each contributor responded fully to suggestions made by the reviewers.

## A State Case: Louisiana Social Science Resources for Climate Change Adaptation

No different from any other resource, how a state fares in providing human resources able to address extreme weather and climate change will contribute to its success doing so. To reinforce this point, I call your attention to the recently created Social Science Extreme Events Research (SSEER) *interactive map*, funded by the National Science Foundation and created by the Natural Hazards Center at the University of Colorado. Into the software is inputted the existence of social scientists engaged in extreme events research, their credentials, and *where they are located*. This GIS product recognizes the importance of the location of social scientists to achieve "communicating ongoing project or research outcomes to affected communities and decision-makers."[2] While just launched, it will be interesting to follow how many Louisiana social scientists register. A recent report on this question did not show Louisiana as having a robust number of such researchers. The comments of a presenter at a recent meeting on the Sacramento Delta who had done a literature search on social science research on deltas reported very limited findings: "Only research on New Orleans and the Mardi Gras culture."[3]

To pursue this part of the Louisiana case, social science resources, the chapter authors were examined. Approximately half of the 21 authors of the chapters have been trained at Louisiana universities and half outside of the state. Fourteen hold academic or nonprofit organization positions within the state, and seven are in academic institutions outside of the state. The latter are affiliated with universities along the Gulf Coast and in the South, Northeast and California.

Three of these scientists outside of the state have left positions within Louisiana to go to these institutions in other states. A possible explanation is that scientists experienced with climate change research are in demand by other areas struggling with similar challenges and thus have out-of-state opportunities. This appears to be the case with the small group examined. Future science human resource research will be important to reveal how scientists respond in their personal careers to

[2] EAGER: Interdisciplinary and Social Science Extreme Events Research, Natural Hazards Center, University of Colorado. https://hazards.colorado.edu/news/research-projects/eager-interdisciplinary-and-social-science-extreme-events-research. Accessed January 27, 2019.

[3] Personal communication, January 30, 2019.

extreme weather challenges and whether these challenges harm or enhance a state-level "pool" of extreme weather scientists focused upon a particular state's adaptation challenges.

What is believed by this editor is that extreme weather adaptation should include consideration by universities within a state and by state management boards of higher education of whether the state is committed to creating and nurturing adequate numbers of scientists specializing in these topics within the state's universities. A recent effort introduced by the Louisiana State University Sea Grant Program reflects well on Louisiana in this matter to overcome the shortfall noted above because it is focused on enhancing state specialists among the young faculty. The program is called *La D-I-A*:

> (It) supports scholars interested in applying their work to real world problems and sharing these discoveries with diverse audiences. By encouraging multidisciplinary research efforts and strong links with coastal communities, the Louisiana Discovery-Integration-Application Program (La D-I-A) connects science, communication and coastal residents. In essence, the program provides a two-way conduit between university research results and community needs.[4]

This is a very encouraging development in Louisiana's need for very extensive research on extreme weather and climate change and adaptation to it and the ability of scientists to collaborate with the state's communities supporting them to adapt better to extreme weather and climate change.

It is with gratitude that I thank the chapter contributors for working with me on their chapters and their refinements. The book would also not have been possible without the assistance of two key individuals—both recent graduates of Louisiana universities—who prepared the chapters for submission (Olivia Porter, MURP, JD) and who created detailed maps or refined the many figures necessary to enhance such topics (Julie Torres, MS ES). I am honored to have had such a team.

More authors than would be expected had to complete their chapters while moving to new positions or being occupied by family responsibilities such as the arrival of three babies and one grandbaby. They have asked me to dedicate our book to these very young twenty-first-century recipients of the adaptation successes that their parents, grandparent, and other authors of this book are contributing in support of their futures.

*To Willa, Elias, James, and Pia.*

New Orleans, LA, USA — Shirley Laska

[4] https://www.laseagrant.org/outreach/ladia/

# Contents

# Chapter 1
# Introduction to the Book: "Ahead of the Curve"

Shirley Laska

## 1.1 Louisiana: A *Whole State* with Extreme Weather Challenges

The coastal areas of Louisiana have been subject to extreme weather ever since the Mississippi River began to create the Delta land 7000+ years ago (Roberts 1997). The extreme weather first impacted the indigenous population that has lived here for millennia and, over the last 300+ years, multi-ethnic immigrants, refugees and enslaved peoples who settled among them (Owens 2015). While inland floods have occurred intermittently over the decades, the two very extreme rain events in the spring and summer of 2016 began to change the framing of the state's extreme weather experiences: No longer was Louisiana two "states" – the coast subject to extreme weather effects . . . and a safer inland. The image of the state in the eyes of government entities, of inland riverine residents, was changing, and likely observers from outside of the state were also seeing the state differently: It is now a state subject to extreme weather *throughout* – urban, rural, coastal, and inland watersheds. And that extreme weather is exacerbated by climate change. Weather specialists declared that in Louisiana, both Hurricane Isaac in 2012 and the summer 2016 inland storms were enhanced by it – stalling the storms' advance and thus generating their extreme precipitation effects (Wiel et al. 2017; Peterson et al. 2016; Kossin 2018).

S. Laska (✉)
University of New Orleans Emerita and Lowlander Center, New Orleans, LA, USA
e-mail: slaska@uno.edu



S. Laska (ed.), *Louisiana's Response to Extreme Weather*,
Extreme Weather and Society, https://doi.org/10.1007/978-3-030-27205-0_1

When these inland floods occurred, it became evident to this book's editor – a researcher and adaptation[1] practitioner studying Louisiana flood risk for about 35 years and having begun research on an inland flood of 1983 – that sadly, the whole state had now moved into an extreme weather state category. I believe this change warrants study of Louisiana *as a unit*: examining and understanding better the level of experience and response embodied within the geographic/political entity of an American state and its government bureaucracies, its communities and citizens. It is hoped that this research will benefit both Louisiana and – as an exemplar – other states, whose extreme weather risk is increasing like Louisiana's has done or is likely to increase in the near future as the new US Global Change Research Committee Report (Jay et al. 2018) portends.

The timeliness of a *statewide* "extreme weather" book became quickly obvious when state and local officials initiated after the 2016 floods adaptation programs additional to those focused only on the coast. As the book took form, some eight state extreme weather adaptation programs and one city program began to take shape. Also, adaptation as reflected in the new programs was no longer siloed in one or another relevant state agency. The watershed program was created and implemented by five state agencies, and the coastal restoration's community/parish capacity and capability study approached the project so that the findings could be shared with the watershed program and possibly expanded statewide (see Sect. 1.3.1). The evolving programs demonstrated an appreciation of the interconnectivity organizationally and experientially of the challenges for the whole state. The "connection" had been made between coastal adaptation and inland adaptation to begin to grow *statewide* adaptation as the goal.

This chapter has the following aims:

1.2 Introduce the key Louisiana extreme weather and climate-change induced conditions that prompted the preparation of this book as well as demonstrate *adaptation* and *adaptation risk* to Louisiana's most current extreme weather and disastrous flood levels of the Mississippi River, which are likely climate change-enhanced threat.
1.3 Summarize the recent adaptation programs that have been created within the state. Then argue that an effective way in which climate-induced extreme weather adaptation can occur in the earlier stages of this climate adaptation phenomenon is through existing federal disaster recovery programs – as the new Louisiana programs are currently being funded. How to achieve climate adaptation through this means is diagrammed, and the details are presented. The process of *exceptional recovery* from disasters if successful results in *essential resiliency* – resiliency embracing "just," comprehensive qualities – and "grows" climate change adaptation.

[1] Climate change *adaptation* is adjustment in ecological, social, or economic systems in response to actual or expected climatic stimuli and their effects or impacts. It refers to changes in processes, practices, and structures to moderate potential damages or to benefit from opportunities associated with climate change (Burton et al. 2001). Climate change *mitigation* refers to actions that reduce the human contribution to the planetary greenhouse effect (National Climate Assessment 2014).

1.4 "Place" the book's chapters about Louisiana as expanding the exceptional recovery/essential resilience model; about one-half of the chapters directly address one of the two very related concepts. In addition, four chapters describe/expand the topic of Louisiana's new adaptation programs and are identified in Table 1.1. Finally, the chapters which analyze challenges of *traditional disaster recovery* will be identified because these challenges will only be exacerbated by climate-induced extreme weather events and thus must be appreciated and their solutions addressed. It is critical not to split the challenges and solutions to traditional disaster recovery from those that are emerging from climate-induced extreme weather events.

1.5 End the chapter with a challenge to the readers: Will the recommendations offered by the chapter authors for extreme weather adaptation be more successfully accomplished and achieve their goals if the states/communities – *subnational* units of government – are the true managers of the adaptation process instead of the federal government which is emphasized at present? While the consideration of community adaptation has been embraced in the twenty-first century (Burton et al. 2001; NASEM 2018; Rockefeller n.d.), it is argued that the technical aspects and especially the financial aspects of turning from federal management/control to state and local units have not been adequately considered. Therefore, I propose that shifting to subnational levels is not adequately informed if these real issues are not considered and addressed. This section will offer considerations –both pro and con– for this change to subnational adaptation emphasis and encourage robust research to achieve the critical "essential" adaptation success needed.

## 1.2 The Louisiana Case: Extreme Weather and Climate Change Experiences

The state of Louisiana's twenty-first-century flood inundation is represented by nine individual extreme storm maps that were combined to create the composite map (Fig. 1.1). In addition to the 9 storms represented in the composite map, another 14 storms occurred that caused additional flooding, mostly within the same areas as the 9 storms represented (see Appendix 1 for the full list). No such map as the composite in Fig. 1.1 was available for use in this book; it had to be constructed. Not to have state-level composite maps that are kept up to date and readily available for researchers, practitioners, and government officials, especially for states subject to so much extreme weather flooding, is a significant gap of needed information for addressing extreme weather events and adaptation to them. (See Appendix 1 for expanded description of the types of data used for the included maps, the methods of map creation, and the recommendations for enhancing map resources.)

The majority of Louisiana floods can be traced to 54 tropical weather events, the third largest number that impacted a US state within the last 166 years. Texas (64) and Florida (117) are the only two states with more (National Hurricane Center

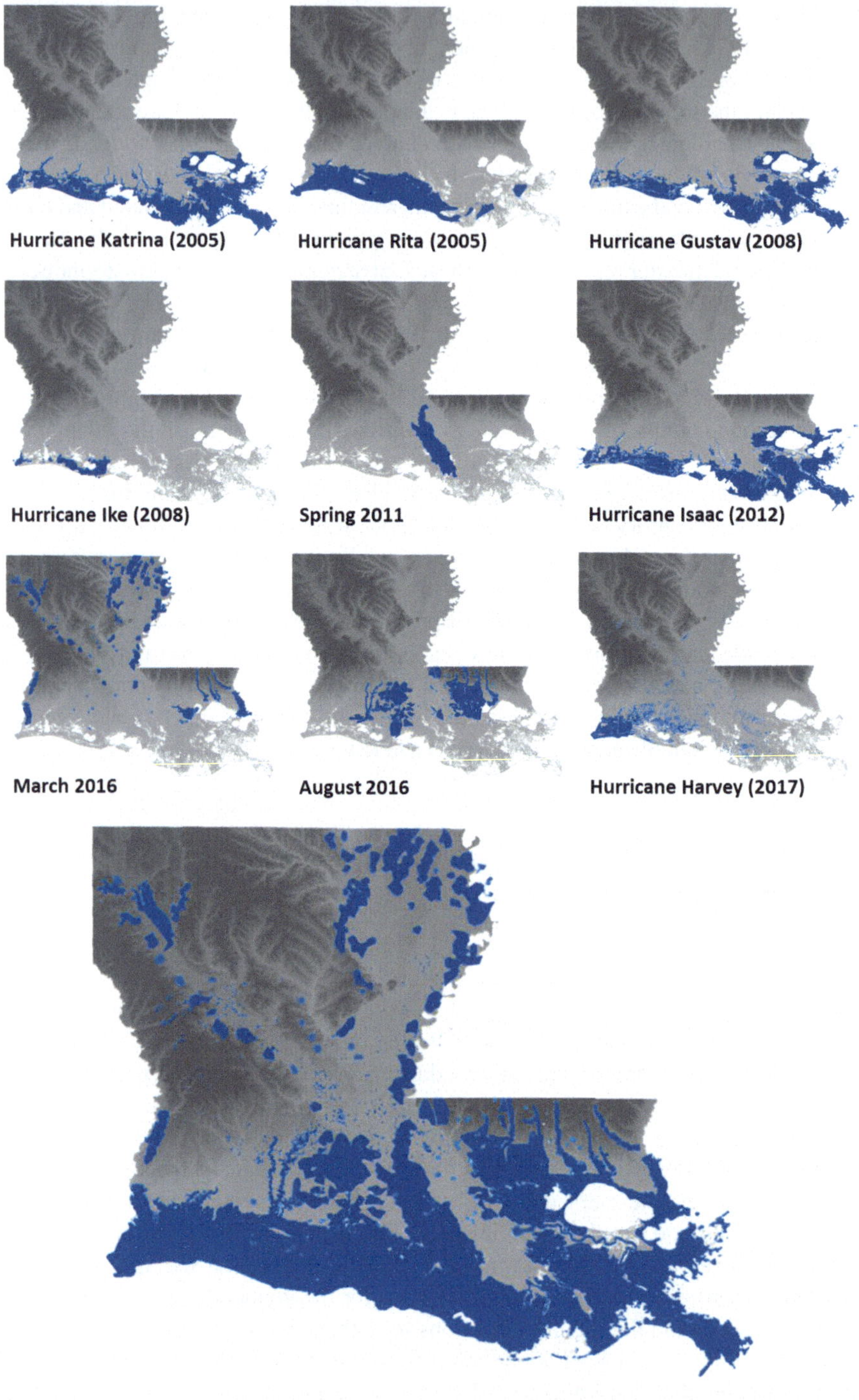


**Fig. 1.1** Flood inundation coverage maps for selected hurricanes and intense rainfall events in Louisiana during the twenty-first century (top) and composite flood map (bottom)

2018). The inland flooding that prompted the creation of this volume (spring and summer 2016 and December 2018) set river records for flood levels that ranged from 15% to 30% higher than previous records that were set as far back as up to 66 years ago (Schleifstein 2019).

The most serious hurricane experience for Louisiana was Hurricane Katrina in 2005, the deadliest and costliest of mainland US storms ever (NHC 2018). Climate change-related qualities of very warm Gulf of Mexico waters and increasing sea level rise (Union of Concerned Scientists 2017a) exacerbated the storm's impacts, especially the significant storm surge that the public was not warned of adequately because it was not included in the Saffir-Simpson scale.[2] Unfortunately Katrina became a "perfect storm" due to these storm and atmospheric qualities combined with weak storm protection (Day et al. 2007), and an ill-conceived ship navigation channel cut through the marsh directly targeted to New Orleans (Freudenburg et al. 2009).

Louisiana is also ranked number one in two other extreme conditions – coastal land loss and sea level rise. Coastal Louisiana was built from sediment suspended in Mississippi River waters flowing down the continent that was deposited in fanlike patterns as the river swished back and forth on the Louisiana coastline that began at the border with the Pleistocene uplands (see Fig. 2.2 in Boesch). However, levees built by early settlers and strengthened after the massive 1927 flood restricted the continued capacity of these depositions to replenish the marshland created by the earlier flows (Couvillion et al. 2011). Resulting marsh subsidence and erosion are exacerbated by saltwater intrusion into the marshes via the thousands of canals created over more than a century of oil and gas exploration (Turner and McClenachan 2018; DeLaune and Pezeshki 1994). Saltwater intrusion via these canals kills the grasses and other plants that hold the marsh soil, resulting in more soil eroding into the water and being washed out into the Gulf, thus causing land loss. Since 1932, approximately 2000 square miles of coastal Louisiana have been lost (Couvillion et al. 2011).

*Relative* sea level rise is predicted to be higher in the coastal Mississippi Delta than anywhere in the world by the end of this century (Parris et al. 2012). This is the result of the subsidence just discussed combined with *eustatic* (global) sea level rise (Rovere et al. 2016) caused by increasing freshwater input and temperature-induced thermal expansion of the world's oceans (Pahl 2016). New measurement techniques identify even more sea level rise on the Louisiana coast than earlier described (Keogh and Tornqvist 2019). The ramifications of such an increase in elevation of the Gulf of Mexico waters along coastal Louisiana are demonstrated by the dramatic number of Louisiana communities expected to be chronically inundated by 2035 in the intermediate sea level rise scenario: some 59 Louisiana communities, including New Orleans, comprise two-thirds of the 91 US communities coast-wide predicted to be so harmfully affected (Union of Concerned Scientists 2017b) if the state's

[2] The Saffir-Simpson scale failed to adequately consider the powerful storm surge caused by Katrina. Seven years later surge measurement was added to National Weather Service forecasting tools (National Weather Service 2012).

coastal restoration efforts are not adequately funded (Davis and Boyer 2016), implemented, and effective.

Finally, as this manuscript is being submitted to Springer in early summer, 2019, the Mississippi River is in the midst of the longest period of high-flood stage in the history of its flood measurement and has surpassed the duration of the 1927 flood (from midwinter expected through midsummer). This event has been caused by extreme storm patterns in the upper Midwest resulting in record-breaking flood elevations all throughout midcontinent rivers that feed the Mississippi. The extremeness of the patterns is suspected to be caused by climate change (Stott 2016). Discussing this particular example of Louisiana extreme weather risk provides the bridge between the just-described presentation of Louisiana's frequent and severe storms via the twenty-first-century storms and *adaptation*, including *adaptation risk*, the two concepts that are the focus of this book.

There are three foci of the protection, i.e., adaptation, which the Corps of Engineers provides to Baton Rouge and New Orleans beginning upriver: the Old River Control Structure(s), the Morganza Spillway, and the Bonnet Carre Spillway. The first maintains the percentage flow between the Atchafalaya Basin and the current Mississippi River channel of 30/70%. Without it, the percentages would reverse, and the two Mississippi River cities would no longer be able to support ocean-going ships and river commerce as it is known today. Due to the extreme importance of this structure remaining intact and providing that service, improvements were made to it in 1973. The Morganza Spillway reduces flood levels for Baton Rouge and relieves flood level pressures on the Old Structure. Finally, the Bonnet Carre permits the Corps to maintain the river below flood stage at New Orleans. Thus, these adaptations serve the goals for which they were designed.

However, the diversions pose *adaptation risks* to the floodways through which the diverted water flows; and it appears very clearly that with frequent and extreme weather events, the risks which these adaptions pose to other communities and livelihoods become disasters in their own right. The harmful impact of the two diversions is on (1) the Atchafalaya Basin (the Morganza Spillway) and (2) Lake Pontchartrain and the Mississippi Sound (the Bonnet Carre) as released water from the latter passes through the lake to the Gulf of Mexico.

Both spillway paths have been "threatened" by this year's high water. While the Corps of Engineers announced a date for the Morganza Spillway opening, it postponed it three times and then postponed it "indefinitely" when this manuscript was going to the publisher; but it is known how the opening would have impacted the spillway from the opening in 2011, only the second in the spillway's history. What is at risk if the Morganza is opened are parts of western Terrebonne Parish, parts of Morgan City, and the Atchafalaya crawfish harvest, rice, sugarcane, cattle farming, and wildlife. Similarly, the Bonnet Carre silted and freshened (added freshwater to brackish) Lake Pontchartrain and, especially harmful, freshened the Mississippi Sound where the freshwater infusion has put at extreme risk the oyster harvest, dolphin and sea turtle populations, and tourism through these wildlife mortalities. Public meetings of community leaders and commercial fishing organizations seeking redress of this harm occurred in late May when the impacts became clearly evident. Newspaper headlines tell the story: "Waveland (MS) 'first in line' for damage from opening of the Bonnet Carre," and "Fisherman outraged by freshwater impact on the Mississippi Sound."

At a public meeting, May 29, 2019, near Biloxi, MS, the attendees wanted first to describe the *layering* (Laska et al. 2015*)* of earlier event impacts that they had experienced – beginning with the BP oil spill and the opening of the Bonnet Carre the earlier time this year (and the third in 4 years) upon which they declared that the current adaptation measures to protect Louisiana river cities posed additional (adaptation) risks to them. They clearly perceived the adaptation event as being so harmful because it came on top of the other events in addition to being a serious event itself, i.e., compounding harm. Although adaptation has been represented as a positive, constructive concept when used in the conversation of climate change, equally important is the need to understand better that adaptation is not a concept free from doing harm and that harm likely increases with the magnitude of climate change. Considering adaptation risk is undertaken in several of this book's chapters: especially Chaps. 3, 5, 6, 7, 8, and 9.

Louisiana's extreme weather history and its predicted future of extraordinary risks – this book does not even include heat waves and droughts that appear also to be in the future of Louisiana – place the state "ahead of the curve" in experiences and thus make it a "poster child" for understanding climate change impacts and for learning about and improving adaptive responses. Extreme weather affects 4.5 million Louisiana residents – rural as well as urban (25/75% population distribution) and coastal as well as inland areas (60/40% population distribution). It impacts a wide variety of communities, cultural groups, and economic activities, both local and national, including farming, ports, fisheries, and tourism, which employ the state's residents. Additional impacts are found in oil and gas extraction/petrochemical production whose normal operations present the state with environmental risks – making their threat even more serious when impacted by extreme weather.

## 1.3 Framing Climate Change-Induced Extreme Weather Adaptation

It would be better to consider and to implement adaptation to climate change-induced extreme weather in a methodical fashion, separate from immediate disaster recovery and separate from uncertainties and fluctuations in funding, with the proposed critical changes from previous extreme weather response clearly identified and included in the societal adjustments to extreme weather. It would be ideal to refine systematically the adaptive responses applying the results of careful climate change social, physical, and atmospheric scientific research and engineering/design and include more than one method in the same effort after considering the compounded benefits (Bailey et al. 2017). What is different with climate change dynamics taking all major vulnerability dimensions into consideration? How is it different? What has caused the differences? What differences will bring the most harm to the social structure, to the social processes, and to the residents as well as the land and ecosystem? How rapidly will these changes occur? What are the differences: magnitude, frequency, and changes from past extreme weather events such as slow-moving storms when they had moved more rapidly in the past? Who is at risk, more than before the extreme weather-induced changes occurred? How can the identified useful adjustments be made economically, with deliberate speed, to protect the largest number of communities and widest array of residents, especially those most vulnerable, and to protect them for the longest time into the future? How can adaptive capacity (Gitz and Meybeck 2012) be enhanced with the new responses? And, of course, how can the adaptation occur while not harming others and other ecosystems?

I believe that those considering what climate change will bring – be they citizens or specialists – are naively assuming that such an adaptive response will likely happen. If a challenge is seen, it is in understanding the likely changes and synching our response successfully with them and in motivating the society to take the necessary adaptive steps. Our society does not raise concerns that the response will not occur as we do about the *mitigation* of $CO_2$ gases, the other response believed to be absolutely necessary in addressing climate change. With the new weather changes, it is believed that new adaptive responses will be determined and implemented. "No problem" as the younger generation likes to say. On the contrary, the chapter content in this book and the new adaptation programs which Louisiana has introduced within the past few years assert that *both* climate adaptation and mitigation pose huge challenges to achieve success.

### *1.3.1 Louisiana Adaptation Innovations and Proposed Early Climate Change Response*[3]

Eight climate change-related adaptation programs came into existence and evolved during the conceiving and preparation of this book (Table 1.1 and Appendix for agency descriptions of each). This was an increase of eight from only two evident programs of significant size before that time and the beginning small steps of the nonstructural aspects of the Coastal Master Plan, which is a substantial amount of growth. Previous programs included the environmental and physical structural efforts undertaken by the Louisiana Coastal Protection and Restoration Authority (LA-CPRA) entitled the *Louisiana's Comprehensive Master Plan for a Sustainable Coast* (Coastal Master Plan) and, one combination of structural and human dimensions, the Community Rating System (CRS) of the National Flood Insurance

[3]Andrea Galinski, Assistant Scholar, Dept. of Landscape Architecture, College of Design, Construction and Planning, University of Florida, had intended to contribute a chapter to this volume on the topic of new adaptation programs but was unable to do so due to a career change from LA-CPRA to Florida. However, she enthusiastically volunteered to offer her assistance with this section. I am appreciative of that willingness.

**Table 1.1** Large, new (since 2016) LA state and local adaptation programs being implemented currently

| Locale emphasis | Description (URL citation) | Govt. unit/program/funding |
|---|---|---|
| *Coastal* | 1. Coast-wide wetlands and barrier islands restoration efforts with *addition of* **Flood Risk and Resilience Program** (CPRA 2017) *(Hemmerling* et al., *in this book)* *(Birch and Carney, in this book)* | LA Coastal Protection and Restoration Authorization (LA-CPRA)/2nd 5-year Master Plan/ multiple federal and state sources |
| | 2. Awarded Purpose (since modified): **Resettlement** of Isle de Jean Tribe of Biloxi-Chitimacha-Choctaw (La. Div. Admin. 2015; 2019, April 23) *(Jessee, in this book)* | LA Office of Community Development (OCD)/Natl. Disaster Resilience Competition (NDRC)/ HUD-CDBG-DR |
| | 3. Gentilly Resilience **District Storm Water Management** Project funded for major project implementation of the Greater New Orleans Urban Water Plan developed during the decade (New Orleans 2019) *(Birch and Carney, in this book)* | City of New Orleans, Office of Resilience and Sustainability (ORS)/ Natl. Disaster Resilience Competition (NDRC)/HUD CDBG-DR |
| *Inland* | 4. **Resettlement** of Pecan Acres subdivision, New Roads, near False River and Silverleaf, City of Gonzales (La. Div. of Admin. 2019, March) *(Peterson, in this book)* | LA Office of Community Dev. (OCD)/Buyout and Resilient Housing Incentive. CDBG-DR from 2016 flood/2017 and Natural Resources Conservation Service, USDA |
| *Both* | 5. LA SAFE. **Determination of 3-tiered coastal areas risk– remove structures, elevate, settle, and parish engagement model projects to respond** (La. Div. of Admin. 2019, April) *(Birch and Carney, in this book)* | LA Office of Community Development (OCD)/Natl. Disaster Resilience Competition (NDRC)/ HUD-CDBG-DR. |
| | 6. **Adaptive capacity for resilience of coupled coastal-inland system** (LSU-CSS 2017) *(Birch and Carney, in this book)* | LSU-Coastal Sustainability Studio/ NAS Gulf Research Program and Robert Wood Johnson |
| | 7. **Framing riverine flood management using watersheds** (Office of the Governor 2018) *(Birch and Carney, in this book)* | Office of the Governor/Watershed Initiative /CDBG-DR from 2016 flood |
| | 8. **Parish flood risk and resilience capability and capacity assessment** (CPRA 2018) | LA CPRA/Flood Risk and Resilience Program evolved to support CPRA nonstructural and the watershed Initiative/multiple federal and state sources |

Program (NFIP), a voluntary program for recognizing and encouraging community floodplain management activities that exceed the NFIP's minimum standards. While, as the title denotes, the Coastal Master Plan focused on the coastal storm and sea level rise, the participating Louisiana CRS communities, some 43 in number, are scattered throughout the state but still mostly in the coastal areas (FEMA 2018b).

There is likely limited data from other states with which to compare the large number of new programs created so close together by Louisiana state agencies and one community. The emergence of such a number within a 4-year time span suggests with little doubt that the different state government agencies, and the largest city, New Orleans,[4] have been recently rapidly growing in their appreciation of the need for extreme weather adaptation. The recent 2016 floods, which damaged 146,000 homes and amounted to approximately $10 billion in economic damage, have likely contributed to reframing adaptation efforts away from a coastal focus alone to a statewide need and effort. The following are (1) a summary of these programs, (2) the description of one for which there is no analysis in chapters contained within this book, and (3) the challenges of such a "surge" of efforts occurring approximately at one time. These topics, it is hoped, will contribute to understanding what we can expect in other areas as they experience more climate change-related impacts or how we can contrast what Louisiana is doing with what other states, communities, and regions of the country are currently undertaking to grow in adaptation knowledge and best practices.

The Louisiana programs are distributed among coastal, inland, and a combined emphasis (Table 1.1). Three of the programs are as would be expected coastal programs, one is specifically inland, and four are both coastal and inland or "coupled" coastal-inland systems. It may be hypothesized that the coupled are likely to be the most fruitful – as is considered in Birch and Carney, Chap. 12, and Peterson, Chap. 7, in this book. However, more research needs to be done on this concept before more knowledgeable assessments of coupling can be described and even better refinements made.

The organizations which created the new adaptation programs range from the state's coastal agency – La Coastal Protection and Restoration Authority (which includes several programs and initiatives – Coastal Master Plan, Flood Risk and Resilience Program, and the Parish Flood Risk and Resilience Capability and Capacity Assessment) – to the agency which manages disaster response, the Office of Community Development (OCD) (three programs – Community Resettlement from the National Disaster Resilience Competition, the LA SAFE program similarly funded to engage and encourage communities and parishes experiencing different gradations of risk from the coast inland to adapt appropriately, and the the inland resettlement of two neighborhoods) to the Governor's Office (Watershed Initiative) and to Louisiana State University's Coastal Sustainability Studio's adaptive capacity for resilience of coupled coastal-inland systems.

This volume was not intended as an analysis of only new adaptation programs (as the programs did not exist/or exist in their current complexity when the book was conceived), but rather it has evolved to describing them and then benefitting from various chapter authors analyzing critical issues of the programs that impact adaptive behavior (see Table 1.1 for the specific chapters that address the various

[4] The work on effective water management began after Hurricane Katrina. The program mentioned herein is an area-wide implementation of the ideas developed since that catastrophic event, especially after 2010.

programs). Further research it is hoped will also be taken up by others, including further research by the contributing authors of this book's chapters. These programs are, in effect, Louisiana's "testing ground" for its climate-induced extreme weather response.

Only the newest program, and one thus not described by the authors of the chapters, will be described as an example of where the state's adaptation efforts stand at publication of this volume after which there will be a beginning analysis of what happens when eight adaptation programs are "gestated" at about the same time. Obviously, the opportunity to follow these programs, to consider if and how they are blending, complementing one another, and defining different areas of need and of professional specialty, is a font of opportunity for those interested in climate change adaptation research.

A number of Louisiana's new programs for responding to climate change-enhanced extreme weather have evolved out of the state's 2017 Coastal Master Plan, which includes recommendations for restoration projects, structural projects (like levees and floodgates), and nonstructural flood mitigation projects (such as home elevations or voluntary acquisitions) across the Louisiana coast. The 2017 Master Plan provides a comprehensive assessment of how coastal storm surge-based flood risk may change over the next 50 years, as well as offers recommendations on where and to what extent nonstructural efforts would most benefit the resiliency of coastal Louisiana. (The plan recommends approximately $6.2 billion in nonstructural mitigation measures in multiple communities over the next 50 years.)

Importantly, the Coastal Master Plan shifts the state's focus from post-disaster storm recovery to planning for proactive flood risk reduction actions. In addition to these mitigation project recommendations, the 2017 plan also lays the groundwork for a state-led program that can better align federal, state, and local funding to advance the implementation of such projects. The aim of CPRA's Flood Risk and Resilience Program is to prioritize areas for nonstructural mitigation and to develop a state-led grant program that supports the implementation of such projects by parishes. The program encourages resiliency actions across a range of state, parish, municipal, and academic/NGO actors through a suite of resiliency policy recommendations. Building more resiliently is also encouraged through funding incentives for increased flood risk standards with up to 100% of a nonstructural mitigation project potentially funded by the state. The program also more broadly aims to advance awareness of current and future flood risk, promote greater inter-agency coordination, and provide resources to build local capacity.

One recent initiative developed under the Flood Risk and Resilience Program (Table 1.1, #8 program) includes the development of a capacity assessment for coastal and near-coastal parishes to better determine their ability to implement nonstructural projects as well as related plans and policies. Andrea Galinski, a former CPRA long-term staff member, explained:

> We wanted to better understand what the current ability is to implement nonstructural projects (and broader resilience/flood risk reduction plans and policies). A capacity assessment was also going to be part of a broader Watershed Initiative across the state, and so this assessment was slightly modified and became framed as a "pilot" that could be used for that effort. (Personal written communication, May 10, 2019)

Galinski also notes that this capacity assessment has helped CPRA to better understand the existing gaps and local needs and has also been informative to other state agencies involved in watershed assessment including the Louisiana Department of Transportation and Development (DOTD), OCD, Governor's Office of Homeland Security and Emergency Preparedness (GOSHSEP), and the Department of Natural Resources (DNR). One important lesson that emerged from the assessment included parishes' concerns with state coordination and actions, which has led to a significant consideration of how state agencies can better coordinate both horizontally (across state agencies) and vertically (between federal and local levels) to reduce flood risk.

What should be the initiation and implementation of a state-level movement toward addressing a "new" problem – in this case extreme weather risk likely associated to climate change and now being experienced by residents throughout the state? Is a flurry of programs the usual way change happens? And then they begin refining their missions, synching their goals, some achieving institutionalization while others fall to the wayside. There is a lack of clarity about if/how these eight human dimension programs will synch. It is not known that they will. Will some be redundant but still stay in existence? Will they expend resources while not achieving the needed integrated framing and implementation of the best adaptive efforts? That the answers to these questions are not known is likely very "normal" for such moments of crisis and a beginning of a broader society push to address serious risk.

However, these and similar questions are being asked about the evolution of this now "macro" adaptation response. Local WWNO public radio reporter Travis Lutz (2019, May 26) queried the director of the La. Office of Community Development with such a question about yet another program, the LA SAFE Program (#5 program in Table 1.1). The director of the Office of Community Development replied: "It is about a new way of thinking about investments we make every day. . ." There is no doubt with all of the energy, new bureaucratic structure, program content, and efforts to identify a path forward that Louisiana is in the midst of a "sea change" in adaptation. Only time and committed research will tell those interested if and how the state succeeded and what can be learned by other states from Louisiana's responses, successes, and failures.

### *1.3.2 Today's Reality of Climate Change Adaptation*

To reiterate, all eight Louisiana significant innovative adaptation projects were, or will be, funded as a result of a specific declared disaster: seven extreme weather and the eighth the Deepwater Horizon oil spill. Given that "tie" to a specific disaster, there is no reason to believe at this time that adaptation innovations will occur regularly, in large numbers and at great investment expense separate from a disaster's recovery funding, its damage legacies. Much effort and resources will have to be invested over a long period of time, and social change occur in major ways so that continual successful adaptation takes place. The reason for such a strong negative assessment by this author to the likelihood of stand-alone adaptation is due to the current resistance to implementing "normal" (not climate change induced) extreme weather resilience efforts. Louisiana is a case in point but unfortunately not one out of the ordinary for US states. It is "nestled" within the norm: there has been very little appetite in Louisiana and in other states for stand-alone extreme weather resilience efforts, traditionally called "pre-disaster mitigation" by FEMA (The Pew Charitable Trust 2018a, b).

Extreme weather adaptation typically occurs during the recovery period *after* major disasters strike (The Pew Charitable Trust 2018b) as in the seven Louisiana examples offered in this chapter. While "pre-disaster flood mitigation" is a term embedded in the vocabulary of the Federal Emergency Management Administration (FEMA), commitment to adapting to an extreme weather threat *before* another weather event occurs in a particular location is currently qualified as "aspirational," as indicated by the modest number of successful awards in FEMA competitions for states, the small amount spent by state governments, their lack of interest in collecting data that would measure such pre-disaster mitigation effects (The Pew Charitable Trust 2018a, b), and the modest interest in the FEMA Pre-disaster Mitigation Program. Louisiana is no exception. For fiscal 2017, the latest year data are available, Louisiana did not receive any funds from FEMA for pre-disaster mitigation (FEMA 2018c).[5]

While there are some new pre-disaster funding opportunities within the new Disaster Recovery Act of 2018,[6] state and local governments are currently challenged to compete successfully and then to implement such systematic adaptation now (The Pew Charitable Trust 2018a). Therefore, if pre-disaster mitigation has not happened much to date with extreme weather disasters, how rapidly and successfully will these funding enhancements lead to significant increases in pre-disaster adaptation in the future? While some of the qualities of the new climate change extreme weather might prompt more pre-disaster adaptation – such as increased magnitude, frequency, clearly observable difference from earlier extreme weather – that link has yet to be studied and, if found to be the case, may not occur repeatedly for quite some time. As the data in the beginning section of this chapter demonstrated, Louisiana has been subject to extreme weather events on the average of more than one a year since the beginning of the twenty-first century. This rate of impact has not motivated the state to successfully compete for pre-disaster mitigation funds as mentioned above (FEMA 2018c). Now there appears to be changes in that response, but the efforts are still being funded by resources tied directly. Research needs to be conducted specifically on this question: Does a significant number of disasters, or continual events, or extreme ones not experienced before or for a lengthy time motivate states to seek pre-disaster mitigation (adaptation) funds more frequently and to successfully qualify for them? If the results of such research lead to the

[5] The disposition of those submitted fell into categories: identified for further review, did not meet hazard mitigation assistance requirements, and not selected.

[6] Improvements in support of pre-disaster implementation for resilience include a reliable stream of 6 percent set aside for public infrastructure pre-disaster hazard mitigation (Section 1234) (FEMA 2018c).

answer "No," then even more concern must be expressed about our culture's assumption that climate change adaptation will be significantly forthcoming. Right now, we should assume that the challenge to adapt prior to disasters will be equivalent to the challenge to mitigate $CO^2$ emissions.

It may be that the effort needed to recover from *particular disasters* or *catastrophes* such as Katrina consumes the citizens' efforts and emotional energy such that interest in continual adaptation is just too much. Or the motivation to adapt is strong while recovery is going on but it fades afterward. Meyer and Kunreuther's recent work (2017) tries to explain these barriers suggesting social psychological causes, i.e., emotions, which discourage commitment to adaptation: myopia, amnesia, optimism, inertia, and simplification of threats and thus responses. The science to explain resistance to constant attention to addressing risk needs considerable additional refinement. And the addition of the qualities of climate change extreme weather – new magnitude, sudden, more frequent, unusual qualities,[7] having both temperature extremes in same event and more media coverage – will also add new dimensions to the needed research. It may be that these qualities will somewhat overcome the impediments to adapt described by Meyer and Kunreuther (2017). They are dramatic, "never have happened before," and extremely damaging, and those qualities beg for attention to a response.

However, given this lack of commitment *now* to continual extreme weather adaptation absent a significant disaster event to draw attention to the topic, I argue that a resiliency framing that focuses on what adaptation is possible in the recovery from a *particular disaster* is the more appropriate focus for this early climate change extreme weather adaptation period, to make these recovery funds as productive as possible in achieving adaptive recovery, rather than merely addressing recovery as putting it back the way it was.

Two new emphases of the Disaster Recovery Reform Act of 2018 (FEMA 2018b) commit to enhancing resilience during the disaster recovery: Section 1235a ensures the Hazard Mitigation grants must "increase resilience to future damage, hardship, loss or suffering" (Section 1235a) and that damaged public facilities be repaired to the latest codes and standards to strive for resilience (Section 1235b). In the spirit of these new federal "commitments," I will now describe a revision of a recovery framing that was first offered in an earlier publication (Laska 2012) to reflect the argument that disaster recovery must be the locus *currently* of much climate change adaptation.

[7]A Louisiana example reinforces the qualities of unusualness of weather events that bring residents' attention to climate change: On December 29, 2018, as this chapter was being prepared, inland flooding occurred in some of the same area flooded by the 2016 spring and summer floods. Television news reporting quoted a victim: "We didn't have any time to prepare for the flooding because what happened in 24 hours in the spring of 2016 happened within a few hours this time."

### *1.3.3 Exceptional Recovery for Essential Resilience*

The extreme weather adaptation frame offered here combines two concepts – *exceptional recovery* and *essential resilience* (Laska 2012).

The *exceptional recovery process* has qualities that have been identified and developed by the authors of this book's chapters. The recovery process must:

- Be based on a robust commitment to citizen participation
- Honor community self-determination of recovery processes and outcomes
- Have a deep commitment to social justice in the recovery processes at all levels of government response
- Expect a sophisticated recognition by government officials of historical experiences that have led to socially constructed vulnerabilities "causing" the current disaster impacts (Tierney 2014; Wisner et al. 2004)
- Appreciate the economics of the recovery process itself that do not support the enablement and adaptation of the entire community to future extreme weather but rather the interest of the corporations that are used to address the damage and of the "growth machine" (Molotch 1976) putting developer interest ahead of community residents
- Have a deep understanding of the institutionally induced harm that manifests itself in the current government-managed recovery including the technocratic framing of disaster funding as dependent upon benefit/cost and to develop recovery processes that are free of such harm

Without such a robust understanding, the recovery process will contribute to reproducing the vulnerabilities that caused the extreme weather event to generate harm in the first place through a disaster or even a catastrophe from which the community or region is now recovering.

Adding the adjective *essential* to the sought-for resilience gives consideration to the qualities of resiliency that must be part of the outcome of the exceptional recovery. The prolific array of publications that have appeared in the last couple of decades speak to the enhancing of the qualities of the society that permit it to "bounce back" or change so that the form the community/region takes after a disaster enables life to go on effectively, e.g., "resiliently." As has been repeatedly affirmed, such resiliency extends way beyond preventing the physical event or modulating generally what the extreme weather event can do to a community physically. The use of essential resiliency in this discussion of climate change adaptation is to encourage the consideration of what qualities of a society, of a community, are *essential* to the robust improvement of the community to withstand future climate change-induced extreme weather impacts. To reiterate, it is the robust, carefully considered *essential* improvements that redound to the benefit of all social classes, races, ethnic groups, and the social organization that supports the full community's ability to function satisfactorily that are the requirements of successful adaptation.

By using such a reasoning, sometimes the improvements that are of focus in disaster recovery can appear to have little to do directly with recovery. However, the work of the chapter authors within this volume and their like-minded researchers reminds us that resilient communities are socially and economically healthy communities with continual efforts to prevent social class, race, and ethnic disparities and discrimination. If, for example, as Andreanecia Morris and Lucas Diaz describe in Chap. 9, lower-income families in a highly hazard risk community like New Orleans are able to improve their income and/or reduce the percentage of their income spent on housing located near good employment, their resilience "when the next storm hits" rises. And if the housing they rent has the ability to be physically resistant to storm winds and water, and to be repaired after the earlier storm, their chances of returning to the community, to their social network, to their employment, and to the contribution which they made before the event are greater, and therefore the entire community is more resilient following the recovery and into the next disaster if one happens. Thus, essential resilience, the outcome of the exceptional recovery process, should be *additive with deliberation and inclusivity*. It should also recognize at its core that much disaster vulnerability is social and economic, not physical (Wisner et al. 2004).

The diagram created for the original concept in 2012 (Laska 2012) has been refined for this book (see Fig. 1.2). It visualizes the difference between the recovery from a disaster undertaken in more traditional ways – support victims in recovery, return their damaged houses to what they were before the event with minimal changes except for elevation for those damaged over 50%, and assist in the repair of the infrastructure of the community back to what it was before the storm other than

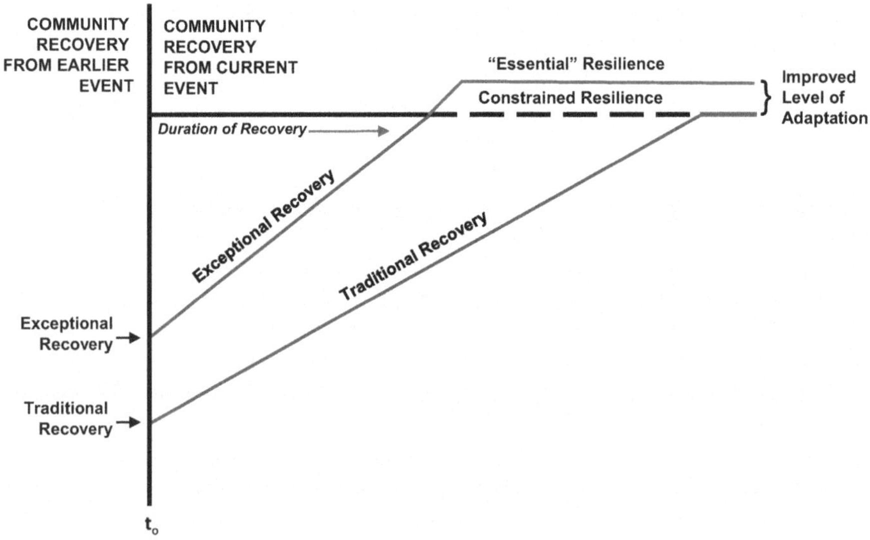


**Fig. 1.2** Resilience deltas when community function is/is not enhanced by pre-event vulnerability reduction measures, i.e., adaptation. (Figure adapted from Laska (2012))

some required improvements due to code improvements. The traditional response is compared to a more adaptive, resilient approach – committing to understand both the physical and especially the social vulnerabilities and undertaking recovery approaches which adhere to the state-of-the-art regulations, community plans, land-use planning, and other resilient qualities including the best scientific knowledge of the anticipated climate change-induced extreme weather effects. Such a process is inclusive of citizens in the learning and decision-making processes.

With the *traditional recovery* being built from the last disaster recovery, the resilience that existed before a disaster will take longer to get back to, and no significant improvement will occur. With *exceptional recovery*, the recovery from each extreme weather disaster event will be more rapid, and the level of resiliency will rise to a better level because the goal will be *essential resilience* (to prepare for the possible next event). Kuhlicke and Steinführer (2010) state that the impact of these adaptations to the phase just before the next disaster, the *new anticipation phase*, would differ from the one described above, in a way that reflects learning and social change, or, to put it differently, a new hazard cycle begins which is not a repetition of the one previously experienced. It will be more resilient. And the community officials and the citizens will be developing a capacity to function adaptively in this more effective way.[8] Diagrams such as the one contained herein help communities, states, and federal government officials and staffs to visualize the simple outline of the process and the outcome. What each level of government might contribute through the exceptional recovery process to the essential resilience goal will be considered in the last section of this chapter.

The irregularities of extreme weather events place the destruction and thus the recovery efforts in different time frames, sometimes the same and sometimes different specific locations and at different levels of disaster – the vagaries of tornado outbreaks in the upper South and Midwest being an example. These uncertain conditions require that the capacity and focused attention of the exceptional recovery be coordinated and documented by state-level or regional agencies within the state rather than only by local communities. One wants the lessons available to all communities and counties subject to a variety of extreme weather disaster events, not just the ones who have experienced a disaster in the past.

Louisiana has the Office of Community Development (LAOCD) which functions primarily as the agency administering the federal disaster response funds. It also sees itself as the state agency responsible for resiliency enhancement:

> OCD-DRU manages the most extensive rebuilding effort in American history and works closely with local, state and federal partners to ensure that Louisiana recovers *safer, stronger and smarter* than before (Louisiana Division of Administration, Office of Community Development n.d.).

[8] Note: Recommending this model of a resilience process is not done ignorant of the fact that disaster occurrence is highly uncertain. Tying adaptation to disaster events may not be the best way to achieve resilience. Just to reiterate, it might produce more resilience at this time than efforts independent of extreme weather events.

Additionally, Louisiana has eight regional planning districts and an umbrella association, Louisiana Association of Planning and Development Districts (LAPDD), which declares interest in recovery and resilience (LAPDD 2018). These organizations can enhance the state's efforts to share the exceptional recovery successes.

## 1.4 "Placing" the Chapters into Exceptional Recovery/Essential Resilience Framework

Can what is learned from research about Louisiana's experience make a contribution to better adaptation by those states growing in extreme weather experiences? This book represents what the American Academy of Arts and Sciences (AAAS) calls "science during crisis" (2019). All of the chapter authors and I hope that this effort adds *long-term adaptation* to the AAAS's goal for scientists and practitioners to "improve crisis response and recovery." The crisis is climate change-enhanced extreme weather impacts.

Part I of the book includes only one chapter that very successfully blends bio/physical/atmospheric analysis with the human/social dimensions of extreme weather response, what is called the "coupled natural-human coastal system" (NASEM 2018). It is written fully in the spirit of the goal that the book honor transdisciplinary research and analysis. Donald Boesch, a native Louisianan with a national reputation for bioecological oceanographic research, academic environmental science management, and environmental policy focused on climate change adaptation, has honored the transdisciplinary goal with a fully integrated discussion of Louisiana's growing extreme weather challenges and their effects on and response by communities and residents of the state.

The chapters in Part II consider methods (exceptional recovery) of achieving successful essential resilience and what challenges are/have been encountered with the efforts undertaken. Chapter 3 by Zachary Lamb is about the force of  in the process of recovery planning, specifically the role of representing seriously flooded neighborhoods as green dots on widely publicized maps that indicate which neighborhoods were proposed "for sacrifice" to recovery redevelopment by turning them into green spaces for holding floodwater. The maps reinforced a politically insensitive representation of class and race privilege in the Katrina recovery planning process contributed to by out-of-town planning consultants and city development leaders. This public memory from Katrina recovery planning taints implementation of the current New Orleans Rockefeller-HUD $141 million resiliency grant for climate change, almost 15 years after Katrina. It demonstrates that exclusion of citizen participation in recovery planning is a mistake and calls into question government legitimacy in response to one disaster and reduces the likelihood of *exceptional recovery* occurring before the next.

Chapter 4 by Kevin Gotham and Megan Faust considers the benefits and drawbacks of national versus state/local responses to extreme weather in a comparison of New Orleans after Hurricane Katrina and New York City after

Hurricane Sandy. *Encompassing cases* merge the combination of state/local and national efforts/policy and consider the benefits and drawbacks of each configuration. Given the expected challenges with relying on federal solutions and aid for extreme weather adaptation, especially for events impacted by climate change, such a comparative consideration and refinement of the *subnational* level are critical for future successes.

Chapter 5 by Scott Hemmerling, Monica Barra, and Rebecca H. Bond offers a very comprehensive description of the evolution of Louisiana's coastal restoration efforts. This chapter offers a picture of one, if not *the* largest, *state*-managed environmental restoration program in the country and its evolution from a project-to-project process to an ecosystem modeling approach. If and how the citizen participation process has improved to support the large, ecosystem-impacting projects is framed in social justice terms. Similar to the green dot example, the risk to rural residents of the project's fisheries impact results in the authors arguing for the importance of creating a planning process that offers an important role to citizen participation and that trusts its importance and contribution to the success of proposed policies and restoration projects.

Part III includes two very different chapters about the issues involved in moving coastal residents inland. Chapter 6 by Nathan Jessee recounts the resettlement process to date of an indigenous tribe that partnered with the state of Louisiana to seek funding for such after two earlier efforts at resettlement failed. The process and challenges to a successful resettlement are presented, and the parallels to earlier treatment of indigenous Americans are described. Resettlement has been romanticized in American culture, while implementation contains parallel structural violence consistent with past experiences of indigenous peoples.

Chapter 7 by Kristina Peterson examines a topic to date rarely found in the social science literature: The dynamics of the relations of climate displaced populations with the receiving communities and their existing challenges. Differences in cultural backgrounds, race and social class are considered as challenges to overcome but may also be seen as opportunities with a firm commitment to make them be so. Peterson proposes approaches that could address these differences and ways migration could be framed to achieve a positive acceptance by the receiving communities including the focus of culture, food, and religion as unifying elements of the "blending" cultures rather than barriers. She also examines issues of identifying environmentally healthy high ground to avoid the repeat victimization that occurred to Katrina migrants during the 2016 floods that hit central Louisiana and flooded them again in their new locales and encouraging receiving community residents and the in-migrants to work together for an improved inland community.

Part IV is organized to include research on adaptation challenges that confront communities of varying sizes, types, and geographic framings, including work on urban, suburban, rural, and watershed communities.

Chapter 8 by Anna Livia Brand and Vern Baxter and Chap. 9 by Andreanecia Morris and Lucas Diaz consider the extreme challenges of achieving *urban* exceptional recovery in the context of economies that understate risks to lower income residents. They examine the way the government and the economy frame recovery in a neoliberal political economy paradigm. The authors demonstrate how social and environmental injustice were manifest in a racialized recovery of New Orleans that stands as an extension of pre-Katrina forces and decisions. They recommend ways to return to citizen participation and expanded citizen influence in the recovery process as ways to deracialize it. Chapter 9 provides a case study of the efforts of a housing coalition to modify the "regime" of decision-making and implementation of adequate affordable housing after Katrina by negotiating a compromise of vested interests. Affordable housing, in crisis mode 14 years after Hurricane Katrina, is seen as a key element to a more resilient and disaster-adaptive populous.

Chapter 10 by Michelle Meyer, Brant Mitchell, Shannon Van Zandt, and Stuart Nolan considers how a climate change-affected extreme weather event presents different requirements for an adaptive recovery. First, the speed of the disaster – multiple inches of rainfall in a short period of time – requires the development of different response assistance as pre-event evacuation is not possible because there is no known severe event predicted early enough to initiate evacuation. This new need is described. Second, the impact is discussed of how a storm with such a deluge of flooded areas outside of the floodplain contributes even more to housing shortages post-event for renters who compete with flood victims in search of rentals, if only as a temporary habitation during rebuilding.

While the population in Louisiana is predominantly urban, coastal rural areas have been subject to frequent destructive storms, while significant inland areas joined this challenge after the 2016 floods. Research on Katrina was severely criticized for its focus on New Orleans to the detriment of learning more about the *exceptional recovery* challenges rural areas experienced. Chapter 11 by Alessandra Jerolleman focuses on the theories of rural extreme weather risk and response with the spring 2016 flood as the case analyzed. Limited resources, distance from the state's power brokers, and possibly being asked to serve as receiver communities while under rural stresses are clearly evident in rural extreme weather challenges.

Design and planning principles explored through a resilience thinking lens can inform a science-based but socially grounded program to increase adaptive capacity, but they are not without their challenges. Chapter 12 by Traci Birch and Jeff Carney offers a review and synthesis of adopted community planning principles and processes that suggest disparate planning frameworks, and agencies are addressing physical and social environmental needs, but a more holistic approach to adaptation is needed.

## 1.5 Subnational Adaptation Management: What Each Level of Government Might Best Contribute to the Exceptional Recovery Process and Essential Resilience Outcome?

The theme of this book – a state's experiences and responses to extreme weather including that which is climate change-induced – will likely be a theme studied over and over again as more states move into the trajectory of such challenges. California is certainly a state that has a near-term robust history of extreme weather/climate experiences and their responses to them like Louisiana, a recent comparison made by Hayden and Cochran (2019). Others could also be mentioned – Texas, Florida, North Carolina, Virginia, New Jersey, and New York.

Undertaking a book about a state also gives me the opportunity to explore if/how more emphasis on state and local adaptation response/control rather than federal would be more effective. I asked specialists their opinion about the question. One replied: "The role of states in hazard mitigation planning was a hot topic up until about a decade ago, but there has been a real drop off."[9] Yet while the interest in the state role has waned, Berke's research itself showed that "federal policies do not make a difference in local land use actions, but state policy exerts a strong influence" (Berke et al. 2014). The increases in climate-induced extreme weather suggest that new research on the combinations and leadership emphases of the government levels are highly warranted. Reinforcing the critical nature of adaptation efforts, former Regional Administrator for the US Department of Housing and Urban Development (HUD) in New York and New Jersey during the region's recovery from Sandy, Irene, and Lee, Holly Leicht, stated: ". . . it is a huge financial and administrative challenge for cities and states to prepare for the ever-widening range of threats the future may hold" (Leicht 2017, p. 2). Note the emphasis on subnational despite her holding a federal agency administrative position.

While the chapter authors were not asked to consider specifically whether/if subnational adaptation would be more successful, they were asked to consider social justice issues about their topics. I ask the reader to consider whether what you learn from the chapter authors may contribute to your assessment of the role of subnational adaptation, including the benefits or not for social justice in extreme weather response. For example, Scott Hemmerling and his co-authors consider the social justice challenges of the state-level coastal restoration plan. That it is at the state level does not seem to have helped the program commit and implement social justice processes and outcomes systematically from the beginning as they have committed say for diversion sediment physical engineering modeling. One might have assumed because the state coastal restoration efforts are closer to the residents and to the communities, that might have been the case. Continuing the thinking, would a federally run coastal restoration program have done any better? Do we as a society know how to fully engage citizens and communities in the critical decision-making process related to climate change extreme weather response that honors residents and communities fully? It is an imperative that we learn how to do so: "Just recovery requires the full harnessing of communities' transformative and adaptive capacity, honoring their definitions of resilience, in order to reduce risks for the future" (Jerolleman 2019, p. 99).

[9] Personal communication with Philip Berke, Director, Institute for Sustainable Communities, Texas A & M University, March 11, 2019

### 1.5.1 Qualities of the Government Levels That Challenge/ Benefit Adaptation

How can the interaction of these three concepts – essential resilience, exceptional recovery, and level of government – be framed for future research on the topic? What has to be considered for each level and the interaction among them to be considered to answer this question? This section will reinforce the need to consider the qualities and challenges of each level of government when deciding whether the federal or the subnational level is best to lead the adaptation. Examples of pros and cons of emphasizing the various levels of government for successful climate change adaptation are offered in Table 1.2. It is not the goal of this introductory chapter to immerse the reader in the details of each of these positive or negative qualities. Rather it is the intent to demonstrate the complexity of the answer to the question: Which level? And, to add to that complexity, the question of which levels serves the most adaptations or the most important adaptations? Or the recovery trajectory, the exceptional recovery, or the utilization of the achieved essential resilience? *MUCH* more research is absolutely necessary.

### 1.5.2 Avoiding Harm While Improving Federal/State/Local Adaptation Configurations

With this deeper exploration of improving extreme weather adaption by reconfiguring the role of the levels of government comes a serious conundrum, and it is flagged by the recommendation I am making to continue to link adaptation with recovery – a *federally* overseen and funded effort – while I am asking you to think about how emphasizing state and local actions might generate more productive climate adaptation than federal. Adaptation innovations must be conceived and implementations attempted and evaluated at different levels of government *while* the current government level in charge of recovery and mitigation is utilized to respond to current disasters and develop improved adaptations. And, there is no time to delay working on both adaptations – within the current federal system *and* adaptations managed by the lower levels of government.

An example of the challenge framed in this conundrum can be seen in the extreme difficulties which the community and the state of Louisiana are having in trying to achieve just resettlement of the Isle de Jean Charles Biloxi-Chitimacha-Choctaw Tribe (Jessee, Chap. 6, this book). The funding came from the innovative Rockefeller/ HUD CDBG-DR NDRC discussed in several chapters of this book. Thus, the funds are *federal funds governed by CDBG requirements*. While the plan that was awarded the $48 million proposes that the Tribe (local) be in charge of the implementation and follow the designs the Tribe created with design/experienced construction implementers, some indigenous, chosen by them, the state has modified that plan to conform to the CDBG implementation process. The Tribe is not in charge; the

**Table 1.2** Pros and cons of different levels of government taking dominant role in climate change adaptation

| | Pros | Cons |
|---|---|---|
| Federal | Knows what practices work for mitigation/adaptation and can give guidance (Leicht 2017)<br>Can hone adaptation standards with nationwide data input and then enforcement<br>Currently where most of the taxes are collected for the country and thus the funding is located (Bullock 2016)<br>Encourages adaptation actions when there is insufficient local support for some such as required elevation for residential structures | Some states/communities may feel that federal requirements are meddling in local and state efforts (Leicht 2017). Freeboard elevation requirements are an example<br>One size fits all – but does it?<br>Mitigation (adaptation) outcomes less flexible and thus may be less useful for specific locales<br>Beginning to be overwhelmed by disaster events and costs and thus threatening to limit recovery funds (FEMA 2018a; Becker 2019)<br>Extremely slow pace of providing recovery response and showing no sign of improving the pace turn locals against federal role in exceptional recovery (Laska et al. 2018)<br>Inadequacy of the federal disaster response staff (GAO 2018; Montjoy et al. 2010). No expectation that federal adaptation management by them would be better<br>Heavy imbalance between recovery efforts for homeowners versus renters/landlords (Hersher and Benincasa 2019). Why expect otherwise for adaptation? Extreme social justice challenges<br>Funding only the most secure protection measures that are appropriate for all flood hazards has been at the expense of explaining and encouraging "less perfect," but much less expensive, efforts that can be effective for shallow, slow moving flood and drainage problems (Wetmore 2019) |
| State | Can easily seek best practices from peers (states) with similar risks (Leicht 2017)<br>Place-based realities and appropriate approaches more evident (Leicht 2017)<br>Sharing adaptation ideas among its constituent communities more proximate, within some similar conditions, more personal sharing experience (State of Louisiana 2018)<br>State is responsible for land-use regulations which will be even more important with adaptation | Possible infrequency of events occurring reduces capacity to use the opportunities for adaptation due to loss of methods in bureaucratic memory loss. Unfortunately, with the prediction of increased climate change disasters expected, this concern may wain<br>Resistance to adaptation due to perceived additional costs of building construction by developers such as resistance by the Louisiana Codes Council to require extra elevation when building residential structures (Smith and Booher 2017)<br>Limited state planning regulations nationwide to address hazard mitigation that would be supportive of climate change adaptation (American Planning Association 2018) |

(continued)

**Table 1.2** (continued)

| | Pros | Cons |
|---|---|---|
| Local | Public/private partnerships may be more doable at local level (Leicht 2017)<br>Better place-based approaches are achievable (Leicht 2017). In the context of "agency," residents can feel and further develop the connection between their knowledge, engagement, and resilience adaptation outcomes to their community's risks (Laska 1986, 1990)<br>Experiencing the climate change-induced extreme weather event with all of its specific extreme and unusual qualities provides a direct link between experience and impetus to adaptive action, overcoming psychological resistance (Meyer and Kunreuther 2017)<br>Achieving adaptation successes or even failures that induce adaptation revisions builds resident capacity and feelings of agency (Laska 1986, 1990) | Neoliberal capitalism encourages benefitting the redevelopment class not necessarily to the benefit of the community, especially poor and racial, ethnic, and Native American minorities (Brand and Baxter, Jessee, both this book)<br>Before citizen capacity to participate is grown, the citizens may not be enable to have full participation in the decision-making, and thus they can be harmed (Lamb, this book)<br>Disparate financial resources among communities may put some at extreme risk even though residents invest in considering adaptation, rural communities, for example (Jerolleman, this book) |
| Across levels | | Multiple administrative layers – at all the different levels of government – hamper efficient, effective, and timely use of disaster recovery funds (Sloan and Fowler 2015). They may hinder climate change adaptation even more due to noninstitutionalized nature of new activities<br>At both federal and state levels, better resourced states and communities and more politically powerful ones – usually co-occurring – likely will achieve most adaptation opportunities |

Note: Appreciation to Alessandra Jerolleman for contributing to refining this list. (Personal communication, May 9, 2019)

implementation process and outcomes are not what the tribe intended (Jessee, again, Chap. 6, this book). Brunner and Nordgren (2016) suggest that past adaptation successes succeeded in making incremental adaptation progress when and where they could adapt their resources to the *circumstances in a community.*

Has this been done adequately with the Tribe? Despite the federal regulations, could it have been done better, like Brunner and Nordgren (2016) propose? Adapt the adaptation resources to the *circumstances in a community* as much as is currently legally possible; and challenge the federal government to adjust their rules and regulations as climate adaptation opportunities emerge in configurations different from the actions current federal programs and rules prescribe. Could climate change innovation have been successfully implemented within the bureaucratic constraints?

Merely coining it a *resilience innovation* was totally inadequate to facilitate an innovation. This example stands as a clear example of the challenges that the country, the states, and the communities are/will have transferring from the earlier recovery model to an adaptation one that is community- and state-based.

The Tribe's cultural and interpersonal existence is being put at extreme risk because the innovation they proposed to reduce their physical and cultural risks from extreme weather, and for which the $48 million was awarded, does not fit the current federal rules and procedures and goals of the state – that being a generic model for resettlement or any CDBG program's implementation. The Tribal members and leadership are caught "dangling" between what they proposed to do – resettle from the physical coastal risk in a manner that would encourage tribal and lifeway survival – *and* what the federal and state governments are prepared to do, which are actually currently being carried out.

Extremely important, this conundrum was not recognized by the Tribe before the competition was implemented by a foundation and a federal agency; to this author's knowledge, it was not even considered adequately and without public communication of the challenge by those who put the competition together and implemented it. Innovation can seriously harm when it is not thoroughly thought through as much as possible *before* the innovation is attempted to be implemented. Careful study of climate adaptation innovation while it is being developed and during its initial/early implementation is an absolute requirement for just, equitable implementation of it. This holds for whichever level of government is the lead as well as the partners at the other levels. The Tribe's innovative plan which got the proposal selected, the $48 million awarded, was caught in a government system that could not handle the innovation.

The tension between innovation and government rules and regulations threatens the most vulnerable more because it is they who need the adaptation the most and the earliest. It is very, very likely that the Tribe will decline in maintaining its cultural practices and tribal interpersonal dynamics that they had before participating in the stressful Rockefeller/NDRC innovation application and ensuing project that has not addressed the vision and the goals the Tribe articulated in their application. We cannot accept this risk to them as the price *they* pay for the society not approaching the Tribe's climate change adaptation very, very carefully. Perhaps the likelihood of increased risk and harm to the most vulnerable should have excluded them from even seeking their resettlement through the competition? No more powerful a conundrum than that.

### *1.5.3 Speed of Recognizing Importance of Subnational Climate Change Response*

When the project of creating this case study book first began, the editor believed that the subnational response to extreme weather was not being adequately considered for climate change-induced extreme weather response rather than just for "normal" disaster recovery. I asked myself: Was there developing a statewide response? Were those responding seeing the differences between previous extreme weather events and what is occurring in the present? What challenges to adaptation were being experienced in specific Louisiana regions and community types of the state? What climate change adaptation efforts were being "birthed"? And their success? It was believed that from such a realistic case combining pre-climate change response with climate change response, recommendations would emerge for the utility or not of subnational adaptation to climate change as it becomes a more powerful driver of extreme weather. The chapter authors contributed their research on specific topics related to this: human-natural system interface, resident engagement requirements, and social justice considerations for those most vulnerable, moving of residents out of coastal risk, and resilience considerations with new climate change risk.

What was not anticipated was how rapidly the recognition of the role of the subnational response was being recognized within Louisiana and also being recognized around the nation. As discussed in Sect. 1.3.1 of this chapter, eight Louisiana adaptation programs were "birthed" during the preparation of this book. And, very clearly from media reports during the same period, subnational leaders – mayors and governors – are stepping forward to assume leadership of climate change adaptation and mitigation without being required to do so (Hersher 2018; Hirji 2019). Media reports of two such responses that were reported during late 2018 confirm this rapidly growing interest in subnational response.

The first example is the response to the release of the *Fourth National Climate Assessment*, Vol. II in late November 2018. Created by government agencies and citizens, it portended a future fraught with *rapidly increasing* climate/weather risks. The next day a media story reported that newly minted US governors recognized that the response to such a threat must include state-level action. The Associated Press headline read: *Natural Disasters Will Be a Priority for Incoming Governors* (Mulvihill 2018).

Similarly, US mayors and governors challenged President Trump's administration in their rejecting climate change by not sending an American representative to the UN Framework Convention on Climate Change (UNFCC) in the fall of 2018. Pittsburgh Mayor Bill Peduto commented:

> There are more than federal governments at stake now, and the sub-national level is really where it's going to get implemented anyway… It's really nice when nations sign documents, but what it really comes down to is *what we do in our own neighborhoods and what we do in our own cities*. (NPR 2018)

It is a critical time to consider how to accomplish the most successful essential adaptation. To that goal it is hoped that this volume about Louisiana and its response to extreme weather at the state and local levels engage other states and their government officials, residents, applied resilience research university and nonprofit researchers and practitioners and college students considering their futures to develop successful, just, equitable adaptations to climate-induced extreme weather, to achieve essential resilience. And the chapter authors and I have the same hope for more successful adaptation to essential resilience for Louisiana, for most of us are natives or "adopted natives" of the state. Finally, in emphasizing the state and local levels of response, we hope to have contributed to the very necessary body of research about which level of government is poised to best lead these adaptation initiatives most successfully. There is no time to spare in appreciating the answer(s) to this question.

## Appendix: Sources of Descriptions of New State and City Adaptation Programs (Numbers Coincide with Numbers on Table 1.1 on Page 9)

1. Coastal Protection and Restoration Authority (CPRA). (2017). Louisiana's Comprehensive Master Plan for a Sustainable Coast. Baton Rouge, LA: Coastal Protection and Restoration Authority. Retrieved from http://coastal.la.gov/wp-content/uploads/2017/04/2017-Coastal-Master-Plan_Web-Book_CFinal-with-Effective-Date-06092017.pdf
2. Louisiana Department of Administration (LDOA). (2015b). *National Disaster Resilience Competition Phase II Application*. Baton Rouge, LA: Louisiana Department of Administration.

   Louisiana Department of Administration (LDOA). (2019b). *Substantial Amendment 5: Introduction of new activities and project narrative clarifications for the utilization of community development block grant funds under the National Disaster Resilience Competition (NDRC) Resettlement of Isle de Jean Charles*. Baton Rouge, LA: Louisiana Department of Administration. Retrieved from https://www.doa.la.gov/OCDDRU/Action%20Plan%20Amendments/NDR/IDJC_Substantial_APA_5_FINAL03272019.pdf
3. City of New Orleans. (2019). Gentilly Resilience District. Retrieved from https://www.nola.gov/resilience/gentilly-resilience-district/
4. Louisiana Division of Administration. Office of Community Development. (2019a). Solution 4: Buyout & Resilient Housing Incentive. Retrieved from https://www.doa.la.gov/OCDDRU/Presentations/CDBG-BootCamp-Restore_Solution4_2019.pdf
5. Louisiana Department of Administration. Office of Community Development. (2019b). Louisiana's strategic adaptations for future environments (LA SAFE). Retrieved from https://www.doa.la.gov/OCDDRU/NDRC/LASAFE_Report_Final.pdf
6. Louisiana State University Coastal Sustainability Lab (LSU-CSS). (2017). Inland from the Coast: A multi-scalar approach to regional climate change responses. Available at https://css.lsu.edu/project/inland-from-the-coast/
7. Office of Gov. John Bel Edwards. (2018). *Louisiana watershed initiative: A long-term vision for statewide sustainability and resilience*. Retrieved from https://www.watershed.la.gov/
8. Coastal Protection and Restoration Authority (CPRA). (2018). *Flood risk and resilience program: Parish flood risk and resilience capability and capacity assessment, executive summary*. Prepared by Foster, C., Sanlee, A. & Cottone, J. Retrieved from http://coastal.la.gov/wp-content/uploads/2017/02/ParishCapabilityCapacityAssessment-9.14.18.pdf

## References

American Planning Association. (2018). Survey of state land use and natural hazards planning laws. Retrieved from https://www.planning.org/nationalcenters/hazards/statesurvey/

Bailey, C., Gramling, R., & Laska, S. B. (2017). Complexities of resilience: Adaptation and change within human communities of coastal Louisiana. In J. W. Day, G. P. Kemp, A. M. Freeman, & D. P. Muth (Eds.), *Perspectives on the restoration of the Misssissippi Delta: The once and future delta* (pp. 125–140). New York: Springer Netherlands.

Becker, I. S. (2019). "Who's going to help me?": Steve King denigrates Hurricane Katrina victims for needing government assistance. *Washington Post.* Retrieved from https://www.washingtonpost.com/nation/2019/03/22/whos-going-help-me-steve-king-denigrates-hurricane-katrina-victims-needing-government-assistance/?noredirect=on&utm_term=.8b36b8c1c828

Berke, P. R., Lyles, W., & Smith, G. (2014). Impacts of federal and state hazard mitigation policies on local land use policy. *Journal of Planning Education and Research, 34*(1), 60–76. https://doi.org/10.1177/0739456X13517004.

Brunner, R. D., & Nordgren, J. R. (2016). Climate adaptation as an evolutionary process: A white paper. In J. A. Bullock, G. D. Haddow, K. S. Haddow, & D. P. Coppola (Eds.), *Living with climate change: How communities are surviving and thriving in a changing climate* (pp. 134–145). Boca Raton: CRC Press.

Bullock, J. A. (2016). Recent actions that federal, state, and local governments are engaged in relative to the issues arising from the impacts of climate change. In J. A. Bullock, G. D. Haddow, K. S. Haddow, & D. P. Coppola (Eds.), *Living with climate change: How communities are surviving and thriving in a changing climate* (pp. 89–113). Boca Raton: CRC Press.

Burton, I., Challenger, B., Huq, S., Klein, R., & Yohe, G. (2001). Adaptation to climate change in the context of sustainable development and equity. In J. McCarthy et al. (Eds.), *Climate change 2001: Impacts, adaptation, and vulnerability* (pp. 887–890). Cambridge: Cambridge University Press. Retrieved from https://library.harvard.edu/collections/ipcc/docs/27_WGIITAR_FINAL.pdf.

Couvillion, B. R., Barras, J. A., Steyer, G. D., Sleavin, W., Fischer, M., Beck, H., et al. (2011). *Land area change in coastal Louisiana from 1932 to 2010 [map]* ((ca. 1:265,000). Scientific Investigations Map 3164). Reston, VA: U.S. Geologic Survey.

Davis, M., & Boyer, D. (2016). *Financing the future-Turning coastal restoration and protection plans into realities: How much is currently funded.* New Orleans: Tulane Institute on Water Resources Law and Policy. Retrieved from http://docs.wixstatic.com/ugd/32079b_300fb856888a4891bcd4e1f226e431d8.pdf.

Day, J. W., Boesch, D. F., Clairain, E. J., Kemp, G. P., Laska, S. B., Mitsch, W. J., et al. (2007). Restoration of the Mississippi Delta: Lessons from hurricanes Katrina and Rita. *Science, 315*(5819), 1679–1684. https://doi.org/10.1126/science.1137030.

DeLaune, R. D., & Pezeshki, S. R. (1994). The influence of subsidence and saltwater intrusion on coastal marsh stability: Louisiana Gulf coast, USA. *Journal of Coastal Research*, (12), 77–89. Retrieved from https://www.jstor.org/stable/25735591.

Federal Emergency Management Agency (FEMA). (2018a). *Disaster declarations by state/tribal government and by year.* Retrieved from https://www.fema.gov/disasters/state-tribal-government

Federal Emergency Management Agency (FEMA). (2018b). Appendix F: Community Rating System, Table 3. Community Rating System Eligible Communities. Retrieved from https://www.fema.gov/media-library-data/1538670889773-81423feb161c06426ac157a409123f3d/app-f_crs_508_oct2018.pdf

Federal Emergency Management Agency (FEMA). (2018c). *Disaster Recovery Reform Act of 2018 transforms field of emergency management* (Release # HQ-18-142). Retrieved from https://www.fema.gov/news-release/2018/10/05/disaster-recovery-reform-act-2018-transforms-field-emergency-management

Freudenburg, W., Gramling, R., Laska, S., & Erikson, K. (2009). *Catastrophe in the making: The engineering of Katrina and the disasters of tomorrow*. Washington, DC: Island Press.

Gitz, V., & Meybeck, A. (2012). *Risks, vulnerabilities and resilience in a context of climate change*. Paper presented at the Building resilience for adaptation to climate change in the agriculture sector [FAO/OECD workshop], Rome. Retrieved from https://www.researchgate.net

Hayden, A. & Cochran, S. (2019). Extreme wet weather in Louisiana and California highlights urgent need for newer, smarter strategies. [web log comment]. Retrieved from http://blogs.edf.org/growingreturns/2019/03/06/extreme-floods-louisiana-california/

Hersher, R. (2018). Mayors and governors rebut Trump Administration position at climate summit. *NPR*. Retrieved from https://www.npr.org/2018/12/12/676001283/mayors-and-governors-rebut-trump-administration-position-at-climate-summit

Hersher, R. & Benincasa, R. (2019). How federal disaster money favors the rich. *National Public Radio*. Retrieved from https://www.npr.org/2019/03/05/688786177/how-federal-disaster-money-favors-the-rich

Hirji, Z. (2019). Climate change is a top priority for the new crop of governors – even one Republican. *BuzzFeed News*. Retrieved from https://www.buzzfeednews.com/article/zahrahirji/governors-climate-change-action

Jay, A., Reidmiller, D. R., Avery, C. W., Barrie, D., DeAngelo, B. J., Dave, M., et al. (2018). Chapter 1: Overview. In D. R. Reidmiller, C. W. Avery, D. R. Easterling, K. E. Kunkel, K. L. M. Lewis, T. K. Maycock, & B. C. Stewart (Eds.), *Impacts, risks, and adaptation in the United States: Fourth National Climate Assessment, Volume II* (pp. 33–71). Washington, DC: U.S. Global Change Research Program. Retrieved from https://nca2018.globalchange.gov/chapter/1/.

Jerolleman, A. (2019). *Disaster recovery through the lens of justice*. New York: Palgrave Macmillan.

Keogh, M., & Tornqvist, T. (2019). Measuring rate of present-day relative sea-level rise in low-elevation coastal zones: A critical evaluation. *Ocean Science, 15*, 61–73. https://doi.org/10.5194/os-15-61-2019.

Kossin, J. (2018). A global slowdown of tropical-cyclone translation speed. *Nature, 558*, 104–107. https://doi.org/10.1038/s41586-018-0158-3.

Kuhlicke, C., & Steinführer, A. (2010). Social capacity building for natural hazards. A Conceptual Frame. CapHaz-Net WP1 Report. Leipzig: Helmholtz Centre for Environmental Research. Retrieved from http://citeseerx.ist.psu.edu/viewdoc/download?doi=10.1.1.476.7009&rep=rep1&type=pdf

Laska, S. (1986). Involving homeowners in flood mitigation. *Journal of the American Planning Association, 52*(4), 452–466. https://doi.org/10.1080/01944368608977119.

Laska, S. (1990). Homeowner adaptation to flooding: An application of the general hazards coping theory. *Environment and Behavior, 22*(3), 320–357. https://doi.org/10.1177/0013916590223002.

Laska, S. (2012). Dimensions of resiliency: Essential, exceptional recovery, and scale. *International Journal of Critical Infrastructure, 8*(1), 246–276. https://doi.org/10.1504/IJCIS.2012.046552.

Laska, S., Peterson, K., Rodrigue, C., Cosse, T., Philippe, R., Burchett, O., et al. (2015). "Layering" of natural and human caused disasters in the context of anticipated climate change disasters: The Coastal Louisiana experience. In Michele Companion (Ed.), *The impact of disasters on livelihoods and cultural survival: Opportunities, losses, and mitigation* (pp. 226–237). Boca Raton/New York: Taylor and Francis (CRC Press).

Laska, S., Howell, S., & Jerolleman, A. (2018). Built-in structural violence and vulnerability: A common threat to resilient disaster recovery. In M. Zakour, N. Mock, & P. Kadetz (Eds.), *Creating Katrina, rebuilding resilience: Lessons from New Orleans on vulnerability & resiliency* (pp. 99–130). Atlanta: Elsevier.

Leicht, H. (2017). *Rebuild the plane now: Recommendations for improving government's approach to disaster recovery and preparedness*. New York: Community Preservation Corporation.

Louisiana Association of Development and Planning Districts (LADPD). (2018). *About LaPDD*. Retrieved from http://www.lapdd.org

Louisiana Office of Community Development. (n.d.). Office of Community Development. Retrieved from https://www.doa.la.gov/Pages/ocd/Index.aspx

Meyer, R., & Kunreuther, H. (2017). *The ostrich paradox: Why we underprepare for disasters*. Philadelphia: Wharton Digital Press.

Molotch, H. (1976). The city as a growth machine: Toward a political economy of place. *American Journal of Sociology, 82*(2), 309–332. Retrieved from https://www.jstor.org/stable/2777096.

Montjoy, R., Farris, M., & Devalcourt, J. (2010). Achieving successful long-term recovery and safety from a catastrophe: Recommendations for public assistance. *CHART Publications*. Paper 4. Retrieved from https://scholarworks.uno.edu/chart_pubs/4

Mulvihill, G. (2018). Natural disasters will be a priority for incoming governors. *AP News*. Retrieved from https://www.apnews.com/75437c1f0aef4cd680f89ff323864fa6

National Academies of Sciences, Engineering, and Medicine (NASEM). (2018). *Understanding the long-term evolution of the coupled natural-human coastal system: The future of the U.S. Gulf Coast*. Washington, DC: The National Academies Press. https://doi.org/10.17226/25108.

National Climate Assessment. (2014). *Mitigation*. Retrieved from https://nca2014.globalchange.gov/report/response-strategies/mitigation

National Hurricane Center (NHC). (2018). *Costliest U.S. tropical cyclones tables updated*. Retrieved from https://www.nhc.noaa.gov/news/UpdatedCostliest.pdf

National Weather Service. (2012). *Inundation mapping* (Version 2.0). Retrieved from https://water.weather.gov/ahps/inundation_mapping_user_guide.pdf

Office of Gov. John Bel Edwards. (2018). *Louisiana watershed initiative: A long-term vision for statewide sustainability and resilience*. Retrieved from https://www.watershed.la.gov/

Owens, M. (2015). Louisiana's traditional cultures: An overview. *Folklife in Louisiana*. Retrieved from http://www.louisianafolklife.org/LT/Maidas_Essay/main_introduction_onepage.html#tab4

Pahl, J. (2016). *2017 Coastal Master Plan: Attachment C-2: Eustatic Sea Level Rise*. Version I. (p. 23). Baton Rouge: Coastal Protection and Restoration Authority. Retrieved from http://coastal.la.gov/wp-content/uploads/2016/04/Attachment-C2-1-Eustatic-Sea-Level-Rise_October-2015.pdf

Parris, A. S., Bromirski, P., Burkett, V., Cayan, D. R., Culver, M. E., Hall, J., et al. (2012). *Global sea level rise scenarios for the United States National Climate Assessment* (NOAA Technical Report OAR CPO-1). Silver Spring: Climate Program Office. Retrieved from https://scenarios.globalchange.gov/sites/default/files/NOAA_SLR_r3_0.pdf

Peterson, K., Laska, S., Philippe, R., Porter, O., Krajeski, R., Steinberg, S., et al. (2016). Refining the process of science support for communities around extreme weather events and climate impacts. In S. L. Steinberg & W. A. Sprigg (Eds.), *Extreme weather, health and communities* (pp. 135–164). New York: Springer.

Roberts, H. H. (1997). Dynamic changes of the Holocene Mississippi River delta plain: The delta cycle. *Journal of Coastal Research*, 605–627. Retrieved from https://www.jstor.org/stable/4298659.

Rockefeller Foundation. (n.d.). National disaster resilience competition. Retrieved from https://www.rockefellerfoundation.org/our-work/initiatives/national-disaster-resilience-competition/

Rovere, A., Stocci, P., & Vacchi, M. (2016). Eustatic and relative sea level changes. *Current Climate Change Reports, 2*(4), 221–231. https://doi.org/10.1007/s40641-016-0045-7.

Schleifstein, M. (2019). Mississippi rising again, could cause 2nd spillway opening. *Nola.com*. Retrieved from https://www.nola.com/environment/2019/04/mississippi-rising-again-could-cause-2nd-spillway-opening.html

Sloan, M., & Fowler, D. (2015). *Lessons from Texas: 10 years of disaster recovery examined*. White Paper. Austin: Texas Appleseed. Retrieved from https://www.texasappleseed.org

Smith, C., & Booher, W. (2017). Guest column: When it comes to Louisiana floods, how much does a foot really matter? *The Advocate*. Retrieved from https://www.theadvocate.com/baton_rouge/opinion/article_e1a4f340-f2dd-11e6-b7a9-138aa486bd71.html

State of Louisiana. (2018). *Louisiana watershed-based floodplain management coordination*. Executive Order # JBE 2018-6. http://gov.louisiana.gov/assets/ExecutiveOrders/JBE%2D%2D18-16-Watershed-Council.pdf

Stott, P. (2016). How climate change affects extreme weather events. *Science, 352*(6293), 1517–1518. https://doi.org/10.1126/science.aaf7271.

(The) Pew Charitable Trusts. (2018a). *What we don't know about state spending on natural disasters could cost us: Data limitations, their implications for policymaking, and strategies for improvement.* Retrieved from https://www.pewtrusts.org/en/research-and-analysis/reports/2018/06/19/what-we-dont-know-about-state-spending-on-natural-disasters-could-cost-us

(The) Pew Charitable Trusts. (2018b). *Natural disaster mitigation spending not comprehensively tracked.* [issue brief]. Retrieved from https://www.pewtrusts.org/en/research-and-analysis/issue-briefs/2018/09/natural-disaster-mitigation-spending%2D%2Dnot-comprehensively-tracked

Tierney, K. (2014). 15. Hazards and Disasters. In *concise encyclopedia of comparative sociology* (pp. 427–436). BRILL.

Turner, R. E., & McClenachan, G. (2018). Reversing wetland death from 35,000 cuts: Opportunities to restore Louisiana's dredged canals. *PloS one, 13*(12), e0207717. https://doi.org/10.1371/journal.pone.0207717.

U.S. Government Accountability Office (GAO). (2018). 2017 *Hurricanes and wildfires: Initial observations on the federal response and key recovery challenges,* GAO-18-472. Retrieved from https://www.gao.gov/products/GAO-18-472

Union of Concerned Scientists. (2017a). *Hurricanes and climate change.* Retrieved from https://www.ucsusa.org/global-warming/science-and-impacts/impacts/hurricanes-and-climate-change.html#.XCpKNy2ZOi4

Union of Concerned Scientists. (2017b). Louisiana faces chronic inundation [fact sheet]. Retrieved from https://www.ucsusa.org/sites/default/files/attach/2017/07/when-rising-seas-hit-home-louisiana-fact-sheet.pdf

Wetmore, F. (2019). Training workshop: Reducing flood risk through nonstuctural floodproofing. Presented at annual meeting of the Association of State Floodplain Managers, Cleveland. Retrieved from https://asfpmconference.org/2019/conference-program/full-conference-program

Wiel, K. V. D., Kapnick, S. B., Oldenborgh, G. J. V., Whan, K., Philip, S., Vecchi, G. A., et al. (2017). Rapid attribution of the August 2016 flood-inducing extreme precipitation in south Louisiana to climate change. *Hydrology and Earth System Sciences, 21*(2), 897–899. https://doi.org/10.5194/hess-21-897-2017.

Wisner, B., Blaikie, P., Cannon, T., & Davis, I. (2004). *At risk: Natural hazards, people's vulnerability and disasters* (2nd ed.). London/New York: Routledge, Taylor and Francis Group.

# Part I
# Louisiana's Risks Anticipating the Future Challenges to Other U.S. Coastal Communities

# Chapter 2
# Managing Risks in Louisiana's Rapidly Changing Coastal Zone

**Donald F. Boesch**

## 2.1 Introduction

While both strategically important to the nation and bountiful in so many ways, Louisiana's coastal zone has always been difficult to access and risky to live in. Much of its landscape consists of wetlands: bottomland forests, swamps, marshes, and mangroves that are continuously, seasonally, tidally, or meteorologically inundated. Most of what passes for dry land is just a few feet above sea level and subject to episodic flooding from the mighty rivers – the Mississippi and the Atchafalaya – that flow through it, locally intense rainfall, and ocean storm surges. Powerful tropical storm winds and associated tornadoes pose additional weather threats to human communities and the built environment.

The complex and dynamic water world that is coastal Louisiana constrains where people live and how they move across the landscape. Early European settlers were confronted by devastating river floods almost immediately after their arrival, and, despite the flood protection systems and elevated infrastructure that were developed over the next 300 years, the threats of rising waters and damaging winds have remained a fact of life for south Louisiana communities and enterprises. Both have moved and adapted in response to extreme weather events in ways that have decreased, but sometimes increased, their vulnerability.

While extreme weather events challenge social resilience, i.e., the ability of communities to cope with and adapt to stresses and disruptions, these transient phenomena are experienced against a background of powerful secular (in the sense of long duration) trends that further test this resilience. Particularly since the mid-twentieth century, the coastal landscapes have been rapidly deteriorating as a net result of geological subsidence, human interference with the processes that build and sustain

D. F. Boesch (✉)
University of Maryland Center for Environmental Science, Annapolis, MD, USA
e-mail: boesch@umces.edu



S. Laska (ed.), *Louisiana's Response to Extreme Weather*,
Extreme Weather and Society, https://doi.org/10.1007/978-3-030-27205-0_2

the coastal landscape, and substantial modification of its hydrology. Moreover, the body of science has made it abundantly clear that human activities are warming Earth's atmosphere and oceans and changing its climate in ways that are enormously consequential for south Louisiana, including accelerated sea-level rise, intensification of precipitation, and more powerful tropical cyclones.

This chapter sets the biophysical stage for the case studies and perspectives on social resilience that follow in this volume. First, I provide an overview of the geomorphic fabric of coastal Louisiana, how it affects human society, and how humans have modified it. I then summarize the kinds of flooding threats, the notable disasters that have occurred, and the flood protection systems that have been created. From there I move to the strategic coastal protection and restoration that is being planned and implemented in Louisiana, before considering global climate change as a threat multiplier that will also have to be addressed. Finally, I conclude with some perspectives on the implications of the rapidly changing coastal landscape for social resilience within these other coastal regions of the United States.

## 2.2 Geological and Human Development

### *2.2.1 Creation and Evolution of Coastal Landscapes*

The people of south Louisiana live on the youngest land in the United States, except for a few small purchases built on barrier islands or filled shallows. As the massive glaciers rapidly melted at the end of the last ice age about 20,000 years ago, the level of the world's oceans rose by about 120 m (400 feet) over 12,000 years (Stanford et al. 2010). Large areas of coastal land were submerged becoming continental shelves, and shorelines retreated until sea level reached a relatively stable point about 7000 years before the present. The level of the world's oceans was nearly constant or slowly declining during the period of European settlement of North America (Kemp et al. 2011). Today, residents of most US coastal areas today live along those same shorelines. But in Louisiana the escarpments marking those 7000-year-old shorelines are now far inland from the Gulf Coast, north of Lake Pontchartrain, and just below Baton Rouge and Lafayette (Saucier 1994).

When the rapid rise in sea level finally slowed, a large marine embayment stood between Baton Rouge and Lafayette into which the Mississippi and other great rivers flowed. With the inland march of the sea finally stalled, sediments discharged by these rivers began to fill up the embayment and then reclaim the shallow Gulf of Mexico by protruding successive delta lobes (Blum and Roberts 2012; Bentley et al. 2016). As a delta lobe grew through the deposition of river-borne sediments, it created branching distributaries, some of which left remnants as today's bayous. Sediments were deposited at river mouths and via overbank flooding, crevasse formation, and infilling of older distributaries (Roberts 1997). As flow gradients diminished, the river sought a quicker path to the sea, breaking out to begin a new delta

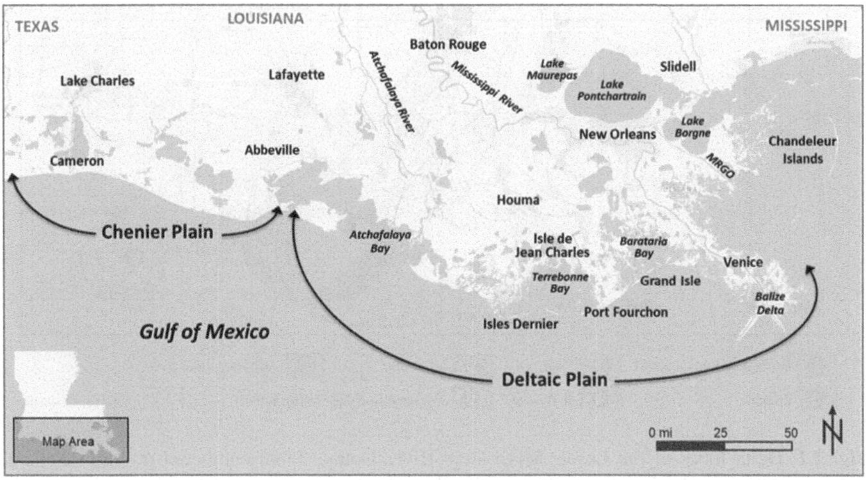


**Fig. 2.1** Map of southern Louisiana showing important cities, water bodies, and geologic provinces. (Base map courtesy of Esri, HERE, Garmin, © OpenStreetMap contributors, and the GIS user community)

lobe. The river's flow did not switch all at once, and the flow was often conveyed down both the old and the new delta.

Eventually, five or six major deltas – depending on how they are distinguished – were formed over the past 4600 years (Roberts 1997; Bentley et al. 2016) with their remnants constituting the landscapes of the Mississippi Deltaic Plain (Fig. 2.1) from Abbeville in the west to the border of the state of Mississippi in the east. The easternmost St. Bernard Delta was active between 2800 and 1000 years ago, extending beyond today's Chandeleur Islands and enclosing large coastal embayments, creating today's lakes Maurepas, Pontchartrain, and Borgne (Fig. 2.2). The earlier Teche Delta (3500–2800 years before present) and Lafourche Delta (1000–300 years ago) filled in the landscapes between the present Atchafalaya and Mississippi rivers. The towns lying along today's bayous Teche and Lafourche sit on natural levee deposits of the past main channels of the great river. The presently active Plaquemines Delta below New Orleans is only 750 years old, and its iconic extension to the edge of the continental shelf in the form of a bird's foot (the Balize Delta) has only existed for about 550 years or since shortly before Columbus discovered America.

A new delta complex began to emerge in Atchafalaya Bay with the 1973 flood (Roberts et al. 2003), more than 20 years after the Atchafalaya River had captured more than 30% of the flow of the Mississippi and Red rivers and its vast swamp basin filled with sediments (Piazza 2014). With the flow since 1963 regulated under law at 30% of water of the lower Mississippi, two delta lobes have been building in the Atchafalaya Bay along central coastal Louisiana.

As the Mississippi river deltas switched back and forth to build southeastern Louisiana, sediments discharged into the Gulf or released from eroding shorelines drifted to the west along the coast under the influence of currents and waves.

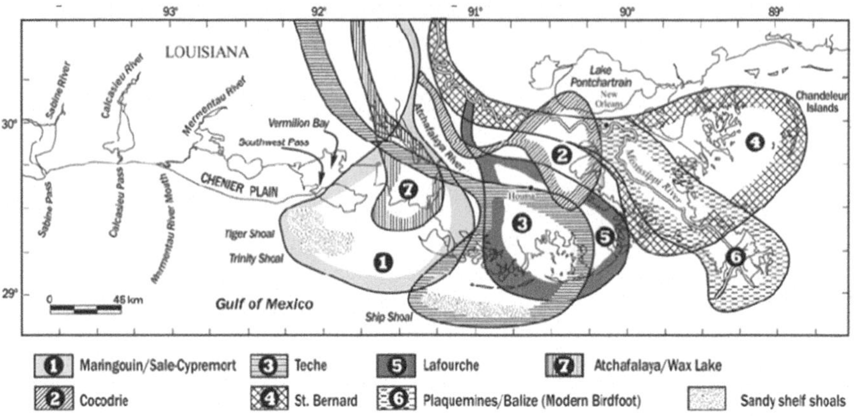


**Fig. 2.2** Delta lobes of the Lower Mississippi River Deltaic Plain numbered in chronological order of formation. (Image source: McBride et al. 2007)

This also resulted in the development of new land in the form of a strandplain characterized by old sandy or shelly beach ridges running parallel to the coast and separated by marshes and swamps (Penland and Suter 1989; Bentley et al. 2016). This Chenier Plain (Fig. 2.1), referring to the oak (chêne in French) trees growing on the ridges, extends for 200 miles along the Louisiana coast from Vermillion Bay to Galveston, Texas. Throughout much of southwestern Louisiana, the Chenier Plain extends inland about 30 miles from the coast.

Once deprived of the river sediments that nourish them, the landforms of both the Mississippi Deltaic Plain and Chenier Plain deteriorate under the effects of geological subsidence caused by compaction of the accumulated sediments and the exposure to forces of the waves, tides, and surges of the Gulf of Mexico. The outer edge of the delta erodes, and the sand deposits remaining are reshaped as flanking barrier islands and the inter-distributary wetlands open up as estuarine bays, such as present-day Barataria and Terrebonne bays (Blum and Roberts 2012). Still, tidal wetlands are sustained for centuries by trapping eroding sediments and growing upward as the soil beneath them subsides (Reed 1989). The skeletal framework of distributary ridges and barrier islands protect interior wetlands from marine forces and saltwater intrusion (Salinas et al. 1986). Coastal ecosystems, consisting of tidal wetlands and channels and shallow bays, are enormously productive of fish and wildlife during this phase (Twilley et al. 2016). Eventually, the barrier island arc becomes detached from land by a broad sound, such as is the present condition for the Chandeleur Islands to the east (Fig. 2.1). Finally, all that remains of the barrier islands are submerged sandy shoals located miles offshore on the inner continental shelf of the Gulf of Mexico. Subsiding and eroding, the deterioration of landforms is exorable until a subsequent delta revisits the area.

The Chenier Plain also underwent periods of land building, when the river delta had moved toward the west, and then erosion, when the delta shifted farther away toward the east (Penland and Suter 1989; Bentley et al. 2016). The beach ridges consisting of coarser sediments were formed during these erosional periods. Deprived of delta sediments, wetlands in the Chenier Plain are currently subsiding at a much faster rate that they are able to vertically accrete soils (Jankowski et al. 2017). In contrast, sediment supplies to the Deltaic Plain wetlands allow them to accrete more soil.

Since human habitation, the expansive coastal zone of Louisiana has always been young, low lying, wet, and highly dynamic, thus posing challenges to human survival, health, prosperity, and social fabric.

### 2.2.2 Human Settlement and Its Risks

Native Americans first occupied the dynamic Mississippi Deltaic Plain about 2000 years ago (McIntire 1958). They left remnants of their occupation in the form of shell middens and earthen mounds located near river channels or distributaries or on barrier ridges. The mounds accommodated their refuge during occasional river and estuarine flooding, providing the community resilience required for living in this bountiful but challenging wet landscape.

Although the establishment of the outpost of Natchitoches preceded it by 4 years, the site of New Orleans was selected for the first French settlement in south Louisiana in 1718 because it controlled the lower Mississippi River and also afforded access via Bayou St. John to Lake Pontchartrain (Colton 2005). In making this decision, Sieur de Bienville was well aware of the frequent risks of river flooding, but, as geographer Peirce Lewis noted, New Orleans was the "inevitable city" in the "impossible" site. The early city was built on the natural levees of the Mississippi River that rose no more than 12 feet above sea level. The colonists did not have to wait long as floods the next spring slowed construction (Campanella 2008). Then, in September 1722, hurricane winds knocked down shoddily built structures, wiping the haphazard slate clean for laying out the street grid that exists in the Vieux Carré today.

Also that year, construction of the first artificial levees to protect from river floods began. Still, frequent floods inundated farms that were spreading along the banks of the river above New Orleans, destroying crops and damaging homes. Moreover, floodwaters reaching the backswamps beyond the natural levees cause backwater flooding of relatively developed areas otherwise protected by river levees. Colonial laws in 1728 and 1743 required landowners to build and maintain levees along their properties fronting the river. By 1763 these stretched 50 miles above the city (Colton 2005). By the time Louisiana became a state in 1812, artificial levees extended from as far north as the Red River to below New Orleans along the west bank and from Baton Rouge to below New Orleans on the east bank. Still, there were occasional urban inundations during the late eighteenth and early nineteenth century due to breaches in the levees fronting the city or its suburbs or resulting from crevasses farther upriver that filled the backswamp and inundated the city from the rear. The most notable example was the 1849 crevasse at Sauvé Plantation that displaced 12,000 of New Orleans' 116,000 residents, the city's worst flood until Hurricane Katrina in 2005 (Campanella 2008).

Nonetheless, the increasing effectiveness of artificial levees along the lower river provided security that allowed expanded development of New Orleans and across the river along the west bank. Paradoxically, it also elevated the threat of river flooding by reducing outlets for floodwaters either over the levees or through natural channels, thus raising the stage of the river for a given flow rate. This realization initiated a nearly century-long debate over whether flood protection should continue to rely on a levee-only strategy or also incorporate floodways to lower the river levels (Barry 1997).

This debate came to a head following the Great Mississippi River Flood of 1927 that inundated 26,000 square miles from Cairo, Illinois, to the Gulf, displacing a half-million people and threatening New Orleans (Barry 1997). The Flood Control Act of 1928 shifted policies from levees- only to include not only massive levees and floodwalls but also control structures and spillways, all under the responsibility of the federal government. Today, high stages in the lower Mississippi are constrained by opening the Bonnet Carré Spillway, sending water to Lake Pontchartrain, or the West Atchafalaya or Morganza floodways, sending water down the Atchafalaya Basin.

As human settlements expanded from along the Mississippi River, across the Atchafalaya Basin to the land of the Attakapas in southwestern Louisiana, and down the bayous of the Mississippi Deltaic Plain, occasional river floods also threatened them. Settlements along Bayou Teche were often flooded, particularly during the 1927 flood (Bernard 2016). Bayou Lafourche carried a portion of the Mississippi flow until it was dammed in 1904. However, there are only modest, if any, artificial levees along these waterways; flooding has been mitigated through various flow control structures.

As development began to extend into the backswamps, canals and levees were constructed to facilitate drainage. Eventually, this required the removal of rainwater by perpetually operated pumps. The dewatering of the highly organic soils of these former swamps resulted in the loss of soil volume due to oxidation and enhanced subsidence (Colten 2005; Dixon et al. 2006; Campanella 2008). Consequently, much of the inhabited area of New Orleans and its suburban parishes lies below sea level, although that land was originally at or slightly above sea level when development began. Similar loss in elevation occurred where there were failed attempts to drain wetlands for conversion to agricultural polders. The resulting urban and agricultural bowls became more susceptible to rainfall-driven flooding and reliant on large-capacity pumps that can keep up with heavy rainfall.

Even before wetland drainage and development, bald cypress and other swamp and bottomland trees were mostly cut down for timber. The loss of tree cover, coupled with drainage and navigation canals (such as the Carondelet and New Basin canals through which commodities were transported into New Orleans), increased the susceptibility of urban areas to winds, tidal incursions, and storm surges. Many of these older canals were filled in or fitted with gates to reduce the risk of flooding resulting from tidal and storm surges; however, massive navigational channels were constructed perpendicular to the coast during the latter half of the twentieth century

(Gulf Intracoastal Waterway, Mississippi River-Gulf Outlet or MRGO, Houma Navigation Canal, and Calcasieu Ship Channel to Lake Charles). They have hastened saltwater intrusion and the resulting loss of cypress swamps and facilitated propagation of tropical storm surges toward population centers distant from the coast (Freudenburg et al. 2009b).

### 2.2.3 *Broader Coastal Deterioration*

The area of land, including wetlands, in the coastal zone of Louisiana more or less continuously expanded after sea level stabilized about 7000 years ago. Surely, abandoned delta lobes subsided and eroded, but new lands created in newly active delta lobes countered the resulting losses. The Chenier Plain lost ground when eastern delta lobes were most active but gained ground when the river switched it course to the west. The multi-millennial trend in slow net land gain was dramatically reversed during the twentieth century, with best estimates of land losses during the late 1970s of 32 square miles per year (83 $km^2/y$), now slowed to 11 square miles per year (28 $km^2/y$). Altogether, over 2000 square miles of land were lost between 1932 and 2016 (Couvillion et al. 2017).

Changes in the Mississippi-Atchafalaya River Basin are responsible for some of the losses. The present Balize Delta is perched on the edge of the continental shelf and deposits much of its terminal load of alluvial sediments into deep waters of the Gulf of Mexico, bypassing the coastal zone where these sediments could be held in wetlands and on shorelines. Erosion associated with land clearing within the Basin during European expansion increased the river's sediment load during the nineteenth century, but then dams constructed throughout the catchment by the middle of the twentieth century trapped sediments upstream. That, coupled with improved soil conservation practices, has resulted in a reduction by half of the suspended sediment of the lower Mississippi since the 1950s (Meade and Moody 2010; Heimann et al. 2011) to loads probably less than those occurring when major delta lobes were being built (Chamberlain et al. 2018). More of the combined river flow began to travel down through the Atchafalaya Basin after Henry Shreve cleared the Great Raft of logs clogging the Red and Atchafalaya rivers in the 1830s. This extensive basin trapped a large share of the riverine sediments transported such that a new delta did not begin to emerge in Atchafalaya Bay until 1973 (Piazza 2014).

Additionally, constraining the flow of the lower Mississippi with its channel by effective flood protection levees and closure of distributary channels almost all the way to its mouth have prevented the broad contribution of riverine sediments to the subsiding wetlands and shallow waters. Indeed, this was foreseen back as far as 1897, when an article on the Mississippi River Delta published in the *National Geographic* (Corthell 1897) stated: "no doubt the great benefit to the present and two or three following generations accruing from a complete system of absolutely protective levees excluding the flood waters entirely from the great areas of the lower delta country, far outweighs the disadvantages to future generations from the subsidence of the Gulf delta lands below the level of the sea and their gradual abandonment due to this cause." Unfortunately, we have already reached the fourth generation without a Plan B.

In addition to changes in the supply and distribution of sediment subsidies required to sustain the coastal plain landscapes, other human activities have resulted in land, and particularly wetland, losses. These include the kind of wetland "reclamation" and dredge and fill activities that caused wetland losses elsewhere, but particularly notable in Louisiana have been the extensive dredging of canals through the coastal wetlands. This includes not only the larger canals constructed for commercial or industrial navigation mentioned earlier but also myriad smaller canals mainly dredged for access to drilling sites and laying pipelines associated with oil and gas production. Dredged canals were seldom backfilled and generally do not fill in naturally by themselves. The spoil banks left interfere with the tidal water-level fluctuations needed for healthy, accreting wetland soils. The wetland losses associated with these indirect hydrological effects may be several times greater than the direct dredge and fill effects, potentially accounting for most of the observed wetland loss (Turner 1997), although this has been questioned (Day et al. 2000). Independent estimates suggest that the net effect of oil and gas canals has been responsible for at least 30% and possibly 50% of the wetland losses during the second half of the twentieth century (Penland et al. 1996). Needless to say, these estimates were strongly contested by the oil and gas industry, and the industry's responsibility has been caught up in political debates and judicial cases concerning liability for the costs of addressing the coastal wetland crisis.

Scientific evidence is also compelling that withdrawals of oil, gas, and associated briny water have increased subsidence rates and thus wetland loss rates in the vicinity of shallow oil and gas fields, such as those in Terrebonne Parish (Morton et al. 2006). The slowdown in fluid withdrawals from these old fields may be the principal cause of the reduction in the rate of subsidence as evidenced in the Grand Isle tide gauge record (Kolker et al. 2011). Similarly, the substantial reduction in new oil and gas canal dredging may have contributed to the lower rates of coastal wetland loss in recent decades (Couvillion et al. 2017).

In aggregate, the multiple consequences of human activities have resulted in deltaic deterioration over less than a century that would take a millennium due to natural processes, such as subsidence, delta lobe abandonment, and erosion due to winds and hurricanes. After the scale and rapidity of coastal wetland loss became apparent in the early 1980s, a succession of plans and programs were developed to slow, if not reverse, the losses. The primary motivation was the restoration of the unique coastal environments and the important natural resources they produce. Protection of coastal communities from flood risks proceeded on a separate, and sometimes competitive, or even antagonistic, track. The disastrous effects of Hurricane Katrina and Rita in 2005 made it clear that deterioration of coastal environments had increased storm surge risks and threatened the very existence of many coastal communities. This realization has required a more integrated and simultaneous approach to planning and implementation of the protection of society and restoration of the environment (Day et al. 2007). Projections of future land losses (Fig. 2.3) and increased flood risks as coastal landscapes continue to degrade (Fig. 2.4) have prompted the integrated planning discussed in Sect. 2.6.

**Fig. 2.3** Predicted land change by 2067 along the Louisiana coast. (Land change data retrieved from the Coastal Restoration & Protection Authority [CPRA]; base map courtesy of Esri, HERE, Garmin, © OpenStreetMap contributors, and the GIS user community)

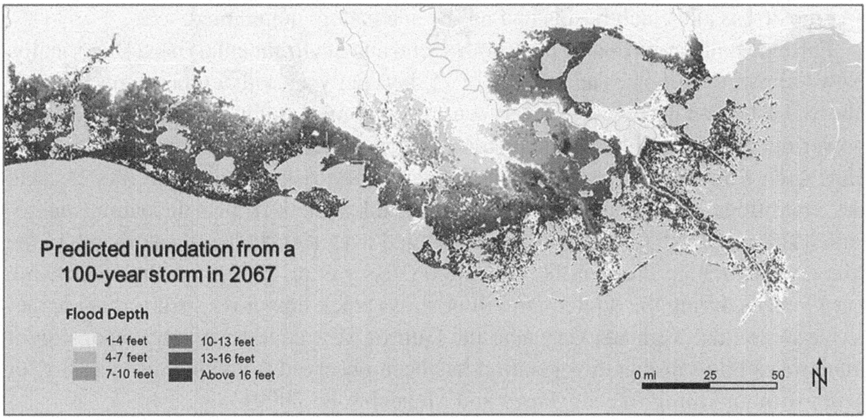


**Fig. 2.4** Predicted inundation depths along the Louisiana coast resulting from a 100-year storm in 2067. (Flood depth data retrieved from CPRA Master Plan Data Viewer)

## 2.3 Extreme Weather Risks

### *2.3.1 South Louisiana's Climate*

Beyond the risks of flooding from the Mississippi and Atchafalaya rivers, there are extreme weather risks associated with coastal Louisiana's climate. South Louisiana has a humid subtropical climate in large part due to the influence of the warm Gulf of Mexico. It has long, hot, humid summers and short, mild winters. Average annual rainfall increases from west to east, from 57 inches (145 cm) in Lake Charles to 64 inches (163 cm) in New Orleans. Rainfall is prevalent during all months, with somewhat higher precipitation in the summer and winter. In summer the prevailing southerly winds provide moist, subtropical weather often favorable for afternoon thunderstorms, sometimes resulting in flooding risks caused by extreme rainfall.

Typically, the most extreme rainfall (as much as 20 inches in a day) has been associated with tropical storms. Even greater rainfall amounts (40–60 inches) occurred in the Houston area when Hurricane Harvey stalled offshore in late summer of 2017 (van Oldenborgh et al. 2017). A similarly stalled depression dumped up to 31 inches of rain in the Amite and Comite river basins near Baton Rouge just a year earlier (van der Wiel et al. 2017). The devastation of these two flood events acted as a wake-up call that, in addition to river floods and tropical storm surges, Gulf Coast communities might be increasingly vulnerable to more extreme rainfall events caused by global warming. The connection with climate change is discussed later but has called into question the adequacy of existing floodplain management and drainage infrastructure for present and future conditions of extreme precipitation. Areas under forced drainage are particularly vulnerable. In August 2017 almost 10 inches of rain resulted in extensive flooding, damage, and inconvenience in New Orleans, which worsened because some of the city's drainage pumps were offline and the drains and catch basins had not be adequately maintained.

Extreme temperatures also pose both social and environmental risks. Historically, New Orleans experiences an average of 75 days per year with temperatures 90 °F or above. Prolonged heat waves or very warm and humid conditions that coincide with power outages caused by tropical storms, such as happened in New Orleans after Hurricane Katrina, pose very serious human health risks. Periods of very hot and dry conditions have been associated with sudden dieback of salt marsh, the so-called Brown Marsh phenomenon that affected over 100,000 ha of salt marsh in the Mississippi Deltaic Plain in the year 2000 (Visser et al. 2002). On the other hand, hard freezes during the winter can kill or stress black mangrove shrubs that characterize some tidal wetlands very near the Gulf of Mexico. Conversely, expansion of mangroves into salt marsh vegetation has been observed following a succession of years without killing freezes (Perry and Mendelssohn 2009).

### *2.3.2 Tropical Cyclones*

Of course, coastal Louisiana is notoriously at risk from the storm surges and damaging winds of tropical cyclones, including depressions, tropical storms, and hurricanes. As was mentioned earlier, the first residents of New Orleans were introduced to the ferocity of a hurricane just 4 years after the city's founding. Over the period of record, an average of about one tropical storm or hurricane per year met landfall along the Louisiana coast (Roth 2010), but there have been periods where there are none (recently 2014, 2015 and 2016) and other years where have been two or more in a year. The occurrence of two powerful storms each in 2005 (Katrina and Rita) and in 2008 (Gustav and Ike) is etched in the memory of many south Louisiana residents.

A strong hurricane hit New Orleans in 1837. While it flooded marshes adjacent to Lake Pontchartrain, the city itself was buffered from the storm surge because of the largely intact marshes and swamps separating it from the lake, except around the two navigation canal basins (Campanella 2008). Another notable hurricane struck the Isles Derniers in Terrebonne Parish in 1856, killing more than 218 vacationers enjoying the relief of beach breezes without any warning of the approaching storm (Dixon 2009). Another hurricane in 1893 killed more than 2000 residents of Cheniere Caminada, between Grand Isle and Port Fourchon. Survivors abandoned that settlement, moving north to other communities farther up Bayou Lafourche (Brasseaux and Davis 2017).

In 1947 a late summer hurricane struck New Orleans with over 100-mile-per-hour winds, pushing modest storm surges inundating outlying areas to the east and in Jefferson Parish (Roth 2010). In response to this storm and one the following year, there was additional levee construction along the Lake Pontchartrain shore and adjacent marshes. In June 1957, Hurricane Audrey came ashore near the Sabine Pass, creating a 12-foot storm surge that destroyed the town of Cameron, causing damage 25 miles inland and killing 526 people in Louisiana alone.

In 1965, Hurricane Betsy had its landfall at Grand Isle with 160-mile-per-hour winds. Facilitated by the Gulf Intracoastal Waterway and the recently completed Mississippi River Gulf Outlet, its large storm surge reached Lake Pontchartrain and breached floodwalls to inundate much of the Gentilly, the Ninth Ward of New Orleans, and the neighboring suburbs in St. Bernard Parish. In response, Congress enacted the Flood Control Act of 1965 that put the federal government in the business of storm protection by raising and constructing levees and strengthening floodwalls to provide Category 3-level storm protection (Campanella 2008). Now protected, areas of New Orleans East subsequently experienced an explosive growth in residences and businesses, in a "levee effect" that paradoxically increases future damages by luring homebuyers into floodplains (Freudenburg et al. 2009a). Despite the protection by levees, the newly developed areas were not protected adequately from interior flooding due to poorly designed drainage (Baxter 2014).

Hurricane Katrina in 2005 had effects that in many ways mirrored those of Betsy, with a massive storm surge on the east side of the river assisted by the navigation canals and meeting little resistance from the by now nearly nonexistent cypresses swamps and deteriorated marshes. Post-Betsy levees in St. Bernard Parish and New Orleans East were overtopped, and floodwalls failed along the Inner Harbor Navigation Canal and the ungated drainage outfall canals penetrating into the city. This inundated not only the Ninth Ward, including post-Betsy developments in New Orleans East, and St. Bernard Parish but also the 80% of the city beyond the high ground along the Mississippi River (McQuaid and Schleifstein 2006). Because of the extent, persistence and devastation of the saltwater flooding and loss of power and other services, most New Orleans residents had to relocate away from the city. Many never returned. Altogether, 1836 people died directly as a result of Hurricane Katrina (Bevan et al. 2008), 1577 of them in Louisiana, and Katrina's total property losses have been estimated at $125 billion (Vigdor 2008).

There is a very voluminous literature on the events, effects, causes, responses, and lingering impacts of the Hurricane Katrina disaster. To the audience of this volume, I recommend books by the veteran reporters McQuaid and Schleifstein (2006) and the deeply experienced social scientists Freudenburg, Grambling, Laska, and Erikson (2009a). Both books emphasize that the disaster was as much human-caused as natural.

Less than a month after Katrina in 2005, a second highly powerful storm struck coastal Louisiana when Hurricane Rita came ashore near the Texas border. It caused major damage to communities in Cameron Parish and elsewhere along the south-west Louisiana coast, damaged freshwater wetlands in the Chenier Plain by inundating them with saltwater, and resulted in storm surge felt along the entire Louisiana coast. Some areas affected by Katrina were flooded again.

During September 2008 Hurricane Gustav came ashore in Terrebonne Parish, and Hurricane Ike had its landfall near the mouth of Galveston Bay just 2 weeks later, flooding and re-flooding many coastal Louisiana communities from Cameron to Plaquemines parishes. Two million people evacuated from south Louisiana in advance of Gustav's arrival, with its storm surge even splashing over newly installed floodwalls in eastern New Orleans.

During the decade of 2000s, Louisiana experienced the effects of a record number of tropical cyclones, including six hurricanes and six tropical storms. These disasters, particularly the Hurricane Katrina disaster, prompted national and regional responses to strengthen storm surge protection and to integrate protection with the rehabilitation of the degrading landscape. These responses are reviewed in the next two sections, starting first with the congenital Louisiana challenge of flood protection.

## 2.4 Flood Protection and Its Limits

### *2.4.1 Mississippi and Atchafalaya Rivers*

The lower Mississippi River flood protection system developed after the Great Mississippi River Flood of 1927 has remained secure and effective despite some challenges. The biggest test came during the 1973 flood when Old River Control Structure was very close to failing when a scour hole developed under the Low Sill structure, causing part of the structure to collapse.

That year the Corps of Engineers opened the nearby Morganza Floodway for the first time since its construction in 1954, and up to 300,000 cubic feet per second (8500 $m^3/s$) of flow was diverted down the Atchafalaya Basin to reduce the flood risks for Baton Rouge and New Orleans. The Morganza Floodway was not opened again until 2011, when up to 173,000 cubic feet per second (4900 $m^3/s$) of flow was diverted. Opening the Morganza Floodway was also seriously considered in 2017. The Corps has had to open the Bonnet Carré Spillway more frequently after Hurricane Katrina than was typical since it was built in 1934: in 2008, 2011, 2016, 2018, and 2019 (twice).

Whether more extreme Mississippi River flows will be experienced with the changing climate remains to be seen, but multiple lines of evidence indicate that artificial channelization upstream has been the predominant cause of the amplification of flood magnitudes over the past century (Munoz et al. 2018). As the Plaquemines-Balize Delta rapidly subsides at rates exceeding 1 cm per year and the level of the Gulf rises, the elevation gradient of the river decreases, slowing flows and inducing sedimentation that further constrains the channel cross-section (Blum and Roberts 2009; Little and Biedenharn 2014). Conversely, because of the diminished elevation gradient, higher storm surges from the Gulf can propagate farther upstream.

During Hurricane Katrina, storm surges overtopped not only the levees intended to protect lower riverside communities in Plaquemines Parish all the way to Venice from hurricane storm surges but also the taller levees protecting from river flooding. With continued subsidence and accelerating sea-level rise, the ability to protect these lower river communities will diminish. The iconic bird-foot distributary system that has characterized the mouth of the Mississippi River over the last 500 years will at some point cease to exist, thus requiring the engineering of a new navigational access to America's great inland waterway. Already, an increasing proportion of the river's flow is being lost above the head of the passes that constitute the toes of the bird's foot, complicating the challenge of maintaining the main navigational entrance by high-velocity flows.

While planning for the eventuality of a new navigational entrance to the river has been put off by the Corps of Engineers and State pending completion of scientific and engineering investigations of lowermost river, a design competition called Changing Course (2016) produced some intriguing concepts, all of which would be expensive and require substantial changes in where and how people live downriver from New Orleans.

### 2.4.2 *Greater New Orleans*

Informed by extensive forensic analyses of Hurricane Katrina, the Corps of Engineers launched an ambitious effort to repair and enhance the flood protection system for greater New Orleans with a network of storm surge levees, strengthened floodwalls, surge barriers, and pumps. Constructed at a cost of $14.5 billion, the system is designed to provide near-complete protection from 100-year storm surge events and to significantly reduce flooding from a 500-year event. The levees were designed to be resilient in that they would not wash away as they did during Katrina, thus overtopping would only last a few hours rather than days. The new system includes a massive barrier east of the city to block storm surges coming from Lake Borgne and the Gulf Intracoastal Waterway and Mississippi River-Gulf Outlet (MRGO). MRGO was also closed to traffic and an armored, earthen dam placed across it.

### *2.4.3 Exurban Coastal Regions*

The exurban areas around greater New Orleans and smaller cities throughout coastal Louisiana have not been afforded that same level of protection. Storm surge from Hurricane Isaac in 2012 raised water levels in Lake Pontchartrain, causing flooding in parts of LaPlace, upriver from New Orleans, and Slidell, across Lake Pontchartrain. Many former residents of New Orleans and St. Bernard Parish had moved to these communities after Hurricane Katrina and were flooded a second time. Extensions of levees, floodwalls, and gates to enhance the protection of communities along the east bank of the Mississippi from Lake Pontchartrain storm surge and communities on the west bank from Barataria Basin storm surge are proposed, but only one has been funded after decades of seeking funding, a $760-million project to protect the east bank of St. John the Baptist Parish and parts of neighboring St. Charles and St. James parishes (Bacon-Blood 2018). Even more expensive are the Morganza-to-the-Gulf system and the Lake Pontchartrain Barrier discussed in the next section on protection and restoration planning.

## 2.5 Coastal Protection and Restoration Planning

### *2.5.1 Evolution of Comprehensive Planning*

Although there had been some earlier legislative or policy efforts to address the degradation of Louisiana's coastal environments, public and political attention to the problem began to be galvanized with the 1980 assessment that the state may be losing as much as 50 square miles per year of its coastal lands (Gagliano et al. 1981). In 1990, Louisiana members of Congress succeeded in enacting the Coastal Wetland Planning, Protection, and Restoration Act (CWPPRA) that produces a relatively modest, but steady, source of dedicated funding for wetland restoration. An implementation plan was developed, but it was clear that a more comprehensive framework was required that takes into account the dynamic geologic realities of the Louisiana coast (Boesch et al. 1994). In 1998 a state task force produced a strategic plan entitled *Coast 2050: Toward a Sustainable Coastal Louisiana* (Louisiana Coastal Wetlands Conservation and Restoration Task Force 1998).

The year prior to Hurricane Katrina, the Corps of Engineers and the State released the Louisiana Coastal Area (LCA) Ecosystem Restoration Study (USACE 2004), and in 2007 the Congress authorized an overarching program that is, much like the Everglades Restoration Program, comprised of an array of separately authorized projects and the first of the intended specific projects. However, the effects of Hurricane Katrina made it clear that coastal restoration and storm surge protection had henceforth to be evaluated, planned, and executed in consort (Day et al. 2007). In response, the Corps of Engineers undertook the Louisiana Coastal Protection and Restoration Study (USACE 2009), and the State formed the Coastal Protection and

Restoration Authority (CPRA). CPRA produced its first Coastal Master Plan in 2007 and refined the plan in 2012 and again in 2017.

### 2.5.2 *Louisiana's Comprehensive Master Plan for a Sustainable Coast*

The latest Louisiana Comprehensive Master Plan for a Sustainable Coast (CPRA 2017; also referred to as the Coastal Master Plan) was approved by the state legislature in June of 2017. The Coastal Master Plan is the product of an extraordinary array of technical and economic analyses that considered varying assumptions about future conditions, resource constraints, and a multitude of project proposals. There was also extensive public consultation throughout its development and after its release prior to its ratification.

The Coastal Master Plan is intended to serve as a blueprint for the State's efforts both in flood protection and ecosystem restoration over the next 50 years. The Plan recognizes the reality of a smaller footprint of coastal lands in the future; thus "restoration" in this context is more of the rehabilitation of functions that sustain the ecosystem and maintain as much land as possible than the return to some previous condition. Implementation of the component projects would require $50 billion both from state resources and through federal appropriations and partnerships. The plan includes some 124 projects that could build or maintain more than 800 square miles of land and reduce expected damages from storm surges and other flooding by a purported $83 billion annually by the year 2067 and by more than $150 billion over the full 50 years. These projects include restoration of barrier islands and headlands, sediment diversions from the two major rivers, hydrological restoration, marsh creation using dredged sediment, ridge restoration, cultivating oyster barrier reefs, shoreline protection, structural protection from floods, and nonstructural risk reduction.

The Coastal Master Plan recognizes that not all needs are addressed by its current array of projects. More will be learned through further investigation and adaptive management of projects that are implemented. In particular, the Plan does not address the challenging questions related to lowermost Mississippi River management and how to maintain navigational access while using more of the river's water and sediment resources for restoration. Nor does it address changes in the allocation of river flow between the Atchafalaya and Mississippi river. These are issues of truly national importance that will have to be resolved.

The 2017 Coastal Master Plan places greater emphasis on coastal communities, incorporating understanding of "the cost of continued land loss and the potential effects of restoration projects on local communities, local businesses, and regional and national economies." In particular, there is a greater focus on flood risk reduction and resilience, including different types of nonstructural options and policies to help communities become more resilient.

### 2.5.3 Nonstructural Adaptation and Relocation

Nonstructural projects included in the Coastal Master Plan have the objective of reducing risks by floodproofing nonresidential structures, raising the elevation of residential structures, and acquisition of residential property. Although it is anticipated that some funding would be provided, all nonstructural projects are considered voluntary. Nonresidential structures in areas with projected 100-year flood depths of 3 feet or less could be renovated so they can be resistant to flood damage. Residential structures located in areas with a projected 100-year flood depth of between 3 and 14 feet could be elevated so that their lowest floors are higher than projected flood depths. Residential acquisition would be offered in areas where projected 100-year flood depths make elevation or floodproofing infeasible. The Coastal Master Plan does not contain specific relocation projects.

Residential acquisition and relocation are obviously very sensitive issues. In coastal Louisiana many residents have multigenerational ties to the places they live and extensive contemporary family and social networks. Still, the reality is that retreat of coastal inhabitants inland has been occurring for a long time, going back at least as far as the relocation of Cheniere Camanada families farther up Bayou Lafourche after the 1893 hurricane. New Orleanians relocated north of Lake Pontchartrain, to upriver communities, or to other parts of the country after Hurricanes Betsy and Katrina. Even less devastating tropical storms prompted movement away from the coast as a result of property damage, insurance settlements, and the cost of complying with new flood insurance requirements. Can this retreat be managed in a more considered manner that maintains the social fabric of communities remaining in the coastal zone or as communities move en masse? In particular, can this be accomplished for marginalized communities that are particularly vulnerable but lack financial resources and political voice?

A current test case is the planned resettlement of a community of Biloxi-Chitimacha-Choctaw people at Isle de Jean Charles, located on a shrinking ridge south of Houma, to a new location 35 miles inland near Shriever. Subsidence attributable to oil and gas withdrawals had hastened the loss of land around Isle de Jean Charles (Morton et al. 2006). The Louisiana Office of Community Development is managing the resettlement with the assistance of a $48 million grant from the US Department of Housing and Urban Development (HUD), with construction beginning in 2019. While the resettlement allows the prospect of keeping the community intact, the residents, while retaining access, will be far removed from the fishing, oyster cultivation, and trapping that have been traditionally the basis of their sustenance.

On a broader front, the State has created, from the same HUD resilience competition as the Isle de Jean resettlement, the Louisiana Strategic Adaptations for Future Environments (LA SAFE) program to assist communities to take proactive steps for adaptation to the rapidly changing coastal environments and risks (Louisiana Office of Community Development Disaster Recovery Unit 2017). The project expressly accepts that some of the most vulnerable communities will need to contemplate resettlement over the next 50 years and that migration is already occurring. Funding thus far is for community engagement and co-design, and sources have not been identified for the significant resources required for residential acquisitions and resettlement nor for the $6.1 billion specified in the Coastal Master Plan for non-structural risk reduction.

### 2.5.4 Implementation and Controversies

Of course, the Coastal Master Plan will require the funding, public acceptance, legal sufficiency, and engineering feasibility of its component projects. After Hurricane Katrina the federal government provided over $14.5 billion to repair and improve structural defenses against storm surge, and the State and local government have invested heavily in improving drainage. At the same time, despite the federal authorization of the LCA Ecosystem Restoration Program, only modest funding has been made available for environmental restoration. However, as a result of payments, penalties, and damage compensation stemming from the BP Deepwater Horizon oil spill that occurred in 2010, the situation has now been reversed. Approximately $8 billion is likely to be provided from these sources for use in restoration in coastal Louisiana over the next decade or so. As a consequence, the State is now advancing planning and implementation of restoration projects without federal appropriations. Meanwhile, even many Congressionally authorized structural protection projects are slowed because of the lack of federal appropriations and limited state and local funding.

Paramount among these authorized but underfunded structural protection projects is the Morganza-to-the-Gulf array of levees, floodwalls, gates, locks and pump stations stretching 98 miles across Terrebonne Parish from to Gibson to Lockport. Intended to protect population center around Houma, the project is proceeding incrementally using State and local funding. At $8.3 billion, the Morganza-to-the-Gulf protection system is the single most expensive project in the $50 billion Coastal Master Plan. However, it confronts significant challenges with regard to the level of risk reduction that would be provided and the sustainability of wetlands enclosed by the levees (Twilley et al. 2008), as well as the engineering feasibility and cost of constructing significant earthen levees across the soft and subsiding substrates of the Terrebonne Basin.

The Lake Pontchartrain Barrier at a cost of $2.4 billion faces its own challenges, including the environmental effects of constraining tidal flows into Lake Pontchartrain and increased storm surge likely to be felt along the Mississippi coast as storm surges are prevented from entering the lake. If structural protection projects are not completed, significant population centers around Houma and Slidell would face increasing risks.

While the concept of coastal restoration enjoys substantial public support, individual projects face opposition from some members of the public or confront issues raised in environmental reviews for permits. Prominent among these projects are diversions of sediment from the rivers to slow the loss and even build new wetlands by recreating the processes that built the Mississippi Deltaic Plain in the first place. Sediment diversions are thought by most coastal scientists to be foundational elements of any credible restoration strategy (Boesch et al. 1994; Day et al. 2007). However, some shrimpers, oyster growers, and sport fishermen have raised opposition because the river flows would freshen brackish estuaries and change the distribution of targeted animals (Muth 2014). Local landowners and residents have raised concerns about increased backwater flooding risks, and shipping interests have objected to potential shoaling of shipping channels as river flows are reduced below diversions.

Federal resource agencies have also raised concerns about the effects of estuarine freshening on essential fish habitat and populations of protected bottlenose dolphins, despite the fact that the present estuarine bays are relatively fleeting features resulting from coastal degradation and may be eventually converted to open Gulf waters without intervention for restoration. Nonetheless, planning and engineering for the Mid-Barataria Sediment Diversion at Myrtle Grove are proceeding, armed with funding from oil spill revenues, state political support, and federal commitments for fast-tracking of environmental permits.

Not only might sediment diversions from the river impact the interests of some stakeholders, but they also will likely take decades to build wetlands. Consequently, there are strong proponents of marsh creation using dredged sediments. The costs of marsh creation projects in the Coastal Master Plan total an estimated $17.9 billion of the $50 billion total costs. Not only will funding be a limiting factor but also will the supply of suitable sediment, at least for marsh creation projects located far from the resources of the Mississippi and Atchafalaya rivers. These will require long-distance pipelines, accessing sand resources from shoals on the continental shelf, or dredging nearby bays, raising questions of the high energy as well as financial costs (Day et al. 2005). Furthermore, marshes created by dredged material require periodic renourishment with dredged sediment to counteract subsidence and relative sea-level rise. Marsh creation may be accomplished more quickly than land building by sediment diversions, but is less sustainable.

## 2.6 Climate Change as a Threat Multiplier

### *2.6.1 Change Is Happening: Human-Caused and Dangerous*

According to an overwhelming scientific consensus, global warming is occurring and is virtually all the result of human activities (USGCRP 2017). The six warmest years on record, in terms of global mean annual temperature, have come in the decade of the 2010s. Natural forces, such as solar activity, have played an inconsequential role in the observed warming since the mid-twentieth century. At current rates of growth in emissions of carbon dioxide and other greenhouse gases, dangerous climate changes would result before the end of this century, threatening the world's biodiversity, acidifying the oceans, amplifying extreme weather events, causing economic hardship, and accelerating sea-level rise to the extent that it would render many low-lying islands and densely populated coastal regions uninhabitable.

These are mainstream scientific assertions (IPCC 2014; USGCRP 2017) that, while widely accepted around the world, are not as widely accepted within Louisiana's political leadership and the south Louisiana citizenry. There are various reasons why this is the case, including perceived economic dependence on the fossil fuel industry, mistrust in government solutions, resentment of the intellectual class, and the fear of cultural eclipse and economic decline about which sociologist Arlie Russell Hochschild (2016) wrote in her book *Strangers in Their Own Lands*. Enigmatically, she argues, those most at risk reject the needed solutions for these reasons. Furthermore, even well-informed Louisianans perceive the current crises as far more the result of natural processes and other human activities than of global climate change in a distant future. In any case, the unwillingness to address the reality and causes of global climate change presents a significant challenge in how its consequences can be brought into planning and action for coastal resilience in coastal Louisiana, both for the environment and for society.

While coastal Louisiana has long had to confront high rates of relative sea-level rise as a result of subsidence, the oceans themselves began to rise beginning only in the late nineteenth century (Kemp et al. 2011). The rise in global mean sea level accelerated through the twentieth century (Dangendorf et al. 2017) and has averaged about 3 mm/year since 1993, when satellite-born altimeters have allowed us to measure the level of entire oceans (Nerem et al. 2018). In addition to the expansion of warming ocean waters and melting of glaciers, the melting of ice sheets perched on Greenland and Antarctica is now contributing to global sea-level rise. Simply projecting the acceleration of rate of rise observed in the satellite record would result in a rise in global sea level of about 65 cm (2.1 feet) by 2100 compared with 2005. On top of subsidence, such a rise would be very challenging for the Louisiana coast but, as will be discussed in the next section, should probably be regarded as the minimum that will likely be experienced.

The scientific consensus at this time is that climate change is unlikely to increase the frequency of tropical cyclones but is very likely to increase the intensity of those that do occur (Knutson et al. 2010). This may particularly be the case on the Gulf Coast as the waters of the Gulf of Mexico continue to warm. A greater percentage of hurricanes are likely to reach Category 4 or Category 5 level on the Safir-Simpson scale. There are many other factors that will influence the trajectories of Atlantic hurricanes, making it impossible to forecast whether the Louisiana coast will experience more or fewer in the coming decades, but those that do impact this coast will probably become stronger.

Climate change also presents risk of increased flooding from extreme rainfall events. Over the last century, precipitation has increased along the northern Gulf Coast, both annually and in the summer (Kunkel et al. 2013). The frequency of rainfall events of 1 inch or more is projected to increase by mid-century and, at the same time, dry spells are likely to become more frequent.

As mentioned earlier, climate change has been implicated in two record-breaking rainfall events and resulting floods, in the Louisiana deluge in the Baton Rouge area in August of 2016 and with Hurricane Harvey around Houston in 2017. Both events occurred when low-pressure systems that developed in the Gulf of Mexico stalled near the coast – consistent with slowdown in tropical storm speeds that has been linked to global warming (Kossin 2018) – allowing them to continue to draw energy and moisture from the anomalously warm waters of the Gulf. Based on observational data and models, researchers found that an event like the Louisiana 2016 deluge is now likely to occur at least 40% more often than prior to the year 1900 and that their precipitation intensity has increased by roughly 10% as a result of human-caused climate change (van der Wiel et al. 2017). For the Houston flood, one study estimated that the chances of observed precipitation accumulations had increased by a factor of 3 and precipitation intensity increased by 15% (van Oldenborgh et al. 2017), while another placed these as a factor of 3.5 and 37%, respectively (Risser and Wehner 2017).

While air temperatures in coastal Louisiana have not increased as much as many other parts of the United States, warmer temperatures later this century are very likely and will pose additional challenges to inhabitants of coastal Louisiana. While there will be fewer killing freezes, an increase in the number of days with temperatures exceeding 95 degrees Fahrenheit (35 degrees Celsius) is projected (Kunkel et al. 2013). Cooling degree days (a measure of how much and for how long outside air temperature is above 65 degrees Fahrenheit) are also projected to increase substantially, placing additional burdens on the poor who may have limited access to air conditioning and on the well-being of the broader population when confronted by power disruptions that result from major storms.

### *2.6.2 Avoiding the Unmanageable*

At the end of 2015, virtually all nations of the world endorsed the United Nations Paris Agreement, the guiding objective of which is to reduce net emissions of greenhouse gases from human activities in order to keep global warming well below an increase of 2 degrees Celsius in global mean temperature above preindustrial levels, with an ambition to limit it to 1.5 degrees Celsius (Rogelj et al. 2016). We are at about 1 degree Celsius above the preindustrial level today. The Paris Agreement recognizes that substantial adaptation to the changing climate will be still required but that as these levels of warming are exceeded, it will be very challenging for human society to adapt. In short, humankind must avoid the unmanageable, while managing the unavoidable.

Limiting global warming to less than 2 degrees Celsius will require the rapid reduction of global greenhouse gas emissions beginning very soon and reaching net zero emissions by mid-century or soon thereafter (Figueres et al. 2017). Absent dramatic breakthroughs in carbon capture and storage technologies, such large and rapid emission reductions would necessitate a transition from a fossil fuel-based economy far more quickly than the citizens and political leadership of south Louisiana may be ready to consider. And yet the fundamental conundrum is that such a global transition is as essential for the future habitation of coastal Louisiana as it is for an imperiled Pacific island nation.

The existential threat to future habitation in coastal Louisiana is global sea-level rise. First, keep in mind that the relative rate of sea-level rise there, half or more due to subsidence, already poses substantial adaptation challenges. To its credit, the 2017 Coastal Master Plan considers three scenarios of environmental changes over the next 50 years, representing sea-level rise (in addition to variable rates of subsidence) of 43, 63 and 83 cm by 2067 for the low, medium, and high scenarios (CPRA 2017). Although the Plan does not link these scenarios to greenhouse emission pathways, it should be obvious that the greater the greenhouse gas concentrations realized, the greater the warming of the atmosphere and oceans and the greater the sea-level rise.

If greenhouse gas emissions continue to grow through the century (the Representative Concentration Pathway 8.5 of the 2014 IPCC assessment), it is increasingly apparent that a very substantial and unstoppable loss of Antarctic ice would probably be triggered with dramatic effects on sea level later in the twenty-first century and beyond (Kopp et al. 2017). This would result in a range of possible sea-level rise by the end of the century that includes the 200 cm (6.6 feet) by 2100 on which Coastal Master Plan's high scenario is based. That would be just the beginning, as the likely sea-level rise during the next century would range between 600 and 900 cm (20–30 feet). The Gulf of Mexico shoreline would retreat to where it was 7000 years ago. Moreover, we would not be able to forecast this with great certainty until it is too late to slow the rate of ice loss by reducing our emissions.

If, on the other hand, global society were to rapidly reduce greenhouse gas emissions to meet the goals of the Paris Climate Agreement to keep the increase in global mean temperature below 2 degrees Celsius (RCP 2.6), catastrophic loss of Antarctic ice mass could be avoided. According to the recent probabilistic projections (Kopp et al. 2017), sea-level rise over the next 50 years would likely be less than what even the low scenario of the Coastal Master Plan assumes and substantially less than the 198 cm by 2100 on which scenario is based. In fact, there would be a 50/50 chance of sea-level rise being less than 100 cm even in 2200, giving the embattled Louisiana coast a fighting chance for adaptation that leads to "essential" social resilience (Laska 2012).

## 2.7 Implications for Social Resilience

### *2.7.1 Transient and Secular Disasters*

The people, families, communities, and institutions of coastal Louisiana will continue to be confronted by transient disasters caused by river flooding, storm surges, and deluges. Within limits, they have been proudly resilient in the past, but many steps can yet be taken to improve social resilience in the future. However, now society is confronted with substantial secular (long duration) changes in the natural environment and their attendant risks in the form of rapid coastal disintegration of this geologically young territory, compounded by global climate change. These "slow motion disasters" require a different kind of approach to social resilience, one that fundamentally takes an intergenerational perspective but with substantial changes even happening fast enough to be experienced within a lifetime.

Enhancing intergenerational social resilience will require that the people of coastal Louisiana have a greater awareness and acceptance of the biophysical changes that will be confronting them. They will have to understand the accommodations and solutions that are possible and their limits in order to effectively participate in civil society. The people of coastal Louisiana can no longer afford to remain "strangers in their own land" as Hochshield (2016) framed the dilemma. While Louisiana's Coastal Protection and Restoration Authority has expended considerable effort to engage the public and has secured political support for the Coastal Master Plan to this point, much more extensive understanding by the public and incorporation of community concerns will be required. Because of the intergenerational nature of the challenge, there should be concerted efforts to raise the socio-environmental literacy of school children about their unique and dynamic coastal landscape and how and why it is being altered, including by climate change.

Enhancing resilience to disasters during an era of rapid change will also require a strategically developed capacity of natural and social scientists, engineers, designers and planners, and social workers. Higher education institutions should focus faculty development and research and training programs with this in mind. New kinds of boundary organizations (Cash et al. 2003) will have to evolve that link knowledge with practice, transcend public and private enterprises, and engage both citizens and decision makers.

### 2.7.2 Role of Natural Systems in Resilience

The concept of ecosystem services (Carpenter et al. 2009) has emerged with the growing recognition of the importance of natural environments to human well-being. The values of coastal wetlands for protection from hurricane waves and storm surges have been specifically assessed (Costanza et al. 2008; Barbier et al. 2013) and are among the many ecosystem services that support the socio-economy of coastal Louisiana. The natural ecosystem resilience of coastal Louisiana is increasingly recognized as an important contributor to social resilience.

Louisiana's consecutive Coastal Master Plans have taken major steps in the right direction by incorporating the benefits of coastal ecosystems in moderating wave and storm surge risks and in integrating protection and restoration. There is clearly much more work to be done on this front for project-specific design and integration. Future efforts will have to navigate the institutional obstacles regarding matching of funding sources, typically restricted to protection or to restoration, and coordination among disparate responsible agencies.

### 2.7.3 *Limiting Climate Change Inseparable from Adaptation*

While not expressly linked to global warming and greenhouse gas emissions, the sea-level rise rates embedded in the future scenarios of the Coastal Master Plan, together with their logical extensions beyond 2067 as discussed earlier, make it clear that the single most effective action to ensure the future well-being of people in coastal Louisiana is the rapid reduction in global greenhouse gas emissions consistent with the Paris Climate Agreement. This is urgent: with each 5-year delay in near-term peaking of carbon dioxide emissions, sea level in 2300 would increase by an estimated 20 cm (Mengel et al. 2018). From the perspective of people desiring to live in coastal Louisiana beyond the next 50 years, it is not an exaggeration to say that effective mitigation to limit climate change is a sine qua non. The benefits of most of the protection and restoration we have undertaken or are planning over the next 50 years would be rendered moot by 2 meters or more of sea-level rise. Climate change adaptation and mitigation are not separate issues but must go hand in hand in order to manage the unavoidable while avoiding the unmanageable.

Recognition of this reality by the public and political leadership in Louisiana is a challenging obstacle. Although there are some exceptions, many of those currently in political leadership at the state and federal level are stationed somewhere between denial (climate change is not happening or is mostly natural) and "lukewarmerism" (it will not be that bad or there is not much we can do about it). Improved public awareness of the scientific realities and the technological possibilities will be required to change this much.

Contributing to this reticence are concerns about impacts on jobs and the regional economy of a phase out in the use of oil and gas as fuels. Production of petroleum hydrocarbons would still be required as feedstocks for chemicals and products that society would use. Existing industrial and technological capacities could be useful in developing renewable energy or in carbon sequestration in the vast deep saline aquifers lying under the northern Gulf of Mexico (DeSilva et al. 2015). For example, the support structures for offshore wind turbines recently installed off Rhode Island were built in an oil platform fabrication yard in Houma, Louisiana. Moving away from energy and transportation systems that rely on fossil fuels also opens up opportunities for creative approaches to coastal restoration and community resilience by the strategic brain trust mentioned above.

### 2.7.4 *Defend, Adapt, or Relocate?*

Difficult decisions are already here today regarding whether to structurally protect, improve resilience where structural protection is infeasible, or relocate vulnerable homes and communities (Bailey et al. 2014). Inclusive efforts that plan for the future such as LA SAFE are critical, and there is much that social scientists can contribute to and learn from these efforts and from planned relocations such as for

Isle de Jean Charles. After all, in coastal Louisiana the challenge is not just resilience to extreme weather events but also rational responses to substantial long-term biophysical changes that ensure human well-being and sustain the sociocultural fabric of communities.

### *2.7.5 Coastal Louisiana as a Harbinger*

The Deltaic and Chenier plains that characterize the Louisiana coastal zone differ in many important ways from other coastal zones of the United States. They are younger, exceptionally low lying, and generally subsiding more rapidly than most coastal landforms. Yet, with relative sea-level rise accelerating and ocean storms and extreme precipitation likely to intensify along most of US coasts, Louisiana serves as a harbinger for the challenges to be faced in risk management for coastal communities elsewhere.

From the increase in the frequency of high tide or so-called nuisance flooding, even on sunny days, in cities such as Atlantic City, Annapolis, Norfolk, Charleston, and Miami (Sweet et al. 2018), to the damages associated with the exceptional storm surge of Superstorm Sandy (Halverson and Rabenhorst 2013), increased risks to communities are more evident, and planning is beginning to take this into account. Even California, which one does not usually think of having a low-lying coast, has updated its sea-level rise guidance (California Natural Resources Agency 2018) based on a rigorous scientific assessment (Griggs et al. 2017). With Louisiana's still massive, if underused, supplies of river-borne sediments, Louisiana might even have some advantages in contending with sea-level rise. South Florida, where huge populations and economies are at risk, has no muddy rivers, and the porous limestone platform that underlies it can render earthen levees ineffective.

What coastal Louisiana is confronting today defines challenges surely to be faced in other coastal regions around the globe. How can cities and towns contend with more regular tidal water flooding, as well as greater storm surges, while at the same time remove precipitation-driven stormwater? How can tidal wetlands be maintained not only for their natural resource values but also as a buffer to storm surges, during the coming period of more rapidly rising seas? How can state governments effectively integrate community hazard protection and coastal ecosystem restoration? How do communities and governments make rational and effective choices among structural storm surge protection, nonstructural adaptation, and relocation?

## References

Bacon-Blood, L. (2018). St. John Parish is getting a new $760 million levee: Here are 10 things you need to know. *The Times-Picayune*. https://www.nola.com/environment/index.ssf/2018/07/10_things_to_know_about_st_joh.html

Bailey, D., Grambling, R., & Laska, S. B. (2014). Complexities of resilience: Adaptation and change within human communities in coastal Louisiana. In J. W. Day, G. P. Kemp, A. M.

Freeman, & D. P. Muth (Eds.), *Perspectives on the restoration of the Mississippi Delta* (pp. 125–140). Dordrecht: Springer.

Barbier, E. B., Georgiou, I. Y., Enchelmeyer, B., & Reed, D. J. (2013). The value of wetlands in protecting southeast Louisiana during hurricane storm surges. *PLoS ONE, 8*(3), e58715. https://doi.org/10.1371/journal.pone.0058715.

Barry, J. M. (1997). *Rising tide: The Great Mississippi River Flood of 1927 and how it changed America*. New York: Simon and Schuster.

Baxter, V. (2014). Rent, real estate, and flood mitigation in New Orleans East. *Antipode, 46*(4), 1014–1031. https://doi.org/10.1111/anti.12080.

Bentley, S. J. S., Blum, M. D., Maloney, J., Pond, L., & Paulsell, R. (2016). The Mississippi River source-to-sink system: Perspectives on tectonic, climatic, and anthropogenic influences, Miocene to Antropocene. *Earth-Science Reviews, 153*, 139–174.

Bernard, S. K. (2016). *Teche: A history of Louisana's most famous bayou*. Oxford: University of Mississippi Press.

Bevan, J. L. I. I., Avila, L. A., Blake, E. S., Brown, D. P., Franklin, J. L., Knabb, R. D., et al. (2008). Atlantic hurricane season of 2005. *Monthly Weather Review*, 1009–1117. https://doi.org/10.1175/2007MWR2074.1.

Blum, M. D., & Roberts, H. H. (2009). Drowning of the Mississippi Delta due to insufficient sediment supply and global sea-level rise. *Nature Geoscience, 2*(7), 488–491.

Blum, M. D., & Roberts, H. H. (2012). The Mississippi delta region: Past, present, and future. *Annual Review of Earth and Planetary Sciences, 40*, 655–683.

Boesch, D. F., Josselyn, M. N., Mehta, A. J., Morris, J. T., Nuttle, W. K., Simenstad, C. A., et al. (1994). Scientific assessment of coastal wetland loss restoration and management in Louisiana. *Journal of Coastal Research, Spec Issue, 20*, 1–103.

Brasseaux, C. A., & Davis, D. W. (2017). *Ain't there no more: Louisiana's disappearing coastal plain*. Oxford: University of Mississippi Press.

California Natural Resources Agency. (2018). *State of California sea-level rise guidance*. Sacramento: California Natural Resources Agency.

Campanella, R. (2008). *Bienville's dilemma: A historical geography of New Orleans*. Lafayette: Center for Louisiana Studies, University of Louisiana, Lafayette.

Carpenter, S. R., Mooney, H. A., Agard, J., Capistrano, D., DeFries, R. S., Diaz, S., et al. (2009). Science for managing ecosystem services: Beyond the millennium ecosystem assessment. *Proceedings of the National Academy of Sciences, 106*(5), 1305–1312.

Cash, D. W., Clark, W. C., Alcock, F., Dickson, N. M., Eckley, N., Guston, D. W., et al. (2003). Knowledge systems for sustainable development. *Proceedings of the National Academy of Sciences, 100*(14), 8086–8091.

Chamberlain, E. L., Törnqvist, R. E., Shen, Z., Mauz, B., & Wallinga, J. (2018). Anatomy of Mississippi Delta growth and its implications for coastal restoration. *Science Advances, 4*, eaar4740. https://doi.org/10.1126/sciadv.aar4740.

Changing Course. (2016). *Changing course: navigating the future of the Lower Mississippi River Delta*. http://changingcourse.us

Coastal Protection and Restoration Authority of Louisiana (CPRA). (2017). *Louisiana's comprehensive master plan for a sustainable coast*. Baton Rouge: Coastal Protection and Restoration Authority of Louisiana.

Colton, C. E. (2005). *Unnatural metropolis: Wresting New Orleans from nature*. Baton Rouge: Louisiana State University Press.

Corthell, E. (1897). The delta of the Mississippi River. *National Geographic, 12*, 351–354.

Costanza, R., Pérez-Maqueo, O., Martinez, M. L., Sutton, P., Anderson, S. J., & Mulder, K. (2008). The value of coastal wetlands for hurricane protection. *Ambio, 37*(4), 241.

Couvillion, B. R., Beck, H., Schoolmaster, D., & Fischer, M. (2017). Land area change in coastal Louisiana. In *Scientific investigations map 3381*. Reston: U.S. Geological Survey.

Dangendorf, S., Maros, M., Woppelmann, G., Conrad, C. P., Frederiske, T., & Riva, R. (2017). Reassessment of 20$^{th}$ century global mean sea level rise. *Proceedings of the National Academy of Sciences, 114*(23), 5946–5951. https://doi.org/10.1038/ncomms14792.

Day, J. W., Britsch, L. D., Hawes, S. R., Schaffer, G. P., Reed, D. J., & Cahoon, D. (2000). Pattern and process of land loss in the Mississippi Delta: A spatial and temporal analysis of wetland habitat change. *Estuaries, 23*(4), 425–438.

Day, J. W., Jr., Barras, J., Clairain, E., Johnston, J., Justic, D., Kemp, G. P., et al. (2005). Implications of global climate change and energy cost and availability for the restoration of the Mississippi delta. *Ecological Engineering, 24*, 253–265.

Day, J. W., Jr., Boesch, D. F., Clairain, E. J., Kemp, G. P., Laska, S. B., Mitsch, W. J., et al. (2007). Restoration of the Mississippi Delta: Lessons from Hurricanes Katrina and Rita. *Science, 315*(5819), 1679–1684.

De Silva, G. P. D., Ranjith, P. G., & Perera, M. S. A. (2015). Geochemical aspects of $CO_2$ sequestration in deep saline aquifers: A review. *Fuel, 155*, 128–143.

Dixon, B. (2009). *Last days of Last Island: The hurricane of 1856, Louisiana's first great storm*. Lafayette: University of Louisiana at Lafayette Press.

Dixon, T. H., Amelung, F., Ferretti, A., Novali, F., Rocca, F., Dokka, R., et al. (2006). Subsidence and flooding in New Orleans. *Nature, 441*, 587–588.

Figueres, C., Schellnhuber, H. J., Whiteman, G., Rockström Hobley, A., & Rahmstorf, S. (2017). Three years to safeguard our climate. *Nature, 546*, 593–595.

Freudenburg, W. R., Gramling, R., Laska, S., & Erikson, K. (2009a). *Catastrophe in the making: The engineering of Katrina and the disasters of tomorrow* (2nd ed.). Washington, D.C.: Island Press.

Freudenburg, W. R., Gramling, R., Laska, S., & Erikson, K. (2009b). Disproportionality and disaster: Hurricane Katrina and the Mississippi River-Gulf Outlet. *Social Science Quarterly, 90*, 497–515.

Gagliano, S. M., Meyer-Arendt, K. J., & Wicker, K. M. (1981). Land loss in the Mississippi River deltaic plain. *Transactions, Gulf Coast Association of Geological Societies, 31*, 295–300.

Griggs, G., Árvai, J., Cayan, D., DeConto, R., Fox, J., Fricker, H. A., et al. (2017). *Rising seas in California: An update on sea-level rise science*. Oakland: California Ocean Science Trust.

Halverson, J. B., & Rabenhorst, T. (2013). Hurricane Sandy: The science and impacts of a superstorm. *Weatherwise, 66*, 14–23.

Heimann, D. C., Sprague, L. A., & Blevins, D. W. (2011). *Trends in suspended-sediment loads and concentrations in the Mississippi River Basin, 1950–2009. Scientific Investigations Report* (pp. 2011–5200). Reston: U.S. Geological Survey.

Hochschield, A. R. (2016). *Strangers in their own land: Anger and mourning on the American right*. New York: The New Press.

Intergovernmental Panel on Climate Change (IPCC). (2014). *Climate change 2014: Synthesis report*. Geneva: Intergovernmental Panel on Climate Change.

Janowski, K. L., Törnqvist, T. E., & Fernandes, A. M. (2017). Vulnerability of Louisiana's coastal wetlands to present-day rates of relative sea-level rise. *Nature Communications, 8*, 14792. https://doi.org/10.1038/ncomms14792.

Kemp, A. C., Horton, B. C., Donnelly, J. P., Mann, M. E., Vermeer, M., & Rahmstorf, S. (2011). Climate related sea-level variations over the past two millennia. *Proceedings of the National Academy of Sciences, 108*, 11017–11022.

Knutson, T. R., McBride, J. L., Chan, J., Emanuel, K., Holland, G., Landsea, C., et al. (2010). Tropical cyclones and climate change. *Nature Geoscience, 157-163*(2010), 3. https://doi.org/10.1038/ngeo779.

Kolker, A. S., Allison, M. A., & Hameed, S. An evaluation of subsidence rates and sea- level variability in the northern Gulf of Mexico. *Geophysical Research Letters, 2011*. https://doi.org/10.1029/2011GL049458.

Kopp, R. E., DeConto, R. M., Bader, D. A., Hay, C. C., Horton, R. M., Kulp, S., et al. (2017). Evolving understanding of Antarctic ice-sheet physics and ambiguity in probabilistic sea-level projections. *Earth's Future*. https://doi.org/10.1002/2017EF000663.

Kunkel, K. E., Stevens, L. E., Stevens, S. E., Sun, L., Janssen, E., Wuebbles, D., et al. (2013). *Regional climate trends and scenarios for the U.S. National Climate Assessment. Part 2. Climate of the Southeast U.S. technical report NESDIS* (pp. 142–142). Silver Spring: National Oceanic and Atmospheric Administration.

Laska, S. (2012). Dimensions of resiliency: Essential resiliency, exceptional recovery and scale. *International Journal of Critical Infrastructures, 8*(1), 47–62.

Louisiana Coastal Wetlands Conservation and Restoration Task Force and the Wetlands Conservation and Restoration Authority. (1998). *Coast 2050: Toward a sustainable coastal Louisiana*. Baton Rouge: Louisiana Department of Natural Resources.

Louisiana Office of Community Development Disaster Recovery Unit. (2017). *LA SAFE: Louisiana's strategic adaptations for future environments*. Baton Rouge: Louisiana Office of Community Development.

McBride, R. A., et al. (2007). A geomorphic process-response model for chenier-plain evolution in southwestern Louisiana, USA. In *Coastal Sediments' 07* (pp. 762–775).

McIntire, W. G. (1958). *Prehistoric indian settlements of the changing Mississippi River delta*. Baton Rouge: Louisiana State University Press.

McQuaid, J., & Schleifstein, M. (2006). *Path of destruction: The devastation of New Orleans and the coming age of superstorms*. New York: Little, Brown.

Meade, R. H., & Moody, J. A. (2010). Causes for the decline of suspended-sediment discharge in the Mississippi River system, 1940–2007. *Hydrological Processes, 24*, 35–49.

Mengel, M., Nauels, A., Rogelj, J., & Schleussner, D. F. (2018). Committed sea-level rise under the Paris agreement and the legacy of delayed mitigation action. *Nature Communications*. https://doi.org/10.1038/s41467-018-02985-8.

Morton, R. A., Bernier, J. C., & Barras, J. A. (2006). Evidence of regional subsidence and associated interior wetland loss induced by hydrocarbon production, Gulf Coast region, USA. *Environmental Geology, 50*, 261. https://doi.org/10.1007/s00254-006-0207-3.

Munoz, S. E., Giosan, L., Therrell, M. S., Remo, J. W. F., Shen, Z., Sullivan, R. M., et al. (2018). Climactic control of Mississippi River flood hazard amplified by river engineering. *Nature, 556*, 95–98.

Muth, D. P. (2014). The once and future delta. In J. W. Day, G. P. Kemp, A. M. Freeman, & D. P. Muth (Eds.), *Perspectives on the restoration of the Mississippi Delta* (pp. 9–27). Dordrecht: Springer.

Nerem, R. S., Buckley, B. D., Fasullo, J. T., Hamlington, B. D., Masters, D., & Mitchum, G. T. (2018). Climate-change-driven accelerated sea-level rise detected in the altimeter data. *Proceedings of the National Academy of Sciences*. https://doi.org/10.1073/pnas.1717312115.

Penland, S., & Suter, J. R. (1989). The geomorphology of the Mississippi River chenier plain. *Marine Geology, 90*, 231–240, 243–258.

Penland, S., Mendelssohn, I., Wayne, L., & Britsch, D. (1996). *Natural and human causes of coastal land loss in Louisiana*. Baton Rouge: Coastal Studies Institute and Wetland Biogeochemistry Institute, Louisiana State University.

Perry, C. L., & Mendelssohn, I. A. (2009). Ecosystem effects of expanding populations of *Avicennia germinans* in a Louisiana salt marsh. *Wetlands, 29*(1), 396–406. https://doi.org/10.1672/08-100.1.

Piazza, B. P. (2014). *The Atchafalaya River Basin: History and ecology of an American wet land*. College Station: Texas A&M University Press.

Reed, D. J. (1989). Patterns of sediment deposition in subsiding coastal salt marshes, Terrebonne Bay, Louisiana: The role of winter storms. *Estuaries and Coasts, 4*, 222–227.

Risser, M. D., & Wehner, M. F. (2017). Attributable human-induced changes in the likelihood and magnitude of the observed extreme precipitation during Hurricane Harvey. *Geophysical Research Letters, 44*, 12457–12464. https://doi.org/10.1002/2017GL075888.

Roberts, H. H. (1997). Dynamic changes of the Holocene Mississippi River Delta Plain: The delta cycle. *Journal of Coastal Research, 13*(3), 605–627.

Roberts, H. H., Coleman, J. M., Bentley, S. J., & Walker, N. (2003). Embryonic major delta lobe: A new generation of delta studies in the Atchafalaya-Wax Lake delta system. *Gulf Coast Association of Geological Societies Transactions, 53*, 690–703.

Rogelj, J., den Elzen, M., Höhne, N., Fransen, T., Fekete, H., Winkler, H., et al. (2016). Paris agreement climate proposals been a boost to keep warming well below 2°C. *Nature*. https://doi.org/10.1038/nature18307.

Roth, D. (2010). *Louisiana hurricane history*. Camp Springs, Maryland: National Weather Service.

Salinas, L. M., DeLaune, R. D., & Patrick, W. H., Jr. (1986). Changes occurring along a rapidly submerging coastal area: Louisiana, USA. *Journal of Coastal Research, 2*(3), 269–284.

Saucier, R. T. (1994). *Geomorphology and Quaternary geologic history of the Lower Mississippi Valley*. Vicksburg: Mississippi River Commission.

Stanford, J. D., Hemingway, R., Rohling, E. J., Challenor, P. G., Medina-Elizalde, M., & Lester, A. J. (2010). Sea-level probability for the last deglaciation: A statistical analysis of far-field records. *Global and Planetary Change, 79*, 193–203.

Sweet, W. V., Dusek, G., Obeysekera, J., & Marra, J. J. (2018). *Patterns and projections of high tide flooding along the U.S. coastline using a common impact threshold. NOAA technical report NOS CO-OPS 086*. Silver Spring: National Oceanic and Atmospheric Administration.

Turner, R. E. (1997). Wetland loss in the northern Gulf of Mexico: Multiple working hypotheses. *Estuaries, 20*(1), 13.

Twilley, R., Allison, M., Capozzoli, L., Laska, S., McKee, L., Meselhe, E., et al. (2008). Morganza-to-the-Gulf Technical Panel review final report. In *Duxbury*. Massachusetts: Battelle.

Twilley, R. R., Bentley, S. J., Sr., Chen, Q., Edmonds, D. A., Hagen, S. C., Lam, N. S. N., et al. (2016). Co-evolution of wetlands landscapes, flooding, and human settlement in the Mississippi River Delta Plain. *Sustainability Science, 11*, 711–731.

U.S. Army Corps of Engineers (USACE). (2004). Louisiana Coastal Area (LCA), Louisiana: Ecosystem restoration study. In *New Orleans*. Louisiana: U.S. Army Corps of Engineers New Orleans District.

U.S. Army Corps of Engineers (USACE). (2009). Louisiana Coastal Protection and Restoration (LACPR): Final technical report. In *New Orleans*. Louisiana: U.S. Army Corps of Engineers New Orleans District.

U.S. Global Change Research Program (USGCRP). (2017). *Climate science special report: Fourth National Climate Assessment, Volume I*. Washington, DC: U.S. Global Change Research Program. https://doi.org/10.7930/J0J964J6.

van der Wiel, K., Kapnick, S. B., van Oldenborgh, G. J., Whan, K., Phillipm, S., Vecchi, G. A., et al. (2017). Rapid attribution of the August 2016 flood-inducing extreme precipitation in south Louisiana to climate change. *Hydrology and Earth System Sciences, 897-92*(2017), 21. https://doi.org/10.5194/hess-21-1631-2017.

van Oldenborgh, G. J., van der Wiel, K., Sebastian, A., Singh, R., Arrighi, J., Otto, F., et al. (2017). Attribution of extreme rainfall from Hurricane Harvey, August 2017. *Environmental Research Letters, 12*, 124009. https://doi.org/10.1088/1748/aa9ef2.

Visser, J. M., Sasser, C. E., Chabreck, R. H., & Linscombe, R. G. (2002). The impact of a severe drought on the vegetation of a subtropical estuary. *Estuaries, 25*(6A), 1184–1195.

# Part II
# Climate Adaptation Challenges and Solutions

# Chapter 3
# Connecting the Dots: The Origins, Evolutions, and Implications of the Map that Changed Post-Katrina Recovery Planning in New Orleans

**Zachary Lamb**

## 3.1 Introduction

On January 11, 2006, a little more than 4 months after the Hurricane Katrina-induced levee failures flooded New Orleans, the city's major newspaper, the *Times-Picayune*, published a front-page story with the arresting headline, "4 MONTHS TO DECIDE". Sub-headlines announced that the "City's Footprint May Shrink" and that "full buyouts proposed for those forced to move if the 'hardest hit areas' could not 'prove viability'" (Donze and Russel 2006). Directly beneath these words lay a map (Fig. 3.1) showing a sea of yellow over much of the city indicating the flood-damaged neighborhoods that would be subject to a proposed building permit moratorium and therefore temporarily off-limits to rebuilding. The map showed six areas of the city overlaid with solid bright green circles indicating areas designated for "future parkland" and 12 red-outlined zones identified for prioritized reconstruction. In text and in words, the map laid out a classification system for the reconstruction of New Orleans and the radical restructuring of land use in the city.

For thousands of displaced New Orleanians, scattered across the country, these headlines and this map represented a graphic manifestation of their worst fears of losing their homes and the right to return to their neighborhoods. In both the popular press and among the urban planning academy, the "green dot map," as the newspaper's map came to be known, occupies a near-mythical status for the role that it played in changing the political landscape of post-Katrina planning in New Orleans. The popularly understood story is that "with the publication of this map, entire neighborhoods were instantly mobilized to protect their homes and communities from environmental expropriation" (Fields 2009). The map was identified as a pivotal moment when "recovery planning power shifted decisively to neighborhood"-based planning and away from a "heavy-handed" technocratic approach (Wooten

Z. Lamb (✉)
Architecture and Urban Planning, Princeton University, Princeton, NJ, USA
e-mail: zlamb@princeton.edu; zlamb@alum.mit.edu



S. Laska (ed.), *Louisiana's Response to Extreme Weather*,
Extreme Weather and Society, https://doi.org/10.1007/978-3-030-27205-0_3

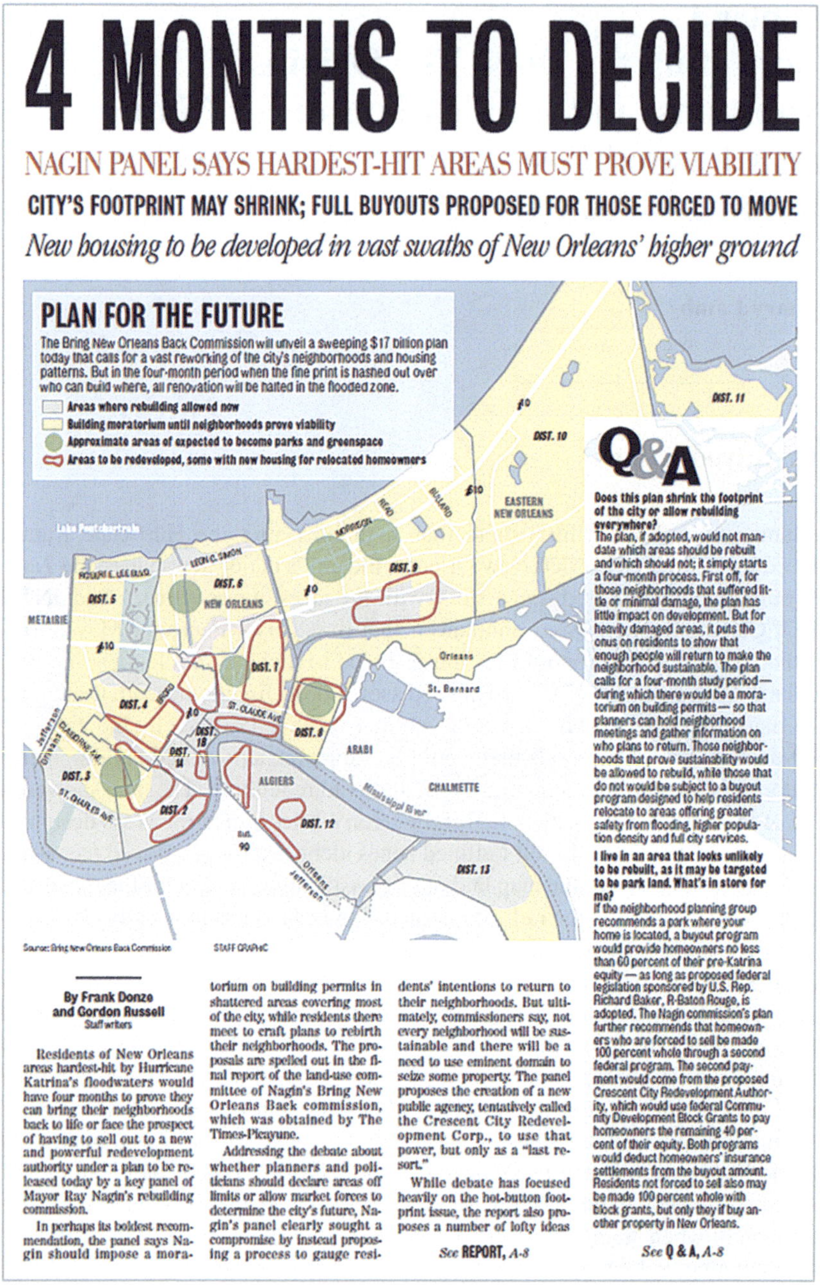

# 4 MONTHS TO DECIDE

## NAGIN PANEL SAYS HARDEST-HIT AREAS MUST PROVE VIABILITY

### CITY'S FOOTPRINT MAY SHRINK; FULL BUYOUTS PROPOSED FOR THOSE FORCED TO MOVE

*New housing to be developed in vast swaths of New Orleans' higher ground*



By Frank Donze and Gordon Russell
Staff writers

Residents of New Orleans areas hardest-hit by Hurricane Katrina's floodwaters would have four months to prove they can bring their neighborhoods back to life or face the prospect of having to sell out to a new and powerful redevelopment authority under a plan to be released today by a key panel of Mayor Ray Nagin's rebuilding commission.

In perhaps its boldest recommendation, the panel says Nagin should impose a moratorium on building permits in shattered areas covering most of the city, while residents there meet to craft plans to rebirth their neighborhoods. The proposals are spelled out in the final report of the land-use committee of Nagin's Bring New Orleans Back commission, which was obtained by The Times-Picayune.

Addressing the debate about whether planners and politicians should declare areas off limits or allow market forces to determine the city's future, Nagin's panel clearly sought a compromise by instead proposing a process to gauge residents' intentions to return to their neighborhoods. But ultimately, commissioners say, not every neighborhood will be sustainable and there will be a need to use eminent domain to seize some property. The panel proposes the creation of a new public agency, tentatively called the Crescent City Redevelopment Corp., to use that power, but only as a "last resort."

While debate has focused heavily on the hot-button footprint issue, the report also proposes a number of lofty ideas

*See* **REPORT**, *A-8*

## Q&A

**Does this plan shrink the footprint of the city or allow rebuilding everywhere?**
The plan, if adopted, would not mandate which areas should be rebuilt and which should not; it simply starts a four-month process. First off, for those neighborhoods that suffered little or minimal damage, the plan has little impact on development. But for heavily damaged areas, it puts the onus on residents to show that enough people will return to make the neighborhood sustainable. The plan calls for a four-month study period — during which there would be a moratorium on building permits — so that planners can hold neighborhood meetings and gather information on who plans to return. Those neighborhoods that prove sustainability would be allowed to rebuild, while those that do not would be subject to a buyout program designed to help residents relocate to areas offering greater safety from flooding, higher population density and full city services.

**I live in an area that looks unlikely to be rebuilt, as it may be targeted to be park land. What's in store for me?**
If the neighborhood planning group recommends a park where your home is located, a buyout program would provide homeowners no less than 60 percent of their pre-Katrina equity — as long as proposed federal legislation sponsored by U.S. Rep. Richard Baker, R-Baton Rouge, is adopted. The Nagin commission's plan further recommends that homeowners who are forced to sell be made 100 percent whole through a second federal program. The second payment would come from the proposed Crescent City Redevelopment Authority, which would use federal Community Development Block Grants to pay homeowners the remaining 40 percent of their equity. Both programs would deduct homeowners' insurance settlements from the buyout amount. Residents not forced to sell also may be made 100 percent whole with block grants, but only they if buy another property in New Orleans.

*See* **Q & A**, *A-8*

**Fig. 3.1** "The Green Dot Map" as published in the *Times-Picayune*, January 11, 2006. ()

2012). In the months and years after its publication, the map provided a unifying enemy around which devastated neighborhoods could organize their resistance. People spoke of their houses and neighborhoods being "green dotted," wore green dots made of paper plates around their necks at neighborhood rallies, and adopted the color of the map's dots in the logos of new community organizations (Wooten 2012; Nelson et al. 2007; Olshansky et al. 2010). Seven years after the map's publication, the symbol retained such potency for the Broadmoor neighborhood (one of the neighborhoods marked for "future parkland") that, when the neighborhood's public library branch reopened in 2012, the coffee shop inside was defiantly christened the "The Green Dot Café." While the intergovernmental power struggles and politics of early post-Katrina planning efforts have been well researched and reported (Olshansky et al. 2010; Nelson et al. 2007; Olshansky 2006), there has been relatively little attention paid to the origins, evolution, and implications of the map that is seen as playing such a pivotal role in this history.

The now infamous green dot map is widely regarded as having been the product of the Urban Planning Committee of the Bring New Orleans Back Commission (BNOBC), a group convened by then Mayor C. Ray Nagin in the fall of 2005 to set the agenda for the city's recovery. In fact, the map that appeared in the *Times-Picayune* that day in January 2006 was the newspaper staff's synthesis and reinterpretation of maps and texts that the BNOBC presented in a press conference earlier that same day. The maps and plans produced for the BNOBC by the landscape and planning firm of Wallace, Roberts, and Todd were, in turn, substantially based on an earlier report produced by a panel of national experts convened in November 2005 by the Urban Land Institute (ULI), a national research and advocacy organization that is closely associated with the real estate development industry. Though the green dot map did become a powerful symbol, most discussions treat the map in the abstract and take its catalytic power for granted. To better understand the role of the map in reshaping the contours of post-Katrina decision-making, this chapter investigates where it came from, what power relations underlay its creation, what values are expressed in its spatial classifications, and, finally, how the map and its reception have shaped planning for water management in New Orleans in the years since.

## 3.2 Literature and Theoretical Context

Though the three primary documents of interest in the story of the green dot map include only a single indirect reference to climate change (a reference to "relative sea level rise" in the ULI report (Urban Land Institute 2005)), this chapter treats the episode as an attempt at de facto climate adaptation. The Intergovernmental Panel on Climate Change (IPCC) defines climate adaptation as "The process of adjustment to actual or expected climate and its effects" by seeking "to moderate or avoid harm"(IPCC 2014). As such, the early planning efforts to restructure New Orleans' post-Katrina urban land uses to reduce flood vulnerability clearly meet the IPCC's definition of adaptation. Viewed as an attempt at climate adaptation, the development and dissemination of the green dot map provide insights into challenges facing more and more cities around the world as they attempt to adapt their historically evolved urban patterns in the face of mounting flood hazards. To develop a fuller understanding of how the green dot map episode might inform future urban climate adaptation, this paper draws from a broad body of literature including work on climate change adaptation, natural hazards research, and critical cartography.

### *3.2.1 Land Use Planning for Hazard Mitigation*

Following pioneering work by geographer Gilbert White, natural hazards scholars from a range of disciplines have produced research on the social and political components of vulnerability to flooding and other natural hazards (White 1945; Wisner et al. 2004; Cutter et al. 2003; Adger 2006). Planning scholars have contributed significantly to exploring how various tools for land use planning and regulation can play a role in reducing hazard exposure (Burby 1998; Burby et al. 2000; Godschalk et al. 1998). Much of this work has advocated for more hazard-informed land use patterns to reduce exposure through a range of federal, state, and municipal policy and planning tools (Olshansky and Kartez 1998; Burby et al. 1999; Beatley 2012). Using Burby et al.'s (1999) schema, the green dot proposal represented a shift in New Orleans' flood hazard mitigation strategy, away from the previous model combining "risk reduction" via levees and building elevation and "risk sharing" via insurance and toward a new model emphasizing "risk elimination," through targeted buyouts and green space preservation in some of the city's most low-lying neighborhoods.

While the natural hazards literature does consider the ways in which planning for natural hazard mitigation can facilitate consensus building (Burby et al. 2000), it largely does not focus on the politics of disaster or land use change or on the distributive implications of such processes. The uneven production and distribution of vulnerability are at the core of much research in human ecology and political ecology (Collins 2008; Pelling 1999, 2003; Gustafson 2015; Hewitt 1983; Wisner et al. 2004). Vale and Campanella's *The Resilient City* explicitly assesses the political and distributive issues that shape how cities recover from disasters (Vale and Campanella 2005).

Of particular interest in this case, authors from a number of different perspectives have addressed the question of to what extent substantial changes in land use are possible or desirable during post-disaster "windows of opportunity"? Much of the early natural hazards planning literature regarded such "windows" as ideal times for "targeting households and business firms to retrofit or relocate"(Olshansky and Kartez 1998). Vale and Campanella, with their emphasis on the politics of resilience, are skeptical of the possibility of substantial change after disasters, observing that it is very rare for post-disaster cities to adopt "visionary new city plans aimed at correcting long-enduring deficiencies or limiting the risk of future destruction" (Vale and Campanella 2005). Like Naomi Klein, who warns against "disaster capitalism" (Klein 2008), Vale and Campanella go further, questioning the desirability of dramatic post-disaster change given the track record of public and private interests "using devastation as a cover for more opportunistic agendas yielding less obvious public benefit" (Vale and Campanella 2005). Writing in the wake of the post-Katrina levee failures, Berke and Campanella (2006) suggest something of a middle ground, arguing that "Hurricane Katrina opened a window of opportunity for creating more resilient communities" but pointing out that taking advantage of such windows may require pre-disaster planning that actively seeks out the viewpoints of often marginalized communities (Berke and Campanella 2006). This tension – between viewing post-disaster planning as a "window of opportunity" for urban restructuring and concerns over post-disaster opportunism and land grabbing – is central to understanding the political conflict arising from the divergent readings of the green dot map.

### *3.2.2 Climate Change Adaptation Through Land Use Planning*

The recent increase in critical attention to climate adaptation has invited renewed attention to the relationship between flood hazards and land use planning. Whereas many previous efforts to reduce flood vulnerability through land use planning were seen as "fighting the last war" by reacting to the most recent disaster event (Godschalk et al. 1998), climate adaptation planning holds promise in inviting approaches to land use planning that are more holistic, forward-looking, and cross-scalar (Adger et al. 2005; Hallegatte and Corfee-Morlot 2011). With little substantial progress from higher levels of government, public and private entities acting at the local and regional levels have taken the lead in local adaptation planning in many areas around the word (Measham et al. 2011; Rosenzweig 2010). Land use planning and regulation are central to many climate adaptation efforts, and they are primarily the responsibility of local government entities in most jurisdictions (Measham et al. 2011). Research on climate adaptation planning has also focused considerable attention on the equity implications of hazard mitigation and other forms of adaptation (Wilson 2006; Hamin and Gurran 2009; Paavola and Adger 2006; Bulkeley et al. 2013). The concerns of poor and marginalized populations of cities are too often underrepresented in climate adaptation planning given the disproportionate vulnerability of such groups to the impacts of climate change (Measham et al. 2011; Bulkeley 2006). Researchers have found that socially and economically vulnerable groups are often directly and indirectly harmed by adaptation efforts (Anguelovski et al. 2016; Sovacool et al. 2015). A range of efforts, including those labeled community-based adaptation, have sought to increase participation and inclusion in climate planning and adaptation (Aylett et al. 2010; Archer et al. 2014; Paavola and Adger 2006).

Adger et al. (2005) proposed four key normative criteria for assessing climate adaptation efforts: *effectiveness*, *efficiency*, *equity*, and *legitimacy* (Adger et al. 2005). These four criteria provide a useful lens through which to understand the conflict which developed surrounding the green dot map, in which different actors placed radically different levels of importance on each of the four criteria.

### *3.2.3 Planning Representation, Maps, and the Shaping of Social and Spatial Reality*

This paper takes as a starting assumption that "maps are knowledge claims that are inherently political"(Kim 2015). This is also the premise at the heart of emerging studies of "critical cartography." Critical cartography includes both analytical and projective tools for questioning the underlying assumptions and power relations behind cartography and developing new ways of using those tools to support alternative claims to knowledge and power (Crampton and Krygier 2005). Arthur Robinson focused early attention on how maps function as *communication* devices that operated through three distinct but related phenomena: sender (mapmaker), medium (map), and receiver (map user) (Robinson and Petchenik 1976; Pickles 2003; Crampton 2001). In "Deconstructing the Map" (1989), J.B. Harley applies the tools of social critique (Foucault and Derrida principally) to cartography (Harley 1989). Harley's work primarily focused on exposing the socially constructed nature of maps and their embeddedness in the power relations of their place and time. While Harley's analysis was largely historical and principally focused on the use of maps as tools of domination, subsequent scholars have gone beyond identifying where maps come from to examine and critique how maps operate to shape social life and power relations (Pickles 2003; Turnbull et al. 1993; Wood and Fels 2009; Crampton and Krygier 2005). This second generation of scholars developed a more complex view of power as "multivocal" and in a constant state of contestation (Pickles 2003). Much of this later critique draws heavily on the critical tools developed by Derrida, Habermas, and Barthes and tends to see maps as narratives or texts that must be read (Crampton 2001). Wood and Fels widened the frame of analysis to consider what they labeled the "paramap," or "everything that surrounds and extends a map" (Wood and Fels 2009). This paramap material includes what they call "perimap," the titles, labels, charts, and borders that frame and situate a map. It also includes "epimap" materials such as any texts or articles to which maps are appended (Wood and Fels 2009).

To date, there has been relatively little attention paid to critical assessment of maps, like the green dot map, which are intended as projective tools for reshaping land use according to changing hazard conditions. While Wood and Fels (2009) do address what they call maps of "threatening nature," they focus more on popular cartography of hazards than on hazard assessment maps or maps for land use planning. In several books, Mark Monmonier has analyzed how the representational tools used in mapping floodplains and coastlines shape perceptions of hazard vulnerability (Monmonier 1997, 2014; Monmonier 2008). However, compared to other critical cartographers, Monmonier tends to deemphasize the political implications of mapping and representation.

Maps are among the most important tools of analysis and communication used by planners and urban designers. John Forester's *Planning in the Face of Power* (Forester 1988) and subsequent work on "communicative planning" (Healey 2012; Innes 1995) highlight the ways in which planners control communication and information to shape power relations. Scholars have long recognized that visual representation used in planning can powerfully shape how practitioners, policy-makers, and the public perceive of planning challenges and proposals. Lisa Peattie analyzed and critiqued the "representations of reality" deployed by planners and other professionals in planning Ciudad Guayana in Venezuela (Peattie 1987). Like Peattie, Raphael Fischler recognized that planners "understand and represent the world in certain ways" that are "necessarily selective and partial" (Fischler 1995). Annette Kim has used the interpretive strategies of critical cartography to analyze how maps and other visual representations created by planners and designer reflected and shaped changing conceptualizations of property in Ho Chi Minh City (Kim 2012, 2015). These and other analyses of visual representation in planning provide useful precedents for interpreting the production and popular reception of the green dot map.

While much of the literature treats critical cartography as an interpretive activity undertaken by scholars to decode the hidden social meaning and politics of maps, some recent work has articulated a variant labeled "lay critical cartography" (Cidell 2008) that shifts the locus of critical analysis to consider explicitly the "social life of maps, the political responses they elicit, and the political possibilities they enable and disable" (Gustafson 2015). Seth Gustafson, a geographer rooted in urban political ecology, has considered the lay critical cartography of landslide hazard maps in North Carolina which ignited intense political opposition from pro-development forces (Gustafson 2015). Such analysis of how a map "provokes new political activities and environmental changes" is a useful precedent in making sense of the response to green dot map.

## 3.3 Methods

The green dot map came to hold tremendous symbolic power because of the immediate context of its production, the chain of interpretation and reinterpretation from which it emerged, and the broader social and historical conditions into which it was projected. In seeking to make sense of this broader context, this chapter traces the creation of the map through three different generations, each of which took different approaches to classifying space and communicating through text and graphics. This chapter uses critical cartography and lay critical cartography to analyze the green dot map and its precedents. It seeks to unearth the "design politics" of the maps to reveal how "social and political preferences are expressed and manipulated" (Vale 2013). As such, it considers how the maps' graphics as well as the "paramap" materials, such as the texts within and surrounding the maps, convey the values of the makers. The analysis of each iteration of the map will address what Bowker and Star call the "practical politics of classifying," by which the maps "arriv(e) at categories" of redevelopment land use and "what (is) visible or invisible" within the categorizations and abstractions of each map (Bowker and Star 1999). The chapter also draws on contemporary media accounts and secondary literature to analyze how the maps were received, both among planners and the public at large. Finally, the chapter includes a brief discussion of the implications and impacts of the green dot map on land use planning and water management in New Orleans. This section is based on interviews with planners, designers, and decision-makers involved in recent and ongoing planning activity in the city.

## 3.4 Analysis

Each of the three generations of plans and maps that lead to the green dot map assumed that New Orleans' post-Katrina population would be significantly reduced and that some degree of "shrinking the footprint" or "neighborhood consolidation" would be necessary to reduce the city's exposure to future flooding (Olshansky et al. 2010). These policies were seen as necessary to ensuring that urban densities in the city would be high enough to allow for efficient and safe delivery of municipal services. While all three presentations called for a smaller New Orleans, they differed in crucial ways, including the composition of the drafting bodies and their claims to authority, their intended audiences, and the graphical and text language deployed to communicate their classification schemes for future land use. The following sections analyze the evolutions and changes through which the green dot map developed.

### *3.4.1 The ULI Plan: "New Orleans, Louisiana: A Strategy for Rebuilding"*

The ULI plan that was released in November of 2005 was the product of a convening of national experts on urban development and planning (Urban Land Institute 2005). Reflecting the professional positioning of the members of the panel and the ULI more generally, the report tends to frame the task of rebuilding in the language of urban real estate development, finance, and administrative efficiency. The report holds that "the city should be rebuilt in a strategic manner" in which the "feasibility of investment" in damaged neighborhoods will be evaluated before public or private funds will be used to rebuild. While the report recognizes that such a strategic reinvestment approach will inevitably impinge on some residents' property rights, they propose a market-based remedy, stating that where property is deemed "unusable," people "should be given fair compensation for their property." Expressing the importance of "government effectiveness," the report calls for the creation of a new body, the Crescent City Rebuilding Corporation (CCRC) that would "provide expeditious compensation for those unable to build."

While the overriding theme of the ULI report is a call for efficient redevelopment that would avoid "scattered, uncoordinated, dysfunctional redevelopment," the report connects these notions of efficient redevelopment to values of security, aesthetic beauty, and environmental balance. The plan calls for reducing the urban footprint of New Orleans in order to "ensure the health and safety of the residents of each neighborhood," to create "functional and aesthetically pleasing neighborhoods," and ultimately to create a city that is "in harmony with the natural environment," particularly with respect to the relationship between urban space and the surrounding waters.

Though the ULI report calls for a radical reconfiguration of the city to accommodate a smaller population on more flood-safe territory, the report does go to great lengths to emphasize the importance of conducting inclusive planning processes to ensure equitable results. Among the report's "Key Findings" are a number of items related to the importance of retaining the city's unique culture. The report goes further to say that "planning for the rebuilding of each neighborhood must involve the citizens from that neighborhood."

Although the panel emphasizes the importance of equity and inclusive planning, they also make clear that, in the cause of reconstruction, diversity and pluralism may have to give way to functional demands. The panel uses the language of equity and security to justify the realignment of residents in saying, "every citizen has a right to return to a *safe* neighborhood" [emphasis added]. Implied in this formulation is that if a neighborhood is deemed unsafe, it may not be rebuilt. Similarly, the report deemphasizes critical conversations about racial justice, an issue of deep resonance in a city and region where planning has long been seen as a tool for enforcing racial hierarchy and systematic prejudice. Though it states that "diversity, equity, and cooperation are of critical importance," the panel's report holds that "the recovery must not be held back by the racial issues that have slowed progress in the past." In this phrase, the panel vaguely blames the contentious history of "racial issues" for impeding progress, ignoring the enormous racial inequities in previous planning actions and arguing for a recovery process freed from the impediments of racial politics.

The only image in the ULI report that puts forward any form of spatial proposal is a single map outlining a "proposed rebuilding framework."[1] The graphics and text of the map, like the larger report, is characterized by a focus on finance and development. The map places the tourist-oriented French Quarter and central busi-

[1] It has not been possible to include the ULI map in this manuscript. It is available at http://uli.org/wp-content/uploads/ULI-Documents/2005NewOrleansPPT.pdf

ness district at its center and is cropped to exclude a sizeable portion of New Orleans East, a primarily African-American suburban neighborhood that sustained heavy flood damages. The map includes very few streets or other landmarks by which one might locate a specific site within the city to see how it might be affected by the proposal. The territory that is included in the map is divided into three "investment zones," zones A, B, and C. The text that accompanies the map states that the investment zone classification should be done according to a broad suite of criteria including the extent of flooding damage, physical vulnerability, infrastructure capacity, historical significance, and housing occupancy and vacancy. In spite of the holistic intention, the ULI map appears to define zones primarily on the basis of flooding depths or topography, ignoring all of the other issues of existing adaptation, infrastructure, and vulnerability that they previously defined as critical. Even if one were to accept topography as a suitable single criterion on which to define investment zones, it is unclear how the panel determined what elevations or depths of flooding constituted logical thresholds for zone classifications.

The panel makes strategic use of the passive voice and technocratic language of urban hydrology to de-emphasize the impacts of their zone classification scheme on residents and neighborhoods. The report indicates that, in Zone A, the most heavily damaged areas, "open space (will) be programmed to reach its greatest capacity to manage storm water retention, treatment, and flow." Though such a statement suggests the replacement of residential areas with functional green space, the panel emphasizes that:

> In these areas, great care must be taken to work closely with residents to determine the exact patterns of reinvestment necessary to restore and create a functional and aesthetically pleasing neighborhood.

Again employing vague, passive, and functionalist language, the report states that, in Zone B, the areas that are moderately damaged and presumably moderately vulnerable to repeated flooding:

> some reprogramming of open space probably will be needed to mitigate the impact of flooding and account for space that may not be rebuilt for any number of reasons.

According to the panel's recommendations, only in Zone C areas, those areas with little or no damage, would building be allowed on a "parcel-by-parcel" basis.

In both text and graphics, the panel employs strategic imprecision to make clear the preliminary nature of their recommendations. The fuzziness and low contrast with which the three zones are rendered in the map are appropriate to the provisional nature of the plan. The accompanying text also expresses uncertainty regarding the spatial classification in saying,

> The precise edges of the respective zones and their transitions cannot be established without detailed on-the-ground surveys, which have yet to occur.

While it is wholly appropriate that such a preliminary report should leave room for revision and refinement, this statement suggests that the missing data necessary to solidify the investment zone classification could be gleaned by a physical survey, without consideration of the range of social, political, and economic factors that would be invisible to such a survey.

In addition to the designation of investment zones, the ULI map indicates "Development Sites" and a network of new proposed "Open Spaces." The map designates sites for "economic development" and "mixed use housing," using red and orange ovoid shapes. While neither the map nor the accompanying report provides insights into how these zones were selected or what their designation would mean for future development, they were intended to be what Monmonier calls "green-lined" zones in which the government would target investment and deploy special incentives for development (Monmonier 2010). The network of designated open space includes new greenways located along the city's canals as well as a major linear green space following the path of Interstate-10. Though the accompanying text provides no clues as to the rationale for the configuration of the open space network, indications elsewhere in the report suggest that the panel advocates the expansion of open space in the city for functional, recreational, and aesthetic purposes.

Though the ULI map and the report in which it is embedded lay out an agenda that would have sweeping implications for reorganizing the city, they also maintain a tone of strategic imprecision and deference to equitable and inclusive planning processes. The plan takes for granted that a radical spatial reorganization of the city will be necessary for reasons of efficiency and security. It uses the language of development, finance, and investment and treats the city as an abstract administrative and financial institution first and foremost. The preliminary map categorizes urban space according to development and investment potential with a mix of appropriate fuzziness and unexplained precision.

### 3.4.2 *The BNOBC Plan: "Action Plan for New Orleans: The New American City"*

Building on the recommendations made in the ULI report, the Urban Planning Committee of the BNOBC issued their proposed plan and maps in a presentation delivered in January 2006. If the ULI map and plan showed a degree of imprecision and deference to inclusive planning processes yet to come, the BNOBC plan and maps were less constrained by such signs of professional humility. From the very name of the presentation, with its emphasis on "action" and "new"-ness, the commission's report took on many of the ULI report's recommendation and stripped away the layers of uncertainty and deference to process. Where the ULI panel was composed of national technical experts, the BNOBC was made up of "knowledgeable community members and professionals" (Ehrenfeucht and Nelson 2011) with a strong bias toward "business and developers"(Allen 2013). The New York Times referred to Joseph Canizaro, the local developer and financier appointed to head the group, as "the mogul who would rebuild New Orleans" (Rivlin 2005a). Both Canizaro himself and many commentators emphasized his ties to both President

George W. Bush and Mayor Nagin (Olshansky et al. 2010; Rivlin 2005a). Where the ULI panel drew its authority from academic and technical credentials, the BNOBC's claim to legitimacy was rooted in financial and political resources.

From early in the formation of the commission, BNOBC members demonstrated the more problematic side of the post-disaster "windows of opportunity," issuing public comments that emphasized the view that the devastation and mass evacuation of New Orleans was an opportunity to reshape the city. Less than a month after Hurricane Katrina made landfall, Mr. Canizaro told the *New York Times*, "I think we have a clean sheet to start again…And with that clean sheet we have some very big opportunities" (Rivlin 2005a). Another commission member, James Reiss, the chairman of the Business Council of New Orleans, went further, explicitly linking the spatial restructuring of the city to the creation of a new social order, telling the *Wall Street Journal* that the rebuilding effort was an opportunity to rebuild the city "in a completely different way: demographically, geographically, and politically" (Wooten 2012). While the ULI report appealed to a sense of unity, smoothing over New Orleans' history of racial divisions and distrust, the final BNOBC presentation makes no mention of race at all. To the extent that issues of racial justice or inclusion were discussed at all, they appear to have taken the form of pragmatic business concern. Canizaro was quoted as expressing the need for the "business community" to work with "our African-American associates" to develop the plan, a phrase that suggests that African-Americans were not a part of the business community (Rivlin 2005a).

Drawing on the ULI plan that had been issued 2 months before, the BNOBC plan frames the reconstruction of the city as, first and foremost, a problem of real estate development and finance. Where the ULI map and report uses the language of investment, the BNOBC presentation focuses on property ownership, site control, and acquisition. The report again assumes that the reconstruction of the city will require a massive reshuffling of land use patterns and establishes categories of redevelopment according to levels of damage, vulnerability, and development potential. Echoing the language of efficiency and equity used in the ULI discussion of planned shrinkage, the BNOBC presentation emphasizes the need to "consolidate neighborhoods with insufficient population to support equitable and efficient service delivery" (Bring New Orleans Back Commission, Urban Planning Committee 2006). Thus, in the BNOBC plan, equity is redefined as a matter of service delivery after a spatial reconfiguration of the city that may or may not be equitable.

The presentation given by the BNOBC on January 11, 2006, included several maps along with accompanying text slides that lay out a four-part categorization of the city's lands. Though the scheme is based on the ULI report's categorization of investment and development zones, it differs in ways that came to be important both substantively and symbolically. Where the ULI panel identified and categorized spaces according to the level of "investment" and "development" that should be targeted for them, the BNOBC presentation added explicit consideration of property acquisition and administrative authority in the form of the city's authority to issue building permits.

The commission identifies "Immediate Opportunity Areas" as those areas with "little or no flood damage." These areas, which roughly correspond to the ULI report's Zone C, are to have "expedite(d) permits for repairs and construction of new housing." The maps call for the "areas contain(ing) deeply flooded and heavily damaged properties," roughly corresponding to the ULI's Zones A and B, to be collapsed into a single category known as "Neighborhood Planning Areas." The name of these zones emphasizes the "neighborhood planning process" that the commission urged be started immediately to "determine the future of the areas." In spite of this emphasis on planning, the recommendation repeated elsewhere in the presentation that the city should "not issue building permits in heavily flooded/damaged areas" led to fear of land grabs and redlining (Olshansky et al. 2010). The BNOBC's focus on *properties* rather than *households* or *people* as the most important unit of analysis for the determination of a neighborhood's fate reinforced the impression that the Commission was primarily concerned with urban land as a legal and financial phenomenon substantially devoid of social importance. When the presentation later gives recommendations for who should be involved in the neighborhood planning processes, "neighborhood residents" constitute only one of the eight named groups identified for participation, with the other seven slots occupied by technical experts of various kinds. Treating people who lived in impacted neighborhoods as just one among several relevant constituencies fits with the broader perception that the planning process was insufficiently attentive to the wishes of residents.

Again emphasizing the commission's focus on development, the BNOBC identified both "Infill Development Areas" and "Targeted Development Areas." The former are defined as those "underutilized sites on high ground" or areas "requiring demolition and clearance that can be developed with houses, commercial, and institutional uses" to accommodate uses relocated from more flood-prone areas. On these sites, marked by bright pink shapes on the maps, the commission recommends an expedited course of development including "consolidat(ion) (of) public and private ownership," "prepar(ation) (of) development plans," and "issu(ing) (of) developer requests for proposals."

Similarly, for the "Targeted Development Areas," the commission recommends that the city, "identify and facilitate financially responsible developers to develop large numbers of houses quickly," suggesting that financial capacity was the primary criteria of importance. While these sites are marked as numbered points on the BNOBC maps, their exact location appears to be somewhat arbitrary; some lie in heavily flooded former residential zones, others in the city's central business district, and still another in a largely industrial zone. The fact that the report does not explain any rationale for the location of these sites raises the question of whether this is a case of "the substitution of precision for validity"(Bowker and Star 1999) or if members of the commission had specific, unstated reasons to target these particular sites.

As in the case of the inexplicably precise locations of the "Targeted Development Areas," the BNOBC employed unexplained spatial and graphical specificity in what would become the group's most infamous map. The "Parks and Open Space" map (Fig. 3.2) drew from the ULI map the idea that the city should develop a network of

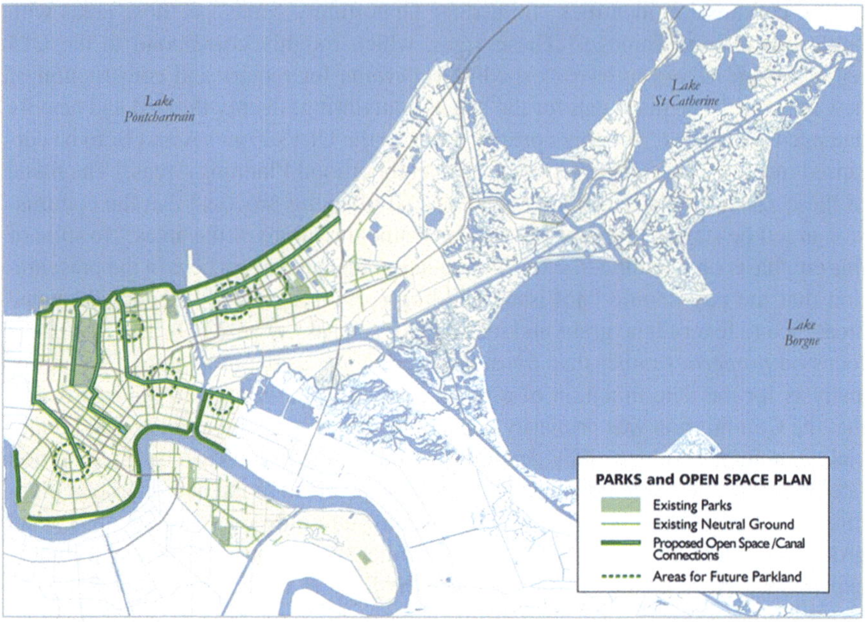


**Fig. 3.2** The "Parks & Open Space Plan" as presented in *Action Plan for New Orleans: The New American City* by the Bring New Orleans Back Commission, Urban Planning Committee in January 2006. (Used with permission of the Bring New Orleans Back Commission)

green spaces that roughly corresponded to the city's drainage canals. In addition to these linear green spaces, the map indicates, with six green dashed, but unfilled, circles, "Areas for Future Parkland."[2] Text elsewhere in the presentation describes the "Parks and Open Space Plan" employing the same functionalist language of systems and real estate acquisition that is favored throughout the document in directing the city to "identify properties that can become part of the system and begin assembly."

To facilitate the assembly of the land necessary for the BNOBC's ambitious green space plan, as in the other development activities included in the plan, the commission recommended the immediate creation of the CCRC. Like the ULI panel, the BNOBC's consideration of homeowners in areas slated for depopulation was largely restricted to the administration of financial transactions. The presentation calls for the city to "aggressively support" legislation to "accommodate buyout of homeowners in heavily flooded and damaged areas" including through the use of eminent domain.

The BNOBC plan, like the ULI plan before it, presumed a radical realignment of the city's population and land use. This plan deemphasized the role of inclusive

[2] Perhaps on account of the furor that this map later generated, the word "potential" was amended to the designation for these circled areas on a later version of the map that appeared in *Architect* Magazine in 2007.

planning processes and called for the immediate enactment of aggressive redlining and greenlining to halt reconstruction in some areas and jumpstart development in others. While the categorization schemes in both plans allude to the need to account for a wide range of criteria, both defaulted to "elevational determinism," wherein topography is the dominant driver of rebuilding decision-making (Wagner and Frisch 2009). The green space map that would become the commission's most recognizable artifact combines the language of systems and functionalism with highly diagrammatic abstract geometries. In all of these classifications, the BNOBC treats the project of urban disaster recovery as a problem of efficiently maximizing real estate finance and public administration virtually devoid of considerations of public consensus or pre-Katrina cultural, social, and economic conditions.

### *3.4.3 The Times-Picayune Map: The Green Dots*

On the same day that the BNOBC presented their plan, the city's largest newspaper ran extensive coverage of the plan along with their own interpretation of the accompanying graphics. The *Times-Picayune*'s coverage, like the BNOBC plan itself, emphasized the plan's sweep and ambition, but it also took several critical steps to reframe the plan in order to address the concerns of readers and residents. The main story's writers, Frank Donze and Gordon Russell, characterize the plan as a "vast reworking of the city's neighborhoods and housing patterns." Rather than presenting this process through the lens of real estate opportunity or civic administration as the previous framings had, the reporters recast the plan as victimizing a traumatized population through powerful new bureaucracy. The story begins:

> Residents of New Orleans areas hardest-hit by Hurricane Katrina's floodwaters would have four months to prove they can bring their neighborhoods back to life or face the prospect of having to sell out to a new and powerful redevelopment authority. (Donze and Russel 2006)

The map that occupied the majority of the front page that morning included several critical reinterpretations of the BNOBC's graphics and text, which shaped the reception of the plan in powerful ways (Fig. 3.1; see the "Introduction" of this chapter). First, the map cropped some of the furthest reaches of the eastern portion of the city in order to zoom in on the more densely populated areas. It also included and labeled major streets, allowing readers to more readily locate specific sites in the city. Though the map includes most of the categorization scheme indicated in the BNOBC maps, it does make some significant changes. Most importantly, the tentatively dash-outlined circles indicating future parkland on the BNOBC map here appear as solid green dots, lending them more graphical prominence and visually associating these areas with the existing parks indicated on the map. The linear green spaces indicated in the ULI and BNOBC maps are omitted in the newspaper's version. Divorced from the "system" of functional greenways, the green dots appear to be arbitrarily sited around the city. Presented in this more solidified form, among the red outlined areas for intensified redevelopment, the green dots came to be seen as a harsh form of "prohibitive cartography" demarking spaces that would be off limits for return and reconstruction (Monmonier 2010).

While the graphical form of the newspaper's map accentuated its prohibitive character, the text that appeared in the map's key and the accompanying story reinforced the conflicting values embedded in the map. With the labeling of each of the plan's land use categories, the writers recast the BNOBC's language of real estate opportunity in terms of the impact on individual homeowners and residents. The BNOBC's "Immediate Opportunity Areas" are rendered as "areas where rebuilding allowed now." "Neighborhood Planning Areas" are labeled as zones where the city will enact a "building moratorium until neighborhoods prove viability," accentuating the prohibition and the proposed process of administrative viability testing. The BNOBC's "Infill Development Areas" are recast as "areas to be redeveloped, some with new housing for relocated homeowners," raising the specter of social reshuffling and intensive infill development.

The story that accompanied the map further reframes the plan, viewing its proposals from the point of view of homeowners. It includes a "Q&A" format that poses and answers such questions as: "I live in an area that looks unlikely to be rebuilt as it may be targeted to be park land. What's in store for me?" In its direct address and use of first-person pronouns, the story lifts the veil of abstraction and technical language that had characterized both the ULI and BNOBC plans to make concrete the impacts of this sweeping proposal for neighborhoods and residents.

## 3.5 Discussion

Proto-critical cartographer Arthur Robinson proposed that maps operate as communication devices through the interaction of three distinct, but related, phenomena: the mapmaker (sender), the map (medium), and the map viewer (receiver) (Pickles 2003). To the extent that the story of the green dot map has been told in the years since its release, it has largely been a story about how the map was received, the transformative impact that it had on the post-Katrina planning process, and, to a lesser extent, the realities and perceptions of who was responsible for the maps. The preceding analysis has clearly shown that, rather than viewing the *Times-Picayune* map as a singular medium of communication that passed from sender to receiver, it is critical to understand the map's creation as an iterative, stepwise process shaped by at least three sets of institutional actors, each with their own vision, values, and priorities for the reconstruction of the city. In the successive reinterpretation of the green dot map, first from the ULI plan to the BNOBC and then from the BNOBC to the *Times-Picayune*, Robinson's three-part schema becomes elongated and shifts from a relatively simple, unidirectional linearity to a series of interpretive tangents. To better understand the impact of the maps, it is important to consider all three of these components: the power structures and institutions that lay behind the creation of the maps, how each set of actors reinterpreted the ideas and images produced by the preceding mapmakers, and finally, the social and political conditions in which the process and products were ultimately received.

### 3.5.1 The Mapmakers

Crampton and Krygier's premise that maps "actively construct knowledge," "exercise power," and "can be a powerful means of promoting social change" highlights the need to interrogate the identities and interests of the people and institutions responsible for mapmaking (Crampton and Krygier 2005). In the case of the green dot map, there are at least three sets of relevant mapmaking actors: the ULI expert panel, the Mayor's BNOB Commission and their consultants, and the reporters and staff of the *Times-Picayune*.

According to the ULI report, in the autumn after Hurricane Katrina:

> ULI assembled expert teams and an advisory panel of economic development, financial, design, redevelopment, land use, and urban planning professionals to work with Mayor C. Ray Nagin's Bring New Orleans Back Commission. (Urban Land Institute 2005)

As this statement indicates, the ULI report and the BNOBC plan lean on one another for their authority and legitimacy. The ULI draws its claims of authority largely from the technical expertise of the assembled panel and then grounds its local legitimacy in the group's political connection to the Mayor of New Orleans and the BNOBC. The 41 experts listed as contributing to the ULI report include a range of respected public and private sector leaders in real estate finance, law, development, construction, and planning, but none of the ULI experts listed New Orleans as their base of operations. While the outsider status of the assembled panel might be seen as beneficial in some settings, in the context of New Orleans, a city whose residents guard their distinctiveness with near-religious zeal, it made the panel and their recommendations immediately suspect.

Where the ULI report and maps drew their authority from the urban real estate, planning, and redevelopment expertise of the assembled group of national experts, the BNOB Commission appointed by the mayor was firmly rooted in New Orleans. The composition of the central committee of the BNOBC was clearly intended to address the city's long-standing racial tensions; it was composed of 17 people, 8 white, 8 black, and 1 Latino. In spite of this superficial diversity, before the group's work had substantially begun, Barbara Major, an African-American activist and the Commission's co-chair expressed skepticism, saying "I think some people don't understand that an equal number of black and white isn't the same as equity" (Rivlin 2005b).

Reflecting Mayor Nagin's long focus on improving the city's business climate, the group's membership was skewed toward business and development interests. The prevalence of business interests on the Commission served as a signal that the city was pursuing a largely privatized reconstruction that would be in line with the neoliberal policy preferences of the federal administration at the time. J. Stephen

Perry, president of the New Orleans Convention and Visitors Bureau at the time, told reporters, "I think the importance of this group is that it will give the federal government the confidence that the city is harnessing the private sector to do a lot of its work" (Rivlin 2005b). Though the group's business affiliations were seen as a strength by some, among many displaced residents and activists, they fueled suspicion that the Commission's recommendations for recovery would not adequately address the concerns of the city's poor citizens (Nelson et al. 2007; Olshansky et al. 2010). Over the course of the fall of 2005, as the Commission undertook its work, quotes from members alluding to the opportunity of the city's post-Katrina "clean sheet" and their desire for demographic restructuring only reinforced these suspicions (Rivlin 2005a; Wooten 2012; Gotham and Greenberg 2014).

While the ULI panel spoke from a position of technical expertise and outside detachment and the BNOBC drew its authority from the wealth, political connections, and business acumen of its members, the *Times-Picayune* maps and accompanying text drew their legitimacy from the position of the newspaper and its reporters as embedded members of the local community. With its urgent headlines and personal tone, the paramap text surrounding the published green dot map reinterpreted the optimism and technical language of the earlier plans to reflect the fears and suspicions of readers. In spite of the technical expertise and superficial diversity of the ULI and BNOBC, the fact that the two groups were widely perceived as not representing the viewpoint of the majority of displaced New Orleanians meant that their plans failed to establish their legitimacy in the eyes of the city's residents.

### 3.5.2 Reinterpretations

The preceding analysis of the three maps considers how each successive generation of maps and plans leading up to the green dot map communicated the values and interests of its makers through graphics and text. The final map reflects a process that began with a deliberately imprecise and highly qualified preliminary planning map produced by the ULI panel. While the ULI plan included recommendations for a planning process that would take account of equity and inclusion, their map and accompanying texts categorized space primarily according to *investment potential*. The ULI recommendations were then reinterpreted by the BNOBC as a real estate development proposal, largely stripped of the ULI's language on race, equity, and participation. The graphics and texts of the BNOBC plan presented a reconstruction process guided by a classification of *real estate acquisition activities*. The BNOBC's plans and maps were then finally recast by journalists and graphic designers at the *Times-Picayune* to focus on the *impacts on people and neighborhoods*.

While the BNOBC proposal draws heavily on the ULI policy and design recommendations, the proposal does not reference the ULI directly anywhere in the text or graphics. The omission of any reference to the ULI panel may have been an attempt by the Mayor's Commission to distance itself from the earlier report, which had drawn local suspicion and resistance (Olshansky et al. 2010). On the other hand, though the map that appeared on the front page of the *Times-Picayune* on January 11, 2006, was substantially different in text and graphics from the BNOBC's maps, it includes a citation in the lower left below the map that reads, "Source: Bring New Orleans Back Commission." By presenting their editorially manipulated map as the product of the Commission, the paper blurred the line between re-presentation and commentary, exacerbating the already widespread distrust of the planning process.

### *3.5.3 Map Receivers*

While the reception of and reaction to the green dot map have been by far the most discussed aspects of the entire episode, it is nonetheless worth considering these reactions systematically through the lenses of lay critical cartography. The map and the categorization system that it represented were roundly rejected and attacked on a number of different fronts. The negative public reactions to the green dot map included critiques of the plan on the basis of all four of the normative criteria for climate adaptation laid out by Adger et al. (2005): *efficiency*, *effectiveness*, *equity*, and *legitimacy*.

Though the BNOBC maps and plan were presented as "a rational path to recovery," they were widely critiqued as both rigid and arbitrary, attacking their claims to *efficiency* and *effectiveness* (Nelson et al. 2007). Many regarded the BNOBC map and plans as overly rigid and formulaic in their use of a logic of "elevational determinism" to condemn low-lying neighborhoods (Wagner and Frisch 2009). Others criticized the BNOBC proposals for arbitrarily condemning some neighborhoods and not others even though virtually the entire metropolitan area is at risk of flooding (Nelson et al. 2007). Calling into question the factual basis on which the plans were based, one Gentilly resident told reporters:

> Unless they could prove to us unequivocally that we were placing ourselves and our children in danger – and they couldn't – then we were not going to allow anyone to unilaterally dictate where we couldn't live. (Krupa 2010)

With its combination of schematic abstraction and unexplained precision, the graphics and paramap text of the *Times-Picayune* map played a significant role in shaping the perception of the plan as both rigid and arbitrary.

While some critiques focused on issues of efficiency and effectiveness, most of the critiques of the green dot map centered on issues of equity and legitimacy. The ULI and BNOBC plans were widely seen as having been the product of "top-down process[es]" (Nelson et al. 2007; Wooten 2012) by "closed-door" committees with "little input from communities" (Irazábal and Neville 2007). Wade Rathke, a leading local activist and founder of ACORN, directly impugned the legitimacy of the process, decrying the "arrogance" of the recommendations and labeling the ULI and BNOBC "unelected and unaccountable"(Rathke 2006). One Ninth Ward resident voiced a distrust of the municipal authorities more broadly in attacking the legitimacy of the city's planning process, saying "I was not going to let a corrupt city government deny my right to return" (Gotham and Greenberg 2014).

Charitable critiques of the green dot map and the plans behind it regarded the plans as naïve to the political realities of the city and inadequate in addressing the interests of the historically victimized low-income and African-American populations of the city. Less charitably minded critiques saw the plans as deliberately hostile to those vulnerable populations. For many critics, the perceived lack of *legitimacy* of the plans went hand in hand with their failure on the *equity* criteria. Political distrust and social division between white and black populations in New Orleans are deeply rooted, and they significantly shaped the response to the early post-Katrina planning processes (Olshansky et al. 2010; Gotham and Greenberg 2014). As in many disasters, the flooding of New Orleans disproportionately harmed African-American and low-income populations in the city due to the heightened physical and social vulnerability of some areas (Tierney 2006).

These same populations and neighborhoods had also suffered disproportionately during previous infrastructure and urban renewal projects in the city (Breunlin and Regis 2006; Nelson et al. 2007). This history of displacement and victimization at the hands of planners and developers led many in New Orleans to harbor a deep distrust of both public and private sector powers seeking to remake the city after the flooding. Given the city's history of racial animus and the racially tinged opportunism of statements from members of the BNOBC, the plan's calls for targeted prohibitions on building permits and buyouts in heavily flooded neighborhoods raised fears that the plan was an elaborate attempt to "keep many African Americans from returning" to the city (Nelson et al. 2007). So charged was the discussion of racial inequality in the proposed land use restructuring that activists labeled it an attempt at "ethnic cleansing" (Nelson et al. 2007) and "class and racial redlining" (Gotham and Greenberg 2014).

## 3.6 The Lasting Impacts of the Green Dot Map

More than a decade after the initial controversy surrounding the green dot map, the episode still looms large in discussions about water management in New Orleans. The swift and overwhelmingly negative response to the proposals presented in the *Times-Picayune* in January 2006 decisively ended official discussions of large-scale reshuffling of land uses to make way for stormwater-absorbing green space. Nonetheless, city leaders and planning and design professionals in New Orleans have continued to pursue "green infrastructure" strategies as one component in the "multiple lines of defense" against urban flooding (City of New Orleans 2015). A series of convenings, plans, and pilot projects have sought to demonstrate the utility of landscape-based stormwater retention and infiltration strategies. Key projects include the Dutch Dialogues series (2006–2009), The Greater New Orleans Water Plan (2013), the New Orleans Resilience Strategy (2015), and the Gentilly Resilience District pilot projects supported by funds from the US Department of Housing and Urban Development's National Disaster Resilience Competition (NDRC) (ongoing). While these projects have made progress in bringing green infrastructure into the mainstream of flood mitigation discussions, many observers cite the green dot map episode as having created an atmosphere of suspicion and distrust, against post-Katrina planning generally and against green infrastructure specifically. This section recounts common themes regarding the lasting implications of the green dot map episode as they emerged from interviews with government officials and planning and design practitioners who have been active in the city's recent green infrastructure efforts. The interviews took place in 2016 and 2017.

While some responded to questions about the ongoing impacts of the green dot map by saying that the city has moved "past it" (New Orleans-based landscape architect 2017) or that "you don't hear much about it as much now" (Senior city official 2017), even those who minimized the ongoing importance of the episode regarded it as having substantially shaped planning over the last decade in New Orleans. One New Orleans-based planning practitioner reported that the episode confirmed the preexisting suspicions that New Orleanians had regarding heavy-handed planning, saying:

> The green dots really just cemented people's skepticism about planning. It was that way before, but it really just cemented it in people's minds. (New Orleans-based planner 2017)

While the episode may have heightened preexisting suspicions of planning generally, it had an especially pronounced impact on efforts to advance green infrastructure flood mitigation strategies. One local designer involved in these efforts reported that "everyone is super conscious of the green dot fiasco" (New Orleans-based designer 2017). A planner working with the city said that when their agency recently initiated several green infrastructure pilot projects, residents asked suspiciously "If you are doing this (green infrastructure), does that mean that you are not going to build more houses in the neighborhood?" (City official 2017). After the early post-Katrina conflicts over green infrastructure, Dutch urban designers who have been involved in water planning in New Orleans reported a wariness of proposals that might get entangled in "local politics" (Dutch water planner 2017). Another Dutch designer said that they avoided becoming involved in discussions of projects that would involve substantial displacement since, "As a Dutch firm, for us to be involved in these society issues ... didn't feel safe... or appropriate" (Dutch urban designer 2017).

While there was widespread agreement that the green dot map episode had negatively colored the public perception of post-Katrina planning generally and green infrastructure specifically, interviewees reported a range of different ways that they perceived the episode as having shaped planning processes and projects in the years since. Some described the failures of the green dot map as failures of communication and translation. One urban designer pointed to the episode as "a good example of the danger of visualizations" and said that the major problem was that a map that "was intended as a discussion... was interpreted as against the lower economic status people" (Dutch water planner 2017). Recognizing that the green dot map emerged from a process of translation as described above, a senior city official pointed out that the harsh reaction was to the map that "the *Times-Picayune* showed in the paper, which was not what the ULI actually showed" (Senior city official 2017).

Reflecting this understanding, that the problems with the green dot map were attributable to failures of communication and representation, one urban designer reported that in their subsequent work, they have "tried to be more specific than dashed circles" in their representation of new green infrastructure. They went on to say that, to avoid the appearance of bias or arbitrariness, they base their recommendations on the "fundamentals of how soil and water interact" so as "to be more defensible" (New Orleans-based designer 2017). Another city planner reported that, in making the case for more green infrastructure investment in a neighborhood, they make a great effort in "connecting [the projects] to economic development and neighborhood revitalization," "forefronting those goals with flood protection subsequent to that" to make the point that these new projects are "very different from the green dot scenario" (City official 2017).

In addition to the ways that the green dot map episode has changed how planners and designers communicate green infrastructure proposals, the experience has also substantively informed planning processes and projects in the years since. A senior city official reported that the BNOBC plan "was done in isolation, with no one in the city." From that experience, they reported that "everyone learned… how to engage with people" and that, "the engagement has gotten much better… partially because of the green dot debacle" (Senior city official 2017).

Several practitioners and officials reported that, after the green dot episode, green infrastructure projects have been more opportunistic and smaller in scale. Designers and planners said that the efforts to institute landscape-based stormwater management have focused on using existing open space rather than advocating for large-scale projects that would require displacement of houses and neighborhoods. A senior planner with the city remarked that:

> Many of the places where we are prioritizing these kinds of projects are in areas that were under the green dots. Without displacing people and without all of the negative connotations of the green dots.

They explained that, "where there is vacancy, you can use that [space]" for green infrastructure. They went on to say, "This is not about taking something away. This is about adding to. We aren't taking houses away, we are adding parks." (Senior city official 2017) These sentiments reflect both the strategic shift to an opportunistic approach to green infrastructure that has characterized recent efforts in New Orleans and the recognition that the legacy of the green dot map has required a reframing of the communication surrounding these projects to address not only their effectiveness and efficiency but also their equity and legitimacy. It is important to note that, in some cases, the open space for this opportunistic green infrastructure approach was made available for those purposes by leaving empty lots on which flooded homes were purchased and demolished by the New Orleans Redevelopment Authority with the decline and movements of New Orleans' population following Hurricane Katrina.

## 3.7 Conclusions

The catalytic impact of the green dot map in shaping post-Katrina planning in New Orleans has been widely recognized. However, scholars and other commentators have paid little attention to where the map came from and how it communicated through graphics and text. With analysis of the maps' makers, the text-based and graphical reinterpretations they employed, and the political and social context of the community they meant to restructure, the story of the green dot map yields deeper insights for future planning and adaptation research and practice. The map was a product of multiple reinterpretations, each undertaken by different groups with distinct values and interests. Each of these iterations reflects a particular design politics colored by the values and interests of its makers. The version of the map presented to the public in the *Times-Picayune* in January of 2006 contains a confounding combination of deliberate abstraction and misplaced precision that, when paired with radical policy prescriptions for reorganizing property and land use, became highly inflammatory.

While public and scholarly critiques of the green dot proposal included concerns related to all four of Adger et al.'s (2005) criteria for successful climate adaptation, issues of the legitimacy and equity were especially central. These equity and legitimacy critiques were rooted in decades of well-earned racialized suspicion and distrust of top-down planning intervention in New Orleans. These suspicions were reinforced by the composition of the planning bodies responsible for the ULI and BNOBC proposals: outsider technocrats and representatives of the city's white-dominated business elite, respectively.

Apologists tend to describe the failure of the BNOBC as a problem of communication. Such a reading of the episode overlooks the fact that the reception of the maps as communication media was deeply shaped by the sociopolitical context into which the maps were released. The communication failings of the green dot map are tightly intertwined with the substantive critiques of the proposals and the preceding process. John Forester points out that the "technical problem-solving" functions of planning are inextricably linked to "planning as a means of processing information and feedback" (Forester 1988). While the ULI and BNOBC experts regarded their plans as reasonable attempts at effective and efficient technical problem-solving, for critics steeped in the history of problematic planning interventions in New Orleans, the green dot map was visual confirmation of their suspicions that the planning process was illegitimate and inequitable. For suspicious residents and critics, the map, with its bewildering combination of hard-edged geometric precision and high levels of abstraction, reflected a top-down process that appeared at once arbitrary and inequitable.

Though the problems with the green dot map run deeper than a benign failure of communication, the particular form of the graphics and text of the map and paramap materials presented in the *Times-Picayune* do matter. In the case of the explosive "green spacing" proposals, the ULI map's fuzzy shapes, the BNOBC's open dotted circles, and the *Times-Picayune*'s solid green dots each communicate different levels of resolution and finality to the plans. Similarly, the shift from the ULI and BNOBC's language of administrative and financial efficiency to the *Times-Picayune's* resident-eye view of radical urban restructuring substantially colored the reception of the proposal. The green dot map episode makes clear that advocates of such urban adaptation projects must be attentive to how their graphics and texts will be reinterpreted, represented, and consumed. These processes are deeply place-specific and historically contingent and, thus, frequently may not be immediately comprehensible to outside technical experts.

Perhaps more than any other episode in the recent history of planning in America, the development and response to the green dot map demonstrate the need for greater sensitivity to the design politics of maps and planning representations among scholars, practitioners, and decision-makers. Skillful graphic communication cannot overcome deep substantive flaws in a planning process or proposal. However, by understanding how planning graphics and texts relate to the specific historically imbedded contexts of a place, planners can communicate with the public and decision-makers in ways that facilitate rather than destroy the potential for effective, efficient, equitable, and legitimate adaptation.

## References

Adger, W. N. (2006, August). Vulnerability. *Global Environmental Change, 16*(3), 268–281. https://doi.org/10.1016/j.gloenvcha.2006.02.006.

Adger, W. N., Arnell, N. W., & Tompkins, E. L. (2005). Successful adaptation to climate change across scales. *JGEC Global Environmental Change, 15*(2), 77–86.

Allen, B. L. (2013). Justice as measure of nongovernmental organization success in postdisaster community assistance. *Science, Technology & Human Values, 38*(2), 224–249. https://doi.org/10.1177/0162243912470726.

Anguelovski, I., Shi, L., Chu, E., Gallagher, D., Goh, K., Lamb, Z., et al. (2016). Equity impacts of urban land use planning for climate adaptation: Critical perspectives from the Global North and South. *Journal of Planning Education and Research, 36*(3), 333–348.

Archer, D., Almansi, F., DiGregorio, M., Roberts, D., Sharma, D., & Syam, D. (2014). Moving towards inclusive urban adaptation: Approaches to integrating community-based adaptation to climate change at city and national scale. *Climate and Development, 6*(4), 345–356. https://doi.org/10.1080/17565529.2014.918868.

Aylett, A., et al. (2010). Participatory planning, justice, and climate change in Durban, South Africa. *Environment and Planning A, 42*(1), 99.

Beatley, T. (2012). *Planning for coastal resilience: Best practices for calamitous times*. Washington, DC: Island Press.

Berke, P. R., & Campanella, T. J. (2006). Planning for postdisaster resiliency. *Annals of the American Academy of Political and Social Science, 604*(March), 192–207.

Bowker, G. C., & Star, S. L. (1999). *Sorting things out: Classification and its consequences*. Cambridge, MA: MIT Press.

Breunlin, R., & Regis, H. A. (2006). Putting the Ninth Ward on the map: Race, place, and transformation in desire, New Orleans. *American Anthropologist, 108*(4), 744–764.

Bring New Orleans Back Commission, Urban Planning Committee. (2006). *Action plan for New Orleans: The new American city*. New Orleans, LA: BNOBC.

Bulkeley, H. (2006). A changing climate for spatial planning. *Planning Theory and Practice, 7*(2), 203–214.

Bulkeley, H., Carmin, J., Broto, V. C., Edwards, G. A. S., & Fuller, S. (2013). Climate justice and global cities: Mapping the emerging discourses. *Global Environmental Change, 23*(5), 914–925. https://doi.org/10.1016/j.gloenvcha.2013.05.010.

Burby, R. J. (1998). Natural hazards and land use: An introduction. In R. J. Burby (Ed.), *Cooperating with nature confronting natural hazards with land use planning for sustainable communities* (pp. 1–26). Washington, DC: Joseph Henry Press.

Burby, R. J., Beatley, T., Berke, P. R., Deyle, R. E., French, S. P., Godschalk, D. R., et al. (1999). Unleashing the power of planning to create disaster-resistant communities. *Journal of the American Planning Association, 65*(3), 247–258.

Burby, R., Deyle, R., Godschalk, D., & Olshansky, R. (2000). Creating hazard resilient communities through land-use planning. *Natural Hazards Review, 1*(2), 99–106. https://doi.org/10.1061/(ASCE)1527-6988(2000)1:2(99).

Cidell, J. L. (2008). Challenging the contours: Critical cartography, local knowledge, and airport noise. *Environment and Planning, A, 40*, 1202–1218.

City of New Orleans. (2015). *Resilient New Orleans: Strategic action to shape our future city*. New Orleans, LA: City of New Orleans.

Collins, T. W. (2008). The political ecology of hazard vulnerability: Marginalization, facilitation and the production of differential risk to urban wildfires in Arizona's White Mountains. *Journal of Political Ecology, 15*(1), 21–43.

Crampton, J. W. (2001). Maps as social constructions: Power, communication and visualization. *Progress in Human Geography, 25*(2), 235–252.

Crampton, J. W., & Krygier, J. (2005). An introduction to critical cartography. *ACME: An International E-Journal for Critical Geographies, 4*(1), 11–33.

Cutter, S. L., Boruff, B. J., & Shirley, W. L. (2003, June). Social vulnerability to environmental hazards. *Social Science Quarterly, 84*(2), 242–261. https://doi.org/10.1111/1540-6237.8402002.

Donze, F., & Russel, G. (2006, January 11). 4 months to decide. *The Times-Picayune, A-8*.

Ehrenfeucht, R., & Nelson, M. (2011). Planning, population loss and equity in New Orleans after Hurricane Katrina. *Planning Practice & Research, 26*(2), 129–146. https://doi.org/10.1080/02697459.2011.560457.

Fields, B. (2009). From green dots to greenways: Planning in the age of climate change in Post-Katrina New Orleans. *Journal of Urban Design, 14*(3), 325–344. https://doi.org/10.1080/13574800903056515.

Fischler, R. (1995). Strategy and history in professional practice: Planning as world making. *Spatial Practices, 1995*, 13–58.

Forester, J. (1988). *Planning in the face of power*. Berkeley and Los Angeles, CA: University of California Press.

Godschalk, D. R., Kaiser, E. J., & Berke, P. R. (1998). Integrating hazard mitigation and local land use planning. In R. J. Burby (Ed.), *Cooperating with nature confronting natural hazards with land use planning for sustainable communities* (pp. 85–118). Washington, DC: Joseph Henry Press.

Gotham, K. F., & Greenberg, M. (2014). *Crisis cities: Disaster and redevelopment in New York and New Orleans*. New York: Oxford University Press.

Gustafson, S. (2015). Maps and contradictions: Urban political ecology and cartographic expertise in southern Appalachia. *Geoforum, 60*(March), 143–152.

Hallegatte, S., & Corfee-Morlot, J. (2011). Understanding climate change impacts, vulnerability and adaptation at city scale: An introduction. *Climatic Change, 104*(1), 1–12.

Hamin, E. M., & Gurran, N. (2009). Urban form and climate change: Balancing adaptation and mitigation in the U.S. and Australia. *HAB Habitat International, 33*(3), 238–245.

Harley, J. B. (1989). Deconstructing the map. *Cartographica: The International Journal for Geographic Information and Geovisualization, 26*(2), 1–20.

Healey, P. (2012). "Communicative Planning: Practices, Concepts, and Rhetorics." *In Planning Ideas That Matter: Livability, Territoriality, Governance, and Reflective Practice, edited by Bishwapriya Sanyal, Lawrence Vale, and C. D. Rosan* (p. 333–55).Cambridge, MA: MIT Press.
Hewitt, K. (1983). *Interpretations of calamity from the viewpoint of human ecology*. Boston, MA: Allen & Unwin.
Innes, J. E. (1995). Planning theory's emerging paradigm: Communicative action and interactive practice. *Journal of Planning Education and Research, 14*(3), 183–189.
International Panel on Climate Change (IPCC). (2014). Summary for policymakers. In *Climate change 2014: Impacts, adaptation, and vulnerability. Part A: Global and sectoral aspects. Contribution of working group II to the fifth assessment report of the intergovernmental panel on climate change*. Cambridge, UK/New York City, NY: Cambridge University Press.
Irazábal, C., & Neville, J. (2007). Neighbourhoods in the lead: Grassroots planning for social transformation in post-Katrina New Orleans? *Planning, Practice & Research, 22*(2), 131–153.
Kim, A. M. (2012). The mixed-use sidewalk. *Journal of the American Planning Association, 78*(3), 225–238. https://doi.org/10.1080/01944363.2012.715504.
Kim, A. M. (2015). *Sidewalk city: Remapping public space in Ho Chi Minh City*. Chicago/London: University of Chicago Press.
Klein, N. (2008). *The shock doctrine: The rise of disaster capitalism* (1st ed.). New York: Picador.
Krupa, M. (2010, August 24). The dreaded dot: After the flood, a map of the city with large green dots representing proposed green spaces sent homeowners into a panic. *The Times-Picayune*
Measham, T. G., Preston, B. L., Smith, T. F., Brooke, C., Gorddard, R., Withycombe, G., et al. (2011). Adapting to climate change through local municipal planning: Barriers and challenges. *Mitigation and Adaptation Strategies for Global Change, 16*(8), 889–909. https://doi.org/10.1007/s11027-011-9301-2.
Monmonier, M. S. (1997). *Cartographies of danger: Mapping hazards in America*. Chicago, IL: University of Chicago Press.
Monmonier, M. S. (2008). *Coast lines: How mapmakers frame the world and chart environmental change*. Chicago, IL: University of Chicago Press.
Monmonier, M. S. (2010). *No dig, no fly, no go: How maps restrict and control*. Chicago, IL/London, UK: The University of Chicago Press.
Monmonier, M. S. (2014). *How to lie with maps*. Chicago, IL: University of Chicago Press.
Nelson, M., Ehrenfeucht, R., & Laska, S. (2007). Planning, plans, and people: Professional expertise, local knowledge, and governmental action in post-Hurricane Katrina New Orleans. *Cityscape*, 23–52.
Olshansky, R. B. (2006). Planning after Hurricane Katrina. *Journal of the American Planning Association, 72*(2), 147–153.
Olshansky, R. B., & Kartez, J. D. (1998). Managing land use to build resilience. In R. J. Burby (Ed.), *Cooperating with nature confronting natural hazards with land use planning for sustainable communities* (pp. 167–200). Washington, DC: Joseph Henry Press.
Olshansky, R. B., Johnson, L., & American Planning Association, & Others. (2010). *Clear as mud: Planning for the rebuilding of New Orleans*. Washington, DC: American Planning Association.
Paavola, J., & Adger, W. N. (2006). Fair adaptation to climate change. *Ecological Economics, 56*(4), 594–609. https://doi.org/10.1016/j.ecolecon.2005.03.015.
Peattie, L. R. (1987). *Rethinking Ciudad Guayana*. Ann Arbor, MI: University of Michigan Press.
Pelling, M. (1999). The political ecology of flood hazard in urban Guyana. *Geoforum, 30*(3), 249–261. https://doi.org/10.1016/S0016-7185(99)00015-9.
Pelling, M. (2003). *The vulnerability of cities: Natural disasters and social resilience*. London/Sterling, VA: Earthscan Publications.
Pickles, J. (2003). *A history of spaces: Cartographic reason, mapping and the geo-coded world*. London/New York: Routledge.
Rathke, W. (2006). A New Orleans for all. (Blog). Tompayne.com, 12 January.
Rivlin, G. (2005a, September 29). A mogul who would rebuild New Orleans. *The New York Times*, sec. Business. http://www.nytimes.com/2005/09/29/business/29mogul.html

Rivlin, G. (2005b, October 1). New Orleans forms a panel on renewal. *The New York Times*, sec. National/National Special. http://www.nytimes.com/2005/10/01/national/nationalspecial/01panel.html
Robinson, A. H., & Petchenik, B. B. (1976). *The nature of maps*. Chicago, IL: University of Chicago Press.
Rosenzweig, C. (2010). Cities lead the way in climate-change action. *Nature, 467*(7318), 909.
Sovacool, B. K., Linnér, B., & Goodsite, M. E. (2015). The political economy of climate adaptation. *Nature Climate Change, 5*(7), 616–618.
Tierney, K. (2006). Social inequality, hazards, and disasters. In R. J. Daniels, D. F. Kettl, & H. Kunreuther (Eds.), *On risk and disaster: Lessons from Hurricane Katrina* (pp. 109–128). Philadelphia, PA: University of Pennsylvania Press.
Turnbull, D., Watson, H., & Deakin University, School of Humanities, and Open Campus Program. (1993). *Maps are territories: Science is an atlas: A portfolio of exhibits*. Chicago, IL: University of Chicago Press.
Urban Land Institute. (2005). *New Orleans, Louisiana: A strategy for rebuilding*. Washington, DC: ULI.
Vale, L. J. (2013). *Purging the poorest: Public housing and the design politics of twice-cleared communities*. Chicago, IL: University Of Chicago Press.
Vale, L. J., & Campanella, T. J. (2005). *The resilient city: How modern cities recover from disaster*. New York: Oxford University Press.
Wagner, J. A., & Frisch, M. (2009). Introduction: New Orleans and the design moment. *Journal of Urban Design, 14*(3), 237–255. https://doi.org/10.1080/13574800903138172.
White, G. F. (1945). *Human adjustment to floods: A geographical approach to the flood problem in the United States*. Chicago, IL: University of Chicago.
Wilson, E. (2006). Adapting to climate change at the local level: The spatial planning response. *Local Environment, 11*(6), 609–625. https://doi.org/10.1080/13549830600853635.
Wisner, B., Blaikie, P., Davis, I., & Cannon, T. (2004). *At risk: natural hazards, people's vulnerability, and disasters*. London/New York, NY: Routledge.
Wood, D., & Fels, J. (2009). *The natures of maps: Cartographic constructions of the natural world*. Chicago, IL: University Of Chicago Press.
Wooten, T. (2012). *We shall not be moved: Rebuilding home in the wake of Katrina*. Boston, MA: Beacon Press.

# Chapter 4
# Antagonisms of Adaptation: Climate Change Adaptation Measures in New Orleans and New York City

**Kevin Fox Gotham and Megan Faust**

## 4.1 Introduction

Scientists increasingly point to the possibility of multiple abrupt negative consequences associated with anthropogenic climate change. Climate change poses risks to many environmental and economic systems—including agriculture, public infrastructure, ecosystems, and human health—and presents a significant financial risk to federal, state, and local governments (US Global Change Research Program 2011; National Research Council 2012a, b). Scientists expect climate change to threaten coastal cities and ecosystems with rising sea levels, elevated tidal inundation, increased storm and flood frequency and intensity, and accelerated erosion and saltwater intrusion (Blum and Roberts 2009; Intergovernmental Panel on Climate Change [IPCC] 2014; Karl et al. 2009). As observed by the US Global Change Research Program (USGCRP), the impacts and costliness of weather disasters resulting from floods, drought, and other events such as tropical cyclones will increase in significance as what are considered "rare" events become more common and intense due to climate change (Karl et al. 2009). Overall, according to the National Research Council and the US Global Change Research Program, changes in Earth's climate—including higher temperatures, changes in precipitation, rising sea levels, and more intense and frequent severe weather events—are underway and expected to grow over time. These risks not only imperil the long-term sustainability of cities and communities, but they could create significant fiscal problems for local, state, and federal governments.

This chapter identifies climate change adaptation measures implemented in post-Sandy New York City and post-Katrina New Orleans and examines their conflictual and contradictory dynamics and impacts. Climate change adaptation measures are an amalgam of government policies, socio-legal regulations, statutes, and laws to

K. F. Gotham (✉) · M. Faust
Sociology and Urban Studies, Tulane University, New Orleans, LA, USA
e-mail: kgotham@tulane.edu; mfaust1@tulane.edu



S. Laska (ed.), *Louisiana's Response to Extreme Weather*,
Extreme Weather and Society, https://doi.org/10.1007/978-3-030-27205-0_4

reduce current and future vulnerability to the negative impacts of climate change (e.g., global warming and sea-level rise) and strengthen social resilience (IPCC 2014). As a risk management strategy, climate change adaptation represents adjustments to natural or human systems in response to actual or expected climate change. The broad goal is to help protect vulnerable sectors and communities that might be affected by changes in the climate (GAO 2013). For example, adaptation measures include raising river or coastal dikes to protect infrastructure from sea-level rise, building higher bridges, and increasing the capacity of storm water systems. State and local authorities are responsible for the planning and implementation of many types of infrastructure projects, and decisions at these levels of government can affect insurance rates for businesses and homeowners as well as influence patterns of economic development. While implementing adaptive measures may be costly, policy-makers and elected leaders are increasingly recognizing that the cost of inaction could be greater as damage from weather-related events becomes more expensive (GAO 2009, 2016).

This paper addresses the ways in which the decentralized and fragmented structure of policy-making and implementation in the United States both constrains the process of formulating and implementing comprehensive climate change adaptation measures and encourages cities to respond to climate change using their own distinctive policy measures. Much social science research has focused on the uneven manner in which climate change adaptation agendas are unfolding in a diverse set of urban contexts (Dunlap and Brulle 2015; Bulkeley and Castán Broto 2013; Bulkeley and Betsill 2013; Burch et al. 2014; Dale et al. 2018; Morrison et al. 2017; McCann 2017). In this paper, we adopt an "encompassing" comparative approach to explain how local climate change developments in New York City and New Orleans reflect, share characteristics with, and contribute to broader socioeconomic and political trends in the United States. Encompassing comparison seeks to understand how local actions and events express the interaction of local-global forces and relations including institutional forms, regulatory strategies, and governance projects. We conceptualize the pairing of climate change adaptation measures as an encompassing comparison, which, according to Charles Tilly (1984, p. 83), "places different instances at various locations within the same system, on the way to explaining their characteristics as a function of their varying relationships to the system as a whole." While our comparison of two cities may lack sufficient scope for statistical generalization, we believe our encompassing mode of comparative analysis provides for breadth of generalization and depth of description that is not possible in quantitative analyses. Our comparison provides an opportunity to reflect upon how decisions surrounding climate change adaptation measures take place in a larger political economy of policy-making that shapes and constraints local actions.

## 4.2 Risk, Resilience, Mitigation, and Adaptation

Over the last decade or so, scholars and policy-makers have debated the steps governments can take to reduce risk of extreme events through climate change adaptation and align such adaptation with broader resilience efforts (Gotham et al. 2011; Gotham and Campanella 2010, 2011). Risk refers to situations or conditions that pose a threat to human health, quality of life, and community well-being (for an overview, see Tierney 2014). Risk is a relational term that is closely connected to the notion of resilience, which the National Research Council (2012b, p. 5) defines as the "ability to prepare and plan for, absorb, recover from, and more successfully adapt to adverse events." In the scholarship on climate change, resilience implies regulatory and policy actions to *reduce* vulnerabilities to the effects of severe weather and to *adapt* to the effects of climate change. Scholars theorize that two related sets of actions—climate change mitigation and climate change adaptation—may be able to enhance resilience by reducing risk. Mitigation refers to human actions to reduce the sources of greenhouse gases (GHGs) that contribute to global warming and, in turn, sea-level rise.

We follow climate change scientists in viewing climate change mitigation and adaptation as conceptually separate and analytically distinct. We recognize that federal agencies such as FEMA and the Army Corps of Engineers have used the term hazard "mitigation" for decades to refer to activities designed to reduce hazard risks. Flood risk reduction, for example, involves a combination of structural—focusing on reducing the *probability* of flooding—and physical and nonphysical nonstructural measures that focus on reducing the *consequences* of flooding (US Army Corps of Engineers 2018). At the individual property level, nonstructural mitigation options include elevating a building to or above the area's base flood elevation, relocating the building to an area with less flood risk, or purchasing and demolishing the building and turning the property into green space. In addition, nonstructural risk reduction measures would include flood insurance, floodplain mapping, improving response capacity, improving post-disaster assessment and communication capacity, and developing more effective strategies to communicate risk and mitigation activities to various stakeholders. Nonstructural mitigation is akin to adaptation. FEMA supports a variety of nonstructural flood mitigation activities to reduce flood risk.

Because mitigation is intended to reduce the harmful effects of climate change, it is part of a broader policy framework that also includes adaptation to climate impacts. Climate change adaptation refers to actions taken by governments, nonprofit organizations, and private firms to reduce the loss of life and property by lessening the impacts of adverse climate change events such as weather-related disasters. Climate change adaptations can also be classed as either *process-oriented* measures—aimed at developing information systems, social structures, and governance needed to support adaptation—or *outcome-oriented* actions, measures taken to reduce vulnerability and exploit opportunities that arise from a changing climate. Climate change adaptation measures can be *effect-oriented* in the sense of building flood protection or *cause-oriented* by adopting approaches such as changing the location of areas for new housing development. Climate change adaptation includes activities such as restoring wetlands and coastal areas to control erosion, improving the quality of road surfaces to withstand hotter temperatures, protecting critical facilities against the negative effects (e.g., inundation) of sea-level rise, and creating permeable surfaces and "green roofs," or roofs partially or completely covered with vegetation, in cities to absorb excess rainfall, provide insulation, and help lower urban air temperatures (Wise et al. 2014; for an overview, see IPCC 2014). Climate change adaptation measures can be proactive and/or reactive. Governments may plan and adopt adaptation measures in advance, establish them in the aftermath of a major disaster, or create them in response to local pressures. In addition to large-scale infrastructure measures to adapt to climate change, governments may also implement policies and regulations to incentivize people to change their behaviors. This approach includes motivating them to use less water, encouraging farmers to plant different crops, and urging more households and businesses to purchase flood insurance.

Many researchers and policy-makers consider climate change a global problem that demands international action and global solutions. But sociologists and others have documented that the effects of global climate change will not be equally distributed around the world, for "many of the countries least responsible for the rise in greenhouse gases will be most likely to feel its impacts in changes in weather, sea-levels, human health costs, and economic hardships" (Nagel et al. 2010, p. 17). The unequal burdens inflicted by climate-related disasters and limited disaster response capacities will exacerbate these inequalities and likely generate unforeseen consequences. Variations in individual, community, and national vulnerability to the impacts of climate change are only part of this global structure of inequality. As the 2007 Intergovernmental Panel on Climate Change (IPCC) report notes, there is an unequal distribution of impacts and vulnerabilities to climate change associated with social class and age in both developed and developing countries: "vulnerability to climate change can be exacerbated by other stresses. These arise from, for example, current climate hazards, poverty and unequal access to resources, food insecurity, trends in economic globalization, conflict and incidence of diseases such as HIV/AIDS" (2007, p. 14). Thus, the impacts of climate change are not spread evenly, and its effects will be felt by different social groups in radically different ways.

The ways in which climate change is closely intertwined with state policy-making, institutional arrangements, and political economy is one of the reasons why it has proven so unique an issue to address internationally as a global problem. Conceptualizing climate change as a global-local issue and using comparative analyses draws attention to different socio-spatial inequalities, local and regional histories and geographies, and their implications for communities. These concerns bring explicit temporal and spatial dimensions to our understanding of the local impacts of global climate change. Climate change adaptation measures are activated and reproduced through the concrete actions taken by state actors, elected leaders, economic elites, and other powerful organized interests. A core assumption of this agent-centric approach is that the adoption of climate change adaptation actions does not develop out of an inevitable and unalterable structural necessity but rather in a contingent manner; it results from the conscious actions taken by individual decision-makers in various institutions, organizations, and communities acting under particular historical circumstances. This emphasis on contingency and agency compels us to examine the actions of human agents, organizations, and interest groups in an effort to grasp how larger climate change dynamics and effects occur at the local level. Underscoring the importance of space and time in climate change research means that any explanation of *why* and *how* climate change policy actions develop will need to take account of *where* and *when* they develop.

In the sections below, we address the obstacles to climate change adaptation by focusing on the dynamic ways in which antagonisms—incongruences and inconsistencies between goals, implementation strategies, and outcomes—develop and persist. Scholars have used terms such as "barriers," "obstacles," "tensions," and "contradictions" to describe the difficulties that hinder the formulation and implementation of climate change adaptation actions (Eakin et al. 2014). Lack of resources to meet the costs of adaptation can be a financial barrier. Lack of administrative and/or political capacity can be an institutional barrier to adaptation. Collective opposition and political mobilization against adaptation can be a social-cultural barrier. Finally, gaps in climate change knowledge and the transmission of information can create an information barrier. Eisenack et al. (2014, p. 867) suggest that the "growing literature on barriers to adaptation reveals not only commonly reported barriers, but also conflicting evidence, and few explanations of why barriers exist and change."

In attempt to move beyond debates over discrete barriers to climate change adaptation, Pelling et al. (2015) have developed the concept of transformation "as an adaptive response to climate change" that can open a range of novel policy options. In their work, transformative adaptation is a multifaceted concept that researchers can use to describe responses that produce nonlinear changes in systems or their host social and ecological environments. The concept also raises distinct ethical and procedural questions for decision-makers and "foregrounds questions of power and preference that have so far been underdeveloped in adaptation theory and practice" (p. 113). The concept of transformative adaptation dovetails with the notion of transformative resilience developed by Gotham and Campanella (2011) and suggests that we view climate change adaptation in a multidimensional fashion, for example, as a political decision point, an opportunity for socio-spatial change, and a prospect for resistance to dominant developmental pathways.

Through a comparison of New Orleans and New York, our research helps to explain the major antagonisms of adaptation and provides insights on how to overcome them to enhance societal resilience to climate change risks. Following the logic of encompassing comparison, we analytically juxtapose policy trajectories and institutional arrangements rather than compare discrete units or fixed variables. The chapter offers a sociological critique of the dominant approaches to adaptation and highlights the institutional and social antagonisms that are shaping the implementation of climate change policy in each city.

Our theorization of the facilitative and discouraging power of state action in the development of climate change adaptation measures focuses on the state as an actor and as an institutional structure. The state comprises many actors that can formulate and implement different policies and socio-legal regulations to respond to climate change. State governments have special charters and can make property rights decisions to alter the organization of firms and corporate hierarchies. Through legislative debate and compromise, the US Congress makes laws, holds hearings to inform the legislative process, conducts investigations to oversee the executive branch, and represents voting constituencies and states in the federal government. Courts can determine the meaning and effect of laws passed by the state legislatures. Over the decades, US judges have played aggressive roles in interpreting policy-making and economic governance (Campbell and Lindberg 1990).

As an institutional structure, the state power and authority are fragmented and restricted to the extent that state and local governments exercise political authority within their own geographical areas. The existence of 50 separate governments combined with hundreds of municipalities in metropolitan areas has played an important role in the development of different markets, real estate financing, and land-use policies and regulatory strategies. For the most part, laws and regulations pertaining to economic activity and investment are locally based. These laws and regulations include, among many others, recording regulations, banking laws, zoning laws, subdivision regulations, private deed restrictions, land-use regulations, building codes, insurance laws, and property tax law (Feagin and Parker 2002; Gotham 2006, 2009). At the same time, local laws and socio-legal regulations establish institutional practices and rules of exchange that coordinate local economic activity among organizations in a particular economic sector (residential, commercial, or industrial activity) and, more importantly, create distinctive locations for policy-making, investment, and economic activity. The implication is that the decentralized and fragmented institutional structure of the state has influenced the development of legal forms that reinforce the place specificity of climate change adaptation measures thereby fostering local uniqueness and innovation (Gotham 2006).

## 4.3 Global Climate Change Risk in New York City and New Orleans

The New York City metropolitan area, with 23 million residents and approximately 3700 miles of tidal coastline, faces a severe social-ecological threat from climate change-driven warming and sea-level rise. The New York City Panel on Climate Change (NPCC), an organization that examines climate change vulnerability and prepares projections for the City and metropolitan region, contends that extreme weather will increase in frequency and severity and that the climate will become more variable. Climate projections encompass a wide range of possible outcomes:

mean annual temperature is projected to increase between 4.1 and 6.6 °F by the 2050s and between 5.3 and 10.3 °F by the 2080s; frequency of heat waves is projected to triple by the 2050s to 5 to 7 heat waves per year; sea level is expected to continue rising by 11 to 21 inches by the 2050s and by 18 to 39 inches by the 2080s, a rate that exceeds the global average in relative sea-level rise. New York City has experienced the devastating effects of coastal storms, most recently during Hurricane Sandy, as well as flooding in low-lying areas during high tides. Sea-level rise is projected to increase the depth, extent, and frequency of flooding from storm surge and during high tides (Horton et al. 2015; New York City Mayor's Office 2015).

Like New York, the New Orleans metropolitan area constitutes a highly cited example of a region experiencing the leading edge of climate-related stresses that are widely anticipated to affect coastal regions worldwide (Hallegatte et al. 2013). Given its low elevation and susceptibility to storm surge, extreme storm events and sea-level rise stand out as two of the most severe consequences of climate change in New Orleans and much of southern Louisiana. González and Törnqvist (2006, 2009) show that the preindustrial millennium (600 to 1600 AD) witnessed a rate of sea-level rise of −0.55 mm $yr.^{-1}$ in coastal Louisiana. In contrast, the past century has seen rates of at least 2 mm $yr.^{-1}$, roughly in line with the global average and a fourfold increase in the rate of relative sea-level rise (IPCC 2013). Moreover, modeling assessments consistently point toward an increase in hurricane intensity with global warming. Hurricanes strike the Louisiana coast with a mean frequency of 2 every 3 years (Kolb and Saucier 1982). In 2005, Hurricane Katrina forced the largest and most abrupt displacement in US history with approximately 1.5 million people evacuated from the Gulf Coast region. Using storm surge models, scientists predict a doubling of Hurricane Katrina-magnitude events over the next century (Grinsted et al. 2013; Holland 2012).

For New Orleans, climate-related environmental change coalesces with other non-climate stressors such as wetland loss and land subsidence. Louisiana harbors approximately 40 percent of the contiguous United States' coastal wetlands yet accounts for almost 80 percent of wetland loss. Louisiana has been losing coastal wetlands since at least the 1930s, but the long-term rate of land loss has slowed since its peak in the 1970s, according to the Department of the Interior's US Geological Survey (Couvillion 2017). In addition to subsidence and accelerated sea-level rise, the dredging of about 15,000 km of canals in the Mississippi River Delta area, primarily for oil and gas infrastructure, has led to widespread saltwater intrusion and ecosystem degradation. However, the fundamental cause of wetland loss is the isolation of the delta plain from sediment input due to the embankment of the Mississippi and Atchafalaya Rivers by artificial levees. Since embankment was completed in the 1930s, the majority of the Mississippi River sediment load has been lost to the deeper waters of the Gulf of Mexico, where the mouth of the river debouches at the shelf edge rather than near shore and inland areas that would replenish the delta plain (Campanella 2017).

Our comparative analysis of climate change adaptation measures in New York and New Orleans suggets two sets of intersecting factors that pose interesting conditions for studying climate change responses. On the one hand, both urban regions represent global climate change's "canaries in the coal mine" in the United States. These highly sensitive regions face a future of rising relative sea level, increased frequency and destructiveness of storm events, extreme vulnerability to flood trauma, and potential for major displacement. Scientists view the two regions as harbingers of climate change impacts to come for coastal ecosystems worldwide (Kent 2012; Reed et al. 2015; Wang et al. 2011). On the other hand, both cities constitute the leading edge of socio-legal experimentation, regulatory inventiveness, and policy innovation that will likely offer new approaches and strategies to help other cities adapt to climate change. Currently, planners and policy-makers debate various policies to reduce coastal risk, and major coastal restoration projects, climate change adaptation, and mitigation efforts are currently underway (Gotham 2016a, b; Gotham and Cheek 2017; Gotham and Cannon 2018; Gotham and Powers 2017; Peyronnin et al. 2013; Fischbach et al. 2012). Yet the level at which elected leaders and policy-makers understand the causes and consequences of climate change, as well as the extent to which they regard climate change as harmful to the ecological and economic sustainability of the two cities and regions, is not known. These concerns underpin the need to examine the local and regional dynamics of climate adaptation policy-making and implementation in detail.

## 4.4 Long-Term Sustainability Challenges Facing New York City and New Orleans

Both New York City and New Orleans face long-term sustainability challenges related to the distinctive and peculiar system of local, state, and federal relationships and financing arrangements in the United States (Gotham and Greenberg 2014). As a distinctive configuration of organizations, the agencies of the different branches of the federal government, as well as agencies at the state and local government levels, are predisposed to struggle and conflict over funding amounts and mechanisms of financing. Insofar as the different parts of local, state, and federal governments have overlapping responsibilities for policy-making, contradictory policy actions and political stalemate are possible. Indeed, the institutionally fragmented nature of US federalism has systematically produced a variety of conflicts and contradictions in many areas of policy-making including economic policy, defense policy, transportation policy, housing policy, regulatory policy, hazard mitigation and disaster response policy, and so on (Campbell and Pedersen 2014; Gotham 2012; Gotham and Wright 2009; Hogan and Howlett 2015). As long as different parts of local, state, and federal governments provide different arenas of access to political actors and organized interests, then political conflict and struggle are possible. Moreover, we can expect conflicting policy actions and policy outcomes to the extent that these actors pursue their interests in different arenas.

Over the last decade, the US federal government has developed a number of climate change adaptation plans that have intersected with local efforts to address urban resilience. In June 2013, President Obama issued the Climate Action Plan, which describes the federal government's existing and planned efforts to prepare for the impacts of climate change and set strategic priorities for the country. For example, the plan directs federal agencies to take appropriate actions to reduce risk to federal investments, specifically calling on them to update their flood risk reduction standards. The plan also established a federal flood risk management standard in January 2015, and implementation guidelines were issued in October 2015 (Executive Order No. 13690). In November 2013, President Obama also issued Executive Order 13653, which directed federal agencies to develop or update comprehensive adaptation plans by describing how they would consider improving climate change adaptation and resilience measures. By 2014, almost 40 federal agencies had created climate change adaptation plans. In addition, several crosscutting interagency plans had been developed to address challenges and improve resilience to climate impacts (Interagency Climate Change Adaptation Task Force 2011).

One example of the antagonistic nature of the formulation and implementation of climate change adaptation measures is the efforts the Trump Administration took to rescind the two Obama Administration executive orders mentioned above. On March 28, 2017, President Trump revoked Executive Order 13653 that aimed to promote (1) engaged and strong partnerships and information sharing at all levels of government; (2) risk-informed decision-making and the tools to facilitate it; (3) adaptive learning, in which experiences serve as opportunities to inform and adjust future actions; and (4) preparedness planning. On August 15, 2017, President Trump revoked Executive Order 13690 in an effort to streamline federal environmental review and approval of major infrastructure projects located in flood-prone areas. The Obama Administration's Executive Order 13690 (2015) required federally funded projects to incorporate flood risk management standards that account for sea-level rise. By revoking this executive order, the Trump Administration adopts a new floodplain risk management strategy that restricts the definition of floodplain, thereby allowing more real estate development in flood-prone areas, rejects determinations of risk based on "a climate-informed science approach," and no longer requires federal agencies to incorporate climate science into the analysis. The decisions of the Trump Administration are not only inconsistent with past decisions of the Obama Administration on climate change adaptation but are at odds with prevailing science-based risk management measures advocated by scientists and policy-makers.

During the Obama Administration, the President's Climate Action Plan directed federal agencies to support climate-resilient investments. For example, the plan directed the Department of Housing and Urban Development (HUD) to require grant recipients funded with supplemental appropriations for Hurricane Sandy response to take sea-level rise into account in their projects and activities. Federal agencies have made other climate-resilient investments. HUD, for example, initiated the Rebuild by Design competition and provided $930 million to fund projects that enhance disaster resilience in areas affected by Hurricane Sandy. One such project proposed building deployable walls attached to the underside of roads that could be used during flood events in Manhattan. In addition, several Federal Emergency Management Authority (FEMA) programs fund hazard mitigation measures to assist states and local governments in their efforts to enhance disaster resilience before disasters occur. Activities eligible for grants include property acquisition, elevation, and retrofitting. Programs such as the Hazard Mitigation Grant Program, Hazard Mitigation Assistance, Pre-disaster Mitigation Grant Program, Flood Mitigation Assistance, Repetitive Flood Claims, and Severe Repetitive Loss undertake these projects (Gotham 2012).

Several federal agencies and funding sources have played major roles in post-Katrina and post-Sandy rebuilding efforts in New Orleans and New York. Following Hurricane Katrina, Congress authorized the US Army Corps of Engineers ("Corps") to design and construct the $14.6 billion Hurricane and Storm Damage Risk Reduction System (HSDRRS) for southeast Louisiana. Over the past 10 years, the Corps has strengthened the levees, floodwalls, gated structures, and pump stations that form the 133-mile Greater New Orleans perimeter system in addition to improving approximately 70 miles of interior risk reduction structures (Gotham 2018). In Louisiana, state officials used $10.5 billion in HUD funds from the Community Development Block Grant (CDBG) program to create the Road Home program, which provided homeowners with up to $150,000 to repair or rebuild damaged homes. Other Road Home funds were used to elevate homes (for a critical overview, see Gotham 2014a). Like FEMA's Hazard Mitigation Grant program, policy-makers designed the Road Home as a resilience-enhancing program, though scholars have raised questions about the overall effectiveness of the Road Home program in achieving resilience goals and outcomes (Gotham 2014b).

Other federal funding sources including the Water Resources Development Act (WRDA) and the Coastal Wetlands Planning, Protection, and Restoration Act (CWPPRA) (authorized by Congress in 1990) have been instrumental in funding coastal restoration and protection efforts in southern Louisiana. Following Hurricanes Gustav and Ike in 2008, the federal government provided $27.4 million to Louisiana for coastal protection and restoration projects that help communities recover from the storms and better withstand future hurricanes. The CDBG funds went to the Louisiana Office of Community Development's Disaster Recovery Unit, which has partnered with the Coastal Protection and Restoration Authority (CPRA) to identify potential flood protection projects such as levee construction or improvements, floodgate installation, critical infrastructure, and shoreline protection. Through Fiscal Year 2015, the CPRA built or improved approximately 250 miles of levees, moved over 150 projects into design and construction phases, approved constructed projects in 20 parishes, and approved construction of 45 miles of barrier islands and berms. These projects represent a total state government investment of nearly $477 million in FY 2015 (CPRA 2017).

In New York, the US Army Corps of Engineers and the State of New York raised $450 million to construct new armored levees and other infrastructure along Midland Beach and Staten Island's East Shore. Funds were also allocated to invest in coastal protection in the Rockaways and the communities surrounding the Jamaica Bay area. In addition, HUD allocated $4.21 billion in disaster aid through the Community Development Block Grant-Disaster Recovery (CDBG-DR) to create to the Business PREP (Preparedness and Resiliency Program). This was a new program to help small businesses better prepare for emergencies and enhance the resiliency of their operations, assets, and physical space. The $4.21 billion in disaster aid helped spur new investment in several urban neighborhoods including the Rockaways, Staten Island, Coney Island, and Hunts Point (New York City Mayor's Office 2014). In 2015, New York City announced that it was using funds from HUD's Rebuild by Design, referred to above, to begin preliminary design work on the Lower East Side to implement a $335 million integrated, neighborhood-sensitive flood protection system to mitigate risk. This project is "intended to be just the first phase of a larger project that will ultimately provide coastal resiliency for all of Lower Manhattan," according to the New York City Mayor's Office (2015).

The generation and distribution of climate change adaptation resources are mediated by the decentralized, fragmented structure of American federalism. This decentralized and fragmented structure localizes the funding mechanisms to generate the monetary resources to support climate change adaptation measures. State governments and localities have historically remained responsible for generating funds for urban revitalization. Consequently, uneven metropolitan development has been a permanent feature throughout US history (Gotham 2014c; Gotham 2001). Moreover, for most people, climate change is distant and remote compared to the everyday concerns of urban health, crime and safety, education, and housing. Governments have typically not included adaptive measures in their planning because they tend to focus their attention and resources on competing, shorter-term priorities like sustaining government services and raising funds to deal with infrastructure upkeep, repair, and long-term challenges. Adaptation is a relatively low priority compared with more traditional and immediate concerns such as managing aging infrastructure systems, sustaining current levels of service, protecting public health and safety, and maintaining service affordability (GAO 2013). Short-term competing priorities make it difficult for decision-makers to address the impacts of climate change, especially since many state and local governments responsible for infrastructure investment face immediate funding and staffing challenges. Given these institutional challenges, the formulation of coherent climate change adaptation policy remains elusive and extraordinarily difficult.

## 4.5 Contradictory Roles of the Federal Government

Recent years have witnessed three ominous developments that threatened to obstruct progress on reducing climate change risks. First, in March 2017, President Donald Trump, as part of his Fiscal Year 2018 budget request, introduced plans to eliminate the National Oceanic and Atmospheric Administration's (NOAA) National Sea Grant College Program, a 50-year old program that funds scientific research focused on climate change adaptation initiatives that will prepare coastal communities for the predicted impacts of climate change. In prior years, the program received $73 million per year. Under the Trump budget, the program would have received no funding at all. The Sea Grant program oversees a network that includes the National Sea Grant Office, 33 university-based state programs, the National Sea Grant Advisory Board, the National Sea Grant Law Center, the National Sea Grant Library, and hundreds of other participating institutions. Also in his budget, Trump announced major cuts to NOAA's Office of Oceanic and Atmospheric Research, where climate research programs are housed. The office would see a $150 million, or 19 percent, budget cut. Other NOAA programs to be zeroed out as part of the administration's budget request included the National Estuarine Research Reserve System, Coastal Zone Management Program grants, and the Pacific Coastal Salmon Recovery Fund. Over the past year, Congress has followed President Trump's lead by pushing for major budget cuts to the Environmental Protection Agency (EPA), such as $513 million in cuts that would effectively terminate climate change adaptation programs such as the Climate Resilience Fund as well as broader environmental programs and management (Meade 2018; Thwaites 2018). While the House ignored the President's budget request, rather increasing funding for Sea Grant and NOAA, the examples continue to be relevant because they reflect a lack of consensus between the legislative and executive branches on funding for climate change mitigation and adaptation.

Second, in President Trump's first budget proposal for Fiscal Year 2018, he proposed major cuts in federal funding for coastal restoration and protection efforts in Louisiana. The proposal upends the Gulf of Mexico Energy Security Act (GOMESA), which is a commitment from Congress to share offshore energy revenues with four of the Gulf states that are impacted by its production. GOMESA is projected to provide approximately $100 million per year beginning in 2017. Congress has dictated that GOMESA funds be used for projects and activities for the purposes of coastal protection, including conservation, coastal restoration, hurricane protection, and infrastructure directly affected by coastal wetland losses. President Trump's proposal follows Obama Administration 2016 and 2017 budgets for the Department of Interior that contained language to redirect GOMESA revenue to broader national programs. GOMESA is a revenue-sharing program that, while estimated to provide substantial long-term funding for climate change adaptation, is not guaranteed across a multi-decade time horizon. A major contradiction in relying on GOMESA to fund climate change adaptation efforts is that GOMESA revenues depend on the continued profitability of offshore oil and gas production, a major producer of GHGs that are the source of global warming and subsequent sea-level rise (Mogensen 2018; Reardon et al. 2017; Selby n.d.).

Third, closely related to recent executive branch efforts to erect roadblocks to climate change adaptation has been a strong political consensus among elected leaders to deny or dismiss the scientific consensus on the extent of climate driven sea-level rise, its significance, and its connection to coastal erosion. In Louisiana, many elected officials have been prominent climate change deniers, calling into question the science behind global warming and adamantly opposing legislation to reduce GHGs that contribute to rising sea levels. Over the years, Louisiana's congressional delegation has voted against legislation that would have factored global warming into federal project planning; voted in favor of barring the EPA from regulating GHGs; voted against enforcing limits on CO2 global warming pollution; voted against tax credits for renewable electricity; voted for tax incentives for fossil fuel energy production; voted against tax incentives for renewable energy; voted against removing oil and gas exploration subsidies and against keeping a moratorium on drilling for oil offshore (outside of Louisiana); voted for authorizing construction of new oil refineries; voted for barring GHGs from Clean Air Act rules; voted for letting the wind energy production tax credit expire; voted against reducing oil usage by 40 percent by 2025 (instead of 5 percent); voted against factoring global warming into federal project planning; voted against implementation of the Kyoto Protocol; voted against including oil and gas smokestacks in mercury regulations (September 2005); and voted to appoint Scott Pruitt—a climate change denier who has sued the EPA over emission regulations—to head that agency (Marshall 2017).

In 2011, US House of Representative member Steve Scalise helped pass an amendment to the Agriculture Appropriations Bill prohibiting funds from being used to implement a new US Department of Agriculture (USDA) regulation requiring agencies to spend taxpayer dollars to study and implement climate change policies and initiatives. The amendment passed by a vote of 238 to 179. Senator Scalise signed the No Climate Tax Pledge and voted for amending the Clean Air Act to prohibit the EPA from promulgating any regulation on the emission of a greenhouse gas to address climate change. He also voted against proposed legislation to require utilities to supply an increasing percentage of their demand from a combination of energy efficiency savings and renewable energy. In 2009, he signed HR 391 to amend the Clean Air Act to declare that nothing in the Act shall be treated as authorizing or requiring the regulation of climate change or global warming. As one journalist put it:

> No landscape in the nation is more threatened by global warming, yet our delegation has consistently voted against carbon legislation at the urging of industry and have voted for bills that would prohibit federal agencies from even studying global warming. Naturally, that has congress people in other states asking the question: Why should we pour billions into rebuilding Louisiana's coast, if their own delegation isn't going to address one of the main causes of its loss? (Marshall 2012)

Climate change denial expresses the mounting political and economic stakes of dealing with the risks of anthropogenic climate change. Conservative think tanks, conservative media, corporations, and industry associations (especially for the fossil fuels industry)—domains dominated by conservative white males—have spearheaded the attacks on climate science and policy from the late 1980s to the present (Dunlap and McCright 2011; Freudenburg and Muselli 2010; Lahsen 2008; McCright and Dunlap 2003, 2010, 2011a, b; Oreskes and Conway 2010). Organized climate change denial has an elective affinity with established conservative think tanks that promote free-market conservatism and front groups promoting industry interests. Sustained climate change denial, promoted largely by the American conservative movement (Dunlap and McCright 2011; McCright and Dunlap 2003, 2010), contributes to political polarization on climate change beliefs (McCright and Dunlap 2011b). This organized "climate change denial movement" has mobilized to undercut public belief in climate science and discourage political support for climate change risk reduction measures and adaptation measures (Dunlap and McCright 2011; Oreskes and Conway 2010; McCright and Dunlap 2010; Powell 2011). The political effect of climate change denial is to off-load the cost of paying for climate change risk reduction to other more vulnerable groups while simultaneously protecting the profiteering interests of the GHG production economy.

## 4.6 Conclusions

Since the establishment of the Intergovernmental Panel on Climate Change in 1988, the potential impacts of global climate change have captured the attention of the natural and physical scientists, the international research community, and the policy-makers around the world. As the human causes and consequences of climate change have become increasingly apparent, scholars and government leaders have called on social scientists to contribute to the scientific understanding of the role of humans in global climate change (for overviews, see Dunlap and Brulle 2015; Nagel et al. 2010). Social scientists debate the short-term and long-term climate impacts on cities and communities around the world. Different nations will have to balance a variety of value-laden considerations related to the impacts of climate change itself, potential costs of mitigation and adaptation, and collective struggles over the appropriate societal response. The difficulty of these tasks is compounded by the need to develop a consensus on fundamental issues such as the level of risk that societies are willing to accept and impose on others, strategies for sharing costs, and planning for unforeseen consequences. These tasks and issues are intertwined with relations of domination and subordination. Different actors and organized interests have varied political-economic prerogatives, and the political system offers differential and unequal access to decision-making power. Thus, any examination of the global social problem of climate change must address the question of adaptation "for whom" and "for what purpose."

In spite of the federal disaster resources directed to New York and New Orleans, the federal government has an inherently limited role in the project-level planning processes central to adapting infrastructure to climate change because these are typically the responsibility of state and local governments. That is, state and local authorities are primarily responsible for prioritizing and supervising the implementation of climate change adaptation measures and projects. Therefore, for the foreseeable future, both New York and New Orleans as well as US cities in general face a constrained situation of poor funding and limited options for raising money. On the one hand, different communities can use a variety of sources to fund capital projects, pay for operations and maintenance costs, and sustain programs. These funding sources can include (1) taxes such as property, sales, and income levies; (2) fees such as charges for inspections and permits; (3) state and federal grants such as those that support improvements to drinking water, wastewater, and storm water systems; (4) bonds which enable communities to borrow money to pay for projects; (5) loans to pay for projects and programs; and (6) public-private partnerships that entail contractual agreements between a public agency and a private sector entity allowing for cooperation and collaboration in the financing, planning, design, construction, and maintenance of water infrastructure.

On the other hand, the particular mix of funding sources and the amount they are capable of generating varies across communities depend on the level of wealth of the tax base, environmental circumstances, and specific community needs. In the United States, the formulation and implementation of climate change adaptation measures work through a complex system of multilevel governance that involves a plethora of statutes, laws, financing mechanisms, administrative capacities, and multiple overlapping and interconnected horizontal and vertical lines of authority. According to a 2010 National Research Council study, no one-size-fits-all adaptation option exists for a particular climate impact because climate change vulnerabilities can vary significantly by infrastructure category, region, community, or institution. Thus, the scale of the urban is of central importance in understanding similarities and differences in the process and patterns of climate change policy-making as well as the outcomes of climate adaptation measures (Bulkeley and Betsill 2013). That is, cities are not only the targets of climate change policy-making but are the locus of political conflicts over the nature and character of climate change adaptation. Cities are creatures of state governments, and state governments are, in turn, constituted by the laws and statutes of the US federal government. Conflicts and struggles over climate change policy and action do not take place in discrete local, national, and international arenas. Rather, climate change politics and policy outcomes express changing cross-scale interactions and complex vertical linkages between local actions, state government institutions, and the higher levels of the US federal government.

Today, many cities in the United States face a situation of chronic fiscal retrenchment, declining federal resources to fund climate change adaptation efforts, and broad hostility to the science of climate change. What the cases of New York and New Orleans show us is that taking steps to implement climate change adaptation measures is difficult for several reasons. Adaptation efforts tend to be expensive and require long-term, concerted planning and consensus building. More important, political and economic elites and elected officials typically oppose adaptation investments until faced with response and recovery expenditures once a disaster has occurred. Additionally, of critical importance is the fact that within the US federal system, the adoption and implementation of many adaptation activities fall within the purview of local governments, and those governmental units are typically influenced by powerful and organized economic development interests that are most likely to oppose adaptation. Moreover, it has only been in the last decade or so that federal funds for adaptation became available to communities, but the future of those funds is uncertain. Consequently, the current US climate change adaptation landscape is a patchwork of differing approaches in which some communities and regions are afforded some protection from some climate risks (but not others), while others remain highly vulnerable to the negative impacts of global climate change.

We conclude with several policy recommendations to facilitate and enable more informed decisions about adaptation. Current efforts include raising public awareness of the adverse impacts of climate change, improving infrastructure decision-makers' access to and use of available climate-related information, providing increased access to local assistance, and considering climate change in existing planning processes (for overviews, see GAO 2013; Dunlap and Brulle 2015). These are important, but we think policy recommendations should embrace an equity-oriented and social justice-based focus. That is, the goal of climate change adaptation cannot lie just in consciousness raising through public awareness campaigns or in the implementation of conventional systems of financing (bond programs, tax incentives, etc.). Rather, climate change adaptation requires an effort to integrate and coordinate housing, infrastructure, and economic development programs with comprehensive, publically financed, and democratically run programs with clear accountability systems. Cross-scale, collaborative governance could enhance the flexibility and spatial targeting of incentives to reduce vulnerability and also provide an institutional foundation for direct participation of community residents in program design, implementation, and oversight responsibilities. The creation of jobs that pay a living wage, adequate benefits for those who cannot work, access to affordable health care, and increased supply of affordable housing might also improve the effectiveness of climate change adaptation programs, especially for low-income and moderate-income communities that face the highest levels of vulnerability to climate change impacts. Developing public and private sector funding criteria to match communities' evolving vulnerability pressures, combined with public works programs, could enhance prospects for achieving equitable climate change adaptation for communities.

## References

Blum, M. D., & Roberts, H. H. (2009). Drowning of the Mississippi Delta due to insufficient sediment supply and global sea-level rise. *Nature Geoscience, 2*, 488–491.

Bulkeley, H., & Betsill, M. M. (2013). Revisiting the urban politics of climate change. *Environmental Politics, 22*(1), 136–154.

Bulkeley, H., & Castán Broto, V. (2013). Government by experiment? Global cities and the governing of climate change. *Transactions of the Institute of British Geographers, 38*(3), 361–375.

Burch, S., Shaw, A., Dale, A., & Robinson, J. (2014). Triggering transformative change: A development path approach to climate change response in communities. *Climate Policy, 14*(4), 467–487.

Campanella, R. (2017). *Delta urbanism: New Orleans*. New York, NY: Routledge.

Campbell, J. L., & Lindberg, L. N. (1990). Property righs and the organization of economic activity by the state. *American Sociological Review, 55*, 634–647.

Campbell, J. L., & Pedersen, O. K. (2014). *The national origins of policy ideas: Knowledge regimes in the United States, France, Germany, and Denmark*. Princeton, NJ: Princeton University Press.

Coastal Protection and Restoration Authority of Louisiana (CPRA). (2017). *Louisiana's comprehensive master plan for a sustainable coast*. Baton Rouge, LA: OTS-State Printing.

Couvillion, B. (2017). USGS: *Louisiana's rate of coastal wetland loss continues to slow*. Washington, DC: U.S. Department of Interior, United States Geological Survey. https://www.usgs.gov/news/usgs-louisiana-s-rate-coastal-wetland-loss-continues-slow

Dale, A., Burch, S., Robinson, J., & Strashok, C. (2018). Multilevel governance of sustainability transitions in Canada: Policy alignment, innovation, and evaluation. In S. Hughes, E. Chu, & S. Mason (Eds.), *Climate change in cities* (pp. 343–358). Cham, Switzerland: Springer.

Dunlap, R. E., & Brulle, R. J. (2015). *Climate change and society: Sociological perspectives*. New York: Oxford University Press.

Dunlap, R. E., & McCright, A. M. (2011). Organized climate change denial. In J. S. Dryzek, R. B. Norgaard, & D. Schlosberg (Eds.), *The Oxford handbook of climate change* (pp. 144–160). London, UK: Oxford.

Eakin, H. C., Lemos, M. C., & Nelson, D. R. (2014). Differentiating capacities as a means to sustainable climate change adaptation. *Global Environmental Change, 27*, 1–8.

Eisenack, K., Moser, S. C., Hoffmann, E., Klein, R. J., Oberlack, C., Pechan, A., et al. (2014). Explaining and overcoming barriers to climate change adaptation. *Nature Climate Change, 4*(10), 867–872.

*Exec. Order No. 13653* (2013), 78 C.F.R. 66819.

*Exec. Order No. 13690* (2015), 80 C.F.R. 6425.

Feagin, J. R., & Parker, R. (2002). *Building American cities: The urban real estate game*. New York: Beard Books.

Fischbach, J. R., Johnson, D. R., Ortiz, D. S., Bryant, B. P., Hoover, M., & Ostwald, J. (2012). *Coastal Louisiana risk assessment model*. Santa Monica, CA: Rand Corporation.

Freudenburg, W. R., & Muselli, V. (2010). Global warming estimates, media expectations, and the asymmetry of scientific challenge. *Global Environmental Change, 20*, 483–491.

González, J. L., & Törnqvist, T. E. (2006). Coastal Louisiana in crisis: Subsidence or sea-level rise? *Eos, 87*, 493–498.

González, J. L., & Törnqvist, T. E. (2009). A new Late Holocene sea-level record from the Mississippi Delta: Evidence for a climate/sea-level connection? *Quaternary Science Reviews, 28*, 1737–1749.

Gotham, K. F. (Ed.). (2001). *Critical perspectives on urban redevelopment: Research in urban sociology* (Vol. 6). New York, NY: Emerald Press.

Gotham, K. F. (2006). The secondary circuit of capital reconsidered: Globalization and the US real estate sector. *American Journal of Sociology, 112*(1), 231–275.

Gotham, K. F. (2009). Creating liquidity out of spatial fixity: The secondary circuit of capital and the subprime mortgage crisis. *International Journal of Urban and Regional Research, 33*(2), 355–371.

Gotham, K. F. (2012). Disaster, Inc.: Privatization and post-Katrina rebuilding in New Orleans. *Perspectives on Politics, 10*, 633–646.

Gotham, K. F. (2014a). Racialization and rescaling: Post-Katrina rebuilding and the Louisiana Road Home Program. *International Journal of Urban and Regional Research, 38*(3), 773–790.

Gotham, K. F. (2014b). Reinforcing inequalities: The impact of the CDBG program on post-Katrina rebuilding. *Housing Policy Debate, 24*(1), 192–212.

Gotham, K. F. (2014c). *Race, real estate, and uneven development: The Kansas City experience, 1900–2010*. Albany, NY: State University of New York Press.

Gotham, K. F. (2016a). Antinomies of risk reduction: Climate change and the contradictions of coastal restoration. *Environmental Sociology, 2*(2), 208–219.

Gotham, K. F. (2016b). Coastal restoration as contested terrain: Climate change and the political economy of risk reduction in Louisiana. *Sociological Forum, 31*(S1), 787–806.

Gotham, K. F. (2018). Katrina is coming to your city: Storm and flood defense infrastructures in risk society. Forthcoming in C. Ermus (Ed.), *Environmental disaster in the gulf south: Two centuries of catastrophe, risk, and resilience* (pp. 161–83). Baton Rouge: Louisiana State University Press.

Gotham, K. F., & Campanella, R. (2010). Toward a research agenda on transformative resilience: Challenges and opportunities for post-trauma urban ecosystems. *Critical Planning, 17*(Summer), 9–23.

Gotham, K. F., & Campanella, R. (2011). Coupled vulnerability and resilience: The dynamics of cross-scale interactions in post-Katrina New Orleans. *Ecology and Society, 16*(3), 12.

Gotham, K. F., & Cannon, C. (2018). Circulating risks: Coastal cities and the specter of climate change risk. In A. Jonas (Ed.), *Routledge handbook on spaces of urban politics* (pp. 383–403). New York, NY: Routledge.

Gotham, K. F., & Cheek, W. (2017). Post-disaster recovery and rebuilding. Forthcoming in S. Hall & R. Burdett (Eds.), *The Sage handbook of the 21st century city* (pp. 279–297). Thousand Oaks, CA: Sage Publications.

Gotham, K. F., & Greenberg, M. (2014). *Crisis cities: Disaster and redevelopment in New York and New Orleans*. New York, NY: Oxford University Press.

Gotham, K. F., & Powers, B. (2017). Constructing and contesting resilience in post-disaster urban communities. In J. Hannigan & G. Richards (Eds.), *Sage handbook of new urban studies* (pp. 139–154). Thousand Oaks, CA: Sage Publications.

Gotham, K. F., & Wright, J. D. (2009). Housing policy. In J. Midgley, M. Livermore, & M. B. Tracy (Eds.), *Handbook of social policy* (2nd ed., pp. 237–255). Thousand Oaks, CA: Sage Publications.

Gotham, K. F., Campanella, R., Lewis, J., Gafford, F., Nance, E., & Avula, M. R. (2011). Reconsidering the new normal: Vulnerability and resilience in post-Katrina New Orleans. *Global Horizons: The Journal of Global Policy and Resilience, 4*, 2.

Government Accountability Office (GAO). (2009). *Climate change adaptation: Strategic federal planning could help government officials make more informed decisions* (GAO-10-113). Washington, DC: Govenrment Accountability Office. https://www.gao.gov/assets/300/296526.pdf

Government Accountability Office (GAO). (2013). *High-risk series: An update* (GAO-13-283). Washington, DC: Government Accountability Office. https://www.gao.gov/products/GAO-13-283

Government Accountability Office (GAO). (2016). *Climate change: Selected governments have approached adaptation through laws and long-term plans* (GAO-16-454). Washington, DC: Government Accountability Office. https://www.gao.gov/assets/680/677075.pdf

Grinsted, A., Moore, J. C., & Jevrejeva, S. (2013). Projected Atlantic hurricane surge threat from rising temperatures. *Proceedings of the National Academy of Sciences, 110*(14), 5369–5373.

Hallegatte, S., Green, C., Nicholls, R. J., & Corfee-Morlot, J. (2013). Future flood losses in major coastal cities. *Nature Climate Change, 3*(9), 802–806.

Hogan, J., & Howlett, M. (Eds.). (2015). *Policy paradigms in theory and practice: Discourses, ideas and anomalies in public policy dynamics*. New York, NY: Springer.

Holland, G. J. (2012). Hurricanes and rising global temperatures. *Proceedings of the National Academy of Sciences, 109*(48), 19513–19514.

Horton, R., Little, C., Gornitz, V., Bader, D., & Oppenheimer, M. (2015). New York City panel on climate change 2015 report: Sea-level rise and coastal storms. *Annals of the New York Academy of Sciences, 1336*(1), 36–44.

Intergovernmental Panel on Climate Change (IPCC). (2007). *Climate change 2007: Synthesis report, summary for policymaker*. Washington, DC: Intergovernmental Panel on Climate Change.

Intergovernmental Panel on Climate Change (IPCC). (2013). *Climate Change 2013: The Physical Science Basis. Contribution of Working Group I to the Fifth Assessment Report of the Intergovernmental Panel on Climate Change*. New York, NY: Cambridge University Press.

Intergovernmental Panel on Climate Change (IPCC). (2014). *Climate change 2014 – Impacts, adaptation and vulnerability: Regional aspects*. New York, NY: Cambridge University Press.

Intergovernmental Panel on Climate Change Adaptation Task Force. (2011). *National action plan: Priorities for managing freshwater resources in a changing climate*. Washington, DC: Intergovernmental Panel on Climate Change.

Karl, T. R., Melillo, J. M., & Peterson, T. C. (Eds.). (2009). *Global climate change impacts in the United States*. New York, NY: Cambridge University Press.

Kent, J. D. (2012). *Assessing the long-term impact of subsidence and global climate change on emergency evacuation routes in coastal Louisiana: A report of findings. Gulf Coast Center for Evacuation and Transportation Resiliency*. Principal Investigator: Joshua D. Kent, Ph.D. Baton Rouge, LA: Center for GeoInformatics, Louisiana State University. http://www.evaccenter.lsu.edu/pub/11-09.pdf

Kolb, C. R., & Saucier, R. T. (1982). Engineering geology of New Orleans. In Geological Society of America. *Reviews in Engineering Geology, 5*, 75–94. https://doi.org/10.1130/REG5-p75.

Lahsen, M. (2008). Experiences of modernity in the greenhouse. *Global Environmental Change, 18*, 204–219.

Marshall, B. (2012, May 29). Approval of 2012 master plan for the Coast is worth celebrating. *The Times-Picayune*. http://www.nola.com/outdoors/index.ssf/2012/05/approval_of_2012_master_plan_f.html

Marshall, B. (2017, May 15). Louisiana keeps voting to drown: Opinion. *The Times-Picayune*. http://www.nola.com/opinions/index.ssf/2017/05/louisiana_eroding_coast.html

McCann, E. (2017). Mobilities, politics, and the future: Critical geographies of green urbanism. *Environment and Planning, 49*(8), 1816–1823.

McCright, A. M., & Dunlap, R. E. (2003). Defeating Kyoto: The conservative movement's impact on U.S. climate change policy. *Social Problems, 50*, 348–373.

McCright, A. M., & Dunlap, R. E. (2010). Anti-reflexivity: The American conservative movement's success in undermining climate science and policy. *Theory, Culture, and Society, 27*(2–3), 100–133.

McCright, A. M., & Dunlap, R. E. (2011a). The politicization of climate change and polarization in the American public's views of global warming, 2001–2010. *The Sociological Quarterly, 52*(2), 155–194.

McCright, A. M., & Dunlap, R. E. (2011b). Cool dudes: The denial of climate change among conservative white males. *Global Environmental Change, 21*, 1163–1172.

Meade, N. (2018, July 7). Trump's cuts in climate-change research spark a global scramble for funds. *New Yorker*. https://www.newyorker.com/news/news-desk/trumps-cuts-in-climate-change-research-spark-a-global-scramble-for-funds

Mogensen, J. L. (2018, September 10). New documents show the Trump Administration cut climate change impacts from its energy plan. *Huffington Post*. https://www.huffingtonpost.com/entry/new-documents-show-the-trump-administration-cut-climate-change-impacts-from-its-energy-planus5b9686fee4b0cf7b004222cb

Morrison, T. H., Adger, W. N., Brown, K., Lemos, M. C., Huitema, D., & Hughes, T. P. (2017). Mitigation and adaptation in polycentric systems: Sources of power in the pursuit of collective goals. *Wiley Interdisciplinary Reviews: Climate Change, 8*, 5, e479.

Nagel, J., Dietz, T., & Broadbent, J. (2010). *Workshop on sociological perspectives on global climate change, May 30–31, 2008*. Sociology Program, Directorate for Social, Behavioral and Economic Sciences, National Science Foundation (MSF). Washington, DC: American Sociological Association. http://www.asanet.org/sites/default/files/savvy/research/NSFClimateChangeWorkshop_120109.pdf

National Research Council. (2010). *Informing an effective response to climate change*. Washington, DC: The National Academies Press. https://doi.org/10.17226/12784.

National Research Council. (2012a). *Climate change: Evidence, impacts, and choices. Answers to common questions about the science of climate change*. Washington, DC: The National Academies Press.

National Research Council. (2012b). *Disaster resilience: A national imperative*. Washington, DC: The National Academies Press. https://doi.org/10.17226/13457.

New York City Mayor's Office. (2014). *Mayor de Blasio announces key resiliency investments to support small businesses and jobs, including new business resiliency program and major upgrades across Sandy-impacted neighborhoods*. Official Website of the City of New York. http://www1.nyc.gov/office-of-the-mayor/news/568-14/mayor-de-blasio-key-resiliency-investments-support-small-businesses-jobs

New York City Mayor's Office. (2015). Mayor de Blasio releases NPCC 2015 report, providing climate projections through 2100 for the first time. *Official Website of the City of New York.* http://www1.nyc.gov/office-of-the-mayor/news/122-15/mayor-de-blasio-releases-npcc-2015-report-providing-climate-projections-2100-the-first
Oreskes, N., & Conway, E. M. (2010). *Merchants of doubt: How a handful of scientists obscured the truth on issues from tobacco smoke to global warming*. New York, NY: Bloomsbury Press.
Pelling, M., O'Brien, K., & Matyas, D. (2015). Adaptation and transformation. *Climatic Change, 133*(1), 113–127.
Peyronnin, N., Green, M., Richards, C. P., Owens, A., Reed, D., Chamberlain, J., et al. (2013). Louisiana's 2012 coastal master plan: Overview of a science-based and publicly informed decision-making process. *Journal of Coastal Research, 67*(1), 1–15. https://doi.org/10.2112/si_67_1.1.
Powell, J. L. (2011). *The inquisition of climate science*. New York, NY: Columbia University Press.
Reardon, S., Tollefson, J., Witze, A., & Ross, E. (2017). US science agencies face deep cuts in trump budget. *Nature News, 543*(7646), 471.
Reed, A. J., Mann, M. E., Emanuel, K. A., Lin, N., Horton, B. P., Kemp, A. C., et al. (2015). Increased threat of tropical cyclones and coastal flooding to New York City during the anthropogenic era. *Proceedings of the National Academy of Sciences, 112*(41), 12610–12615.
Selby, J. (n.d.). The Trump presidency, climate change, and the prospect of a disorderly energy transition. *Review of International Studies*, 1–20. https://doi.org/10.1017/S0260210518000165.
Thwaites, J. (2018). US 2018 budget and climate finance: It's bad, but not as bad as you might think. (Blog). *World Resources Institute*. 23 March. https://www.wri.org/blog/2018/03/us-2018-budget-and-climate-finance-its-bad-not-bad-you-might-think
Tierney, K. (2014). *The social roots of risk: Producing disasters, promoting resilience*. Palo Alto, CA: Stanford University Press.
Tilly, C. (1984). *Big structures, large processes, huge comparisons*. New York: Russell Sage Foundation.
U.S. Army Corps of Engineers (USACE). (2018). Project planning: National nonstructural committee (NIC). U.S. Army Corps of Engineers. https://www.usace.army.mil/Missions/Civil-Works/Project-Planning/nfpc/.vlCo
United States Global Change Research Program (USGCRP). (2011). *America's climate choices*. Washington, DC: U.S. Global Change Research Program. http://www.globalchange.gov/what-we-do/assessment.
Wang, F., Xu, Y. J., & Dean, T. J. (2011). Projecting climate change effects on forest net primary productivity in subtropical Louisiana, USA. *Ambio, 405*, 506–520.
Wise, R. M., Fazey, I., Smith, M. S., Park, S. E., Eakin, H. C., Van Garderen, E. A., et al. (2014). Reconceptualising adaptation to climate change as part of pathways of change and response. *Global Environmental Change, 28*, 325–336.

# Chapter 5
# Adapting to a Smaller Coast: Restoration, Protection, and Social Justice in Coastal Louisiana

**Scott A. Hemmerling, Monica Barra, and Rebecca H. Bond**

## 5.1 Introduction

Louisiana's coastal zone (Fig. 5.1) is a naturally dynamic area that has undergone many changes over the past 8000 years coinciding with the shifting deltaic lobes of the Mississippi River (see Fig. 2.2 in Boesch). Recent decades, however, have seen a rapid acceleration of rates of land loss and transitions of habitat types resulting from increasing sea level rise and land subsidence, saltwater intrusion, reduced sediment flow, increasing eutrophication, large storm events, and habitat clearing and alteration due to infrastructure development along the coast (Carruthers et al. 2017). Over this same time period, coastal residents have become increasingly and disproportionately dependent on the coastal zone for living space and recreation, ports and harbors, oil and gas production, commercial and recreational fisheries, marine construction, ship and boat manufacturing, tourism and recreation, and marine transportation (Hemmerling et al. 2016; Weinstein et al. 2007). Most of these economic activities are based on local renewable and nonrenewable natural resources and are therefore largely immovable and highly sensitive to natural and human-induced changes, including fluctuating global economic conditions, environmental stress, climate change impacts, coastal habitat destruction, and increasing social and economic pressures. Shoreline erosion and coastal land loss also threaten the onshore infrastructure that supports these activities, including the

S. A. Hemmerling (✉)
The Water Institute of the Gulf, Baton Rouge, LA, USA
e-mail: shemmerling@thewaterinstitute.org

M. Barra
University of South Carolina, Columbia, SC, USA
e-mail: mbarra@seoe.sc.edu

R. H. Bond
Independent Researcher, Washington, DC, USA



S. Laska (ed.), *Louisiana's Response to Extreme Weather*,
Extreme Weather and Society, https://doi.org/10.1007/978-3-030-27205-0_5

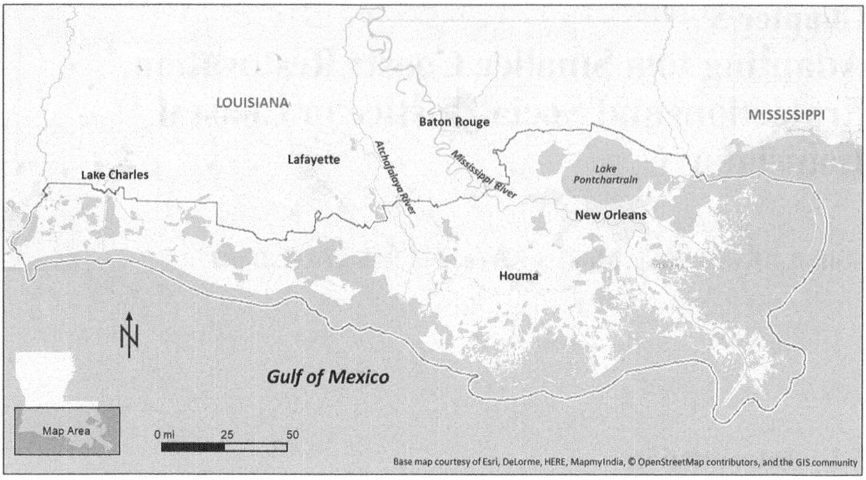


**Fig. 5.1** Map of southern Louisiana showing major cities and the coastal zone. Base map courtesy of Esri, DeLorme, HERE, MapmyIndia, © OpenStreetMap contributors and the GIS community. (Data retrieved from Louisiana Department of Natural Resources)

extensive network of oil and gas infrastructure and pipelines that cross the coastal zone (Hemmerling et al. 2016).

The extent of societal dependency on these activities in the face of increased levels of environmental, economic, and social vulnerability has made managing coastal resources for the public good more challenging than at any time in the past (Weinstein et al. 2007). Recent decades have seen a number of shifts in coastal management priorities: from local problem-solution approaches to broader ecological restoration strategies and, most recently, to large-scale, unified restoration plans that are no longer constrained by political boundaries, embracing multiple jurisdictions and watershed-ocean gradients (Reed 2009). Each of these regimes has a strong science or engineering foundation and seeks to produce predetermined ecological outcomes (Colten 2017) and ultimately results in a shifting of the distribution of the benefits of protection and restoration to greater spatial and temporal scales. In the case of Louisiana's Coastal Master Plan, a science-driven restoration plan that relies heavily on numerical models to optimize project selection and location, the benefits of restoration and protection are anticipated to be widely distributed among a larger, but more abstract, coastal population. Further, numerical model results are able to examine the expected distribution of benefits to future populations.

State policy makers acknowledge that the restoration and protection benefits derived from the numerical model results are spatially variable, owing to both funding and biogeophysical constraints. In some locations of the coast, even with full implementation of the Master Plan, land area is expected to greatly diminish, while in others the land area will be largely maintained. Further, planners acknowledge that implementation of structural protection projects may not be feasible for some coastal communities and areas outside the major levee systems. In some cases, residents of these areas will require nonstructural protection, while in others the adverse future environmental conditions may ultimately displace people, infrastructure, and possibly even the entire communities. With these consequences in mind, coastal planning in Louisiana raises several issues of social justice, many of which arise from the techniques and practices state coastal policy makers use to select restoration and protection projects. The main concern is the extent to which the voices and values of residents bearing the greatest burden of coastal restoration are integrated into planning practices.

This chapter traces the history of coastal planning and the social justice implications of the shift from piecemeal to comprehensive, coastwide planning. It tackles these questions through a close examination of several key topics. First, it traces the emergence and evolution of coastal planning processes, focusing on the shift toward science-driven, numerical models and how and to what extent public engagement contributed to planning processes and the development of conceptual frameworks. Next, the chapter examines three different aspects of social justice – distributive, procedural, and contextual – asking how each might or might not be impacted by the activities of coastal restoration planning in Louisiana. Finally, the chapter ends with a close examination of recent efforts by state policy makers to enhance public participation for the development of the 2017 Coastal Master Plan and analyzes to what extent new techniques for public engagement potentially translate into more socially just selections for future coastal proposals geographically, socially, and economically.

## 5.2 Historical Evolution of Coastal Restoration Planning in Louisiana

To more fully understand the costs and benefits of coastal restoration and how these are distributed among coastal residents, it is important to understand the historical development of restoration polices and the restoration planning process itself. The coastal protection and restoration planning process has continuously evolved over several decades as local, state, and federal agencies developed a number of plans and policy proposals to combat the persistent loss of land that has affected Louisiana's coastline since at least the 1930s. The most expansive plan developed thus far has been the 2017 update to the state's 50-year Coastal Master Plan. Unanimously approved by the Louisiana Coastal Protection and Restoration Authority (CPRA) in April 2017, this plan significantly expanded upon the original 2007 Master Plan as well as the 2012 update. As part of the plan, state officials proposed 124 projects that would maintain or build approximately 800 square miles of land and could save as much as $150 billion in flood damages over the next 50 years. The CPRA estimates that the state would lose another 2250 square miles of land by 2067 if not able to fully implement the Master Plan in the coming decades, resulting in over $12 billion in annual flood damages (Schleifstein 2017). The state acknowledges that, even if fully implemented, the Coastal Master Plan will be unable to protect the entire coast of Louisiana and that the combination of land loss, sea level rise, and subsidence will continue to take a toll on Louisiana's coastal communities for decades to come. Just as the impacts of the state's coastal crisis will be more pronounced in some communities than in others, the degree of protection afforded by the Coastal Master Plan will not be evenly distributed across the coast. The planning process itself will necessarily result in the establishment of winners and losers in coastal protection and restoration, raising the possibility of social and environmental injustices and outcomes.

### 5.2.1 Pre-Katrina: From Piecemeal Projects to Broad-Scale Ecological Planning

Though the 2017 Coastal Master Plan contains a number of grim warnings about the future of the state's coast, Louisianans are not unfamiliar with the risks posed by shoreline erosion. As early as the 1970s, scientists and researchers began warning government officials and the public about the potential impacts of losing the marshes and swamps that make up large sections of the state's coastal topography. In 1972 and 1973, the Louisiana Advisory Commission on Coastal and Marine Resources published three reports examining the loss of coastal wetlands and the potential negative consequences (Louisiana Advisory Commission on Coastal and Marine Resources 1972, 1973a, 1973b). Five years after the commission's reports, legislators passed the State and Local Coastal Resources Management Act of 1978 in an effort to manage development in 19 coastal parishes and help protect the wetlands ("Managing Our Coastal Resources" 1980). Two years later, the federal government approved a coastal management plan (CMP) that included a Coastal Use Permit system to provide additional oversight of activities in the coastal zone (Louisiana Department of Natural Resources 1980).

Although the National Environmental Policy Act of 1969 increased the ability for residents to challenge projects based on environmental and social impacts through the Environmental Impact Statement process, these early coastal policies and practices in Louisiana largely adhered to the more traditional public input process of having comments made on a report instead of public involvement in the design of projects. These initial efforts produced single-purpose project designs that restricted the vision of engineers and scientists to addressing the primary objective, whether it was flood protection, converting wasteland to productive real estate, or species perpetuation (Colten 2017). Initial efforts to manage coastal resources in Louisiana, for example, focused on species-specific habitat enhancements and often involved the manipulation of tidal regimes to maintain water levels to provide access and encourage growth of particular forage vegetation (Reed 2009).

After the CMP's acceptance in 1980, relatively few significant advances in policy implementation or administration of coastal restoration occurred for the next 9 years. However, at the end of the decade, two important policy proposals became law. First, Louisiana voters approved a constitutional amendment in 1989 that established the Wetlands Conservation and Restoration Trust Fund (WCRTF), which was intended to provide a reliable source of money for restoration projects in the state (McMahon 1989). The second notable policy development took place in summer of the following year when Senators John Breaux and J. Bennett Johnston ushered a bill through Congress that finally brought federal dollars to the state's coastal erosion crisis. Signed by President George H.W. Bush in November 1990, the Coastal Wetlands Planning, Protection and Restoration Act (CWPPRA or Breaux Act) provided dedicated funding, meaning the state would not have to request money from Congress every fiscal year. However, the amount of money CWPPRA could deliver was relatively small in relation to estimates for a fully funded restoration program reaching several billion dollars in the early 1990s (McKinney 1989).

While the limited amount of funding was a concern, CWPPRA did make some positive steps in other areas of coastal management and restoration. The legislation created a bureaucratic framework that could bring order to the piecemeal administrative regime that had troubled wetlands policy during the previous two decades. CWPPRA established a task force comprised of officials from the state of Louisiana and five federal agencies, including the US Army Corps of Engineers (USACE), Department of the Interior, Environmental Protection Agency, Department of Agriculture, and Department of Commerce. Each agency's secretary appoints a representative to serve on the task force, and every year the group selects a certain number of restoration projects to be funded under CWPPRA (Louisiana Coastal Wetlands Conservation and Restoration Task Force 1993). Furthermore, the law encouraged Louisiana to develop a conservation plan that included the goal of "no net loss" for wetlands in future developments along the coast. The two plans – conservation to prevent future losses and restoration to address ongoing losses – were intended to be complementary with regulations and allow for more comprehensive management of the coast (Coastal Wetlands Planning, Protection and Restoration Act 1990).

Unlike the majority of previous efforts, the CWPPRA process depends on project nominations from the public, state and federal agencies, coastal parishes, and other coastal entities such as ports. These project nominations are then reviewed by regional planning teams and technical teams and ultimately brought to the task force for funding decisions. This gives local communities the ability to have initial input into project selection before engineering and design even begins. During its first few years, CWPPRA tended to fund small, localized projects that primarily benefitted the immediate vicinity. Those smaller projects were successful, but they could not stop wetlands loss on a large scale over a long period of time (Louisiana Coastal Wetlands Conservation and Restoration Task Force 1996). As the severity of coastal land loss became apparent, broad-scale ecological restoration emerged as the dominant management regime in coastal Louisiana. A primary focus of ecological

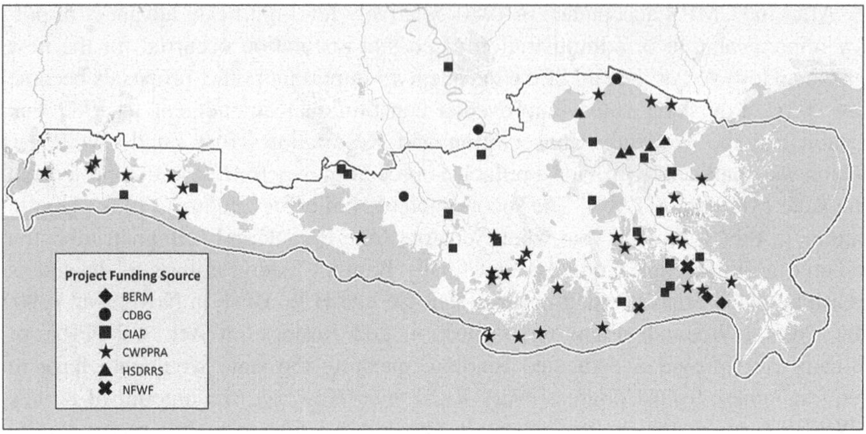


**Fig. 5.2** Map showing existing restoration projects in the Louisiana coastal zone symbolized by funding source. Includes projects completed as of April 2018. (Project data retrieved from the Coastal Protection and Restoration Authority (CPRA) CIMS database)

restoration is to re-establish wetlands to a condition that will protect the region's ecology and major economic interests in the state (Colten 2017). Despite the broadening focus, however, these projects have tended to address single missions such as coastal protection or coastal restoration or were focused on geographically limited locations (Hemmerling 2017). Ultimately, these piecemeal efforts (Fig. 5.2) have had limited success and have not resulted in a net gain of wetlands (Peyronnin et al. 2013). In 1995, groups such as the Coalition to Restore Coastal Louisiana (CRCL) and members of the Louisiana Department of Natural Resources began to advocate for bigger projects that could address losses across the entire coast. In September of that year, the CWPPRA task force agreed that large-scale projects such as freshwater diversions and barrier island restoration should receive priority status on annual project lists. Going forward, the task force would dedicate two-thirds of its yearly project funding to large-scale projects and the remainder for small-scale projects (Anderson 1995a, b).

The shift to prioritizing large-scale projects was an important step toward more effective restoration projects, but officials still needed a single vision for Louisiana's coast to help officials oversee protection and development. In the mid-1990s, dozens of agencies had authority in and around the wetlands, and there was no unifying vision to guide their activities (Schleifstein 1996b). Both Louisiana and the federal government's approach remained piecemeal, even after several years of project development under CWPPRA. If the current approach were to continue, the estimates were that only 22% of future losses could be avoided (Anderson 1997; Louisiana Coastal Wetlands Conservation and Restoration Task Force and Louisiana Wetlands Conservation and Restoration Authority 1998; Schleifstein 1996a, b).

In response to such predictions, political officials in Louisiana's government and representatives from federal agencies initiated a series of meetings in 1997. The CWPPRA task force and the state's Wetlands Authority in the Governor's Office led the process of developing a unifying strategy including inviting agencies such as the National Marine Fisheries Service and the Louisiana Department of Environmental Quality to participate in the process. The group's goals were to build a consensus about what Louisiana's coast should look like in the year 2050, ensure cooperation among the variety of agencies involved in coastal management, and determine how to administer a unified restoration plan (Horst 1997a, b; Louisiana Coastal Wetlands Conservation and Restoration Task Force and Louisiana Wetlands Conservation and Restoration Authority 1998). Over the course of 18 months, public officials met with concerned citizens 65 times to determine what coastal users wanted to see in a restoration plan (Louisiana Coastal Wetlands Conservation and Restoration Task Force 1999).

The Coast 2050 task force built on previous plans but also made sure that the best available science guided their decision-making for future restoration policies. The Coast 2050 Plan was largely a vision document that pointed out challenges and potential solutions. However, it stopped short of listing specific projects and instead focused on strategies such as "maximize land building in Atchafalaya Bay" or "lower water levels in upper Penchant marshes." The final proposal emphasized striking a balance between wetlands protection and economic development and recognized there were multiple interests invested in using Louisiana's coast for a variety of purposes (Dunne 1998). Participating agencies agreed that there was no way to return Louisiana's coast back to the way it had been prior to the 1930s, but there was a consensus that a smaller, sustainable wetlands ecosystem was possible (Gagliano 1994).

The official response to the *Coast 2050* report was largely positive. All 20 coastal parishes formally endorsed the plan, and Governor Mike Foster voiced his support for the adoption of *Coast 2050* as a unified coastal management strategy (Schon 1998). In 2002, the state partnered with the US Army Corps of Engineers to further refine the *Coast 2050* plan and develop a coastwide comprehensive restoration plan under the existing Louisiana Coastal Area (LCA) planning authority (Day et al. 2007; Reed 2009). The resulting LCA Study produced detailed analyses of the costs and benefits of various groupings of restoration projects and included a list of critical restoration projects, many of which had already undergone planning through the CWPPRA process (Reed 2009).

Despite the scientific advances made in these early unified plans, there was still the issue of who would pay the billions of dollars necessary to implement a coastwide restoration and management plan. Some progress was made in funding Louisiana's coastal restoration efforts in summer 2005 when Congress approved the passage of a Coastal Impact Assistance Plan (CIAP). The program was designed to provide revenues to states that contributed to oil development in the Outer Continental Shelf (OCS), and nearly $1 billion in expected revenues would be split among six states. Louisiana's share was estimated to be around $540 million. The funds would be distributed starting in 2007 and could only be used for projects related to coastal erosion and mitigation of the impacts of Outer Continental Shelf (OCS) oil and gas development (Alpert 2005; Radtke Russel 2007).

However, just a few weeks after Congress approved the law containing CIAP, the devastation caused by Hurricanes Katrina and Rita changed the trajectory of coastal restoration in the state. The hurricanes led to the loss of approximately 1800 lives in Louisiana and resulted in $200 billion in damages along the Gulf Coast (Louisiana Governor's Office of Homeland Security and Emergency Preparedness 2015). New Orleans was submerged under water when the city's hurricane protection levees failed during Katrina, and removing the floodwaters took over 40 days. More than one million Louisianans were displaced from their homes in the aftermath of the hurricane (Knabb et al. 2005). This displacement was not equally distributed across the population, nor was recovery. In New Orleans, for example, it was found that black residents were less likely to return to their homes after the storm than white residents primarily because the storm did the most damage in those low-lying areas of the city disproportionately populated by black residents (Groen & Polivka 2010). This in turn reflects historical environmental inequities as black residents were relegated to the low-lying and more vulnerable areas long before the storm ever hit, highlighting that social justice for the future depends on decisions that are made in the present. Katrina thus brought issues of social vulnerability and justice to the forefront of coastal protection and restoration science and marked a dramatic shift in the state approach to coastal planning and urgency of generating more effective policies.

### *5.2.2 Post-Katrina: Establishing CPRA and Louisiana's Coastal Master Plan*

While residents fled to other locations or struggled to rebuild in the wake of the storms, state officials took action to address some of long-standing administrative issues regarding coastal restoration. Governor Kathleen Blanco signed Act 8 into law in November 2005, which created the Louisiana Coastal Protection and Restoration Authority to replace the Wetlands Restoration and Conservation Authority. The new body was directed to coordinate "the efforts of local, state, and federal agencies to achieve long-term comprehensive coastal restoration and hurricane protection." Act 8 also charged the CPRA with creating a "Master Plan that presents a conceptual vision of a sustainable coast based on the best available science and engineering." Projects related to CWPPRA, the LCA near-term plan, and funds from the CIAP were forthwith to be organized "toward a common goal" (Louisiana Coastal Protection and Restoration Authority 2007). Legislators asserted that "the state must have a single agency with authority to articulate a clear statement of priorities," and that "without this authority, the safety of citizens, the viability of state and local economies, and the long-term recovery from disasters such as Hurricanes Katrina and Rita remain in jeopardy" (Louisiana Act No. 8 2005).

In fall 2006, voters approved a constitutional amendment that replaced the Wetlands Conservation and Restoration Trust Fund with the Coastal Protection and

Restoration Fund. They also approved a measure that directed all potential income from any OCS revenue-sharing scheme into wetlands conservation, coastal restoration, and hurricane protection (Sentell 2006). The revenue-sharing amendment was passed in anticipation of Congress authorizing the Gulf of Mexico Energy Security Act (GOMESA) in 2006. In contrast to CWPPRA or the 2005 CIAP, GOMESA was designed to provide a much larger scale of funding over a longer period of time (Walsh 2006). Though GOMESA promised another important source of funding for restoration in Louisiana, coastal advocates warned even that large amount of money was insufficient for the scope of the problem. Mark Davis, then with CRCL, praised the new revenue from GOMESA and said it was "hugely important, but it's also hugely important to know that it's only a down payment" (Shields 2006).

Another significant development happened in 2006. The state legislature had charged the CPRA with developing a Coastal Master Plan to be updated every 5 years, and the committee worked throughout the year to accomplish that goal. To maximize the benefits of coastal projects and comprehensively address both restoration and protection, the CPRA began to focus on the development of unified restoration plans that acknowledge the systematic complexity of interrelated issues in Louisiana's coastal zone and developed more coordinated, integrative frameworks. These frameworks utilize a multiple lines of defense strategy that incorporates a broad suite of structural, nonstructural, and coastal restoration features, including the Gulf of Mexico shelf, the barrier islands, the sounds, marshland bridges, natural ridges, manmade ridges, flood gates, flood levees, pump stations, home and building elevations, and evacuation routes (Lopez 2009). The unified restoration plan approach has culminated with the development of Louisiana's *Coastal Master Plan for a Sustainable Coast*, a numerical model-driven plan built on previous efforts and based upon a theoretically unbiased evaluation of hundreds of previously proposed projects, including nonstructural measures, under both current and future conditions (Fig. 5.3; Peyronnin et al. 2013). While this plan involved extensive public comment periods, public meetings, stakeholder meetings, and presentations, the science-based numerical models do not incorporate these comments. Instead, the comments are made after the plan is drafted and adjustments are made at that time. The difference between outreach, which the planning effort does well, and community engagement, which is still lacking, can make it difficult to meaningfully include social justice issues into the decision-making process.

A preliminary draft of the 2007 Master Plan was released in November 2006 and included a tentative vision for merging coastal restoration and hurricane protection. The draft also proposed some recommendations that had been previously seen as politically toxic such as closing the Mississippi River Gulf Outlet (MRGO). The state had requested that the Corps close the MRGO before the 2005 hurricane season, but the federal agency had been reluctant. A small number of shippers still used the channel; however, after Katrina, decommissioning the MRGO seemed more feasible. There were concerns that the navigation corridor had acted as a "super highway" for storm surge, and officials in St. Bernard Parish welcomed the closure (Committee on Homeland Security and Governmental Affairs 2006; Freudenburg et al. 2009; Schleifstein 2006). Other proposals in the initial draft report faced

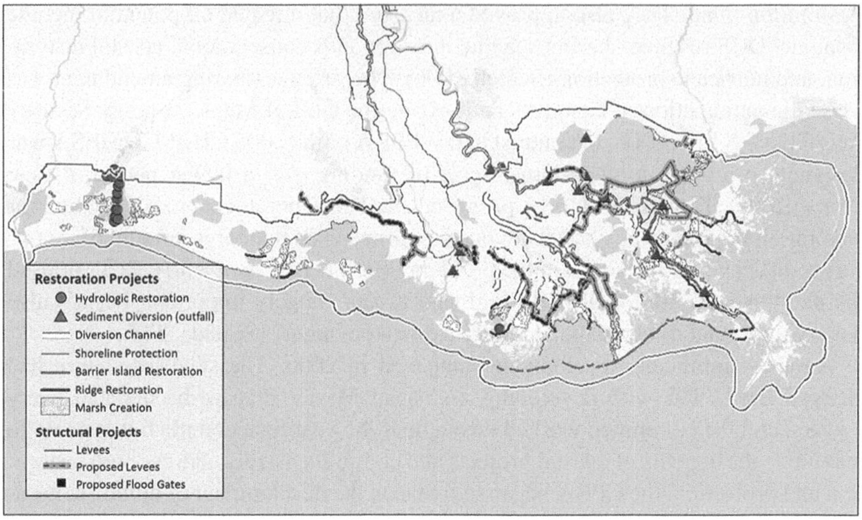


**Fig. 5.3** Map showing Master Plan 2017 projects symbolized by project type. (Project data retrieved from the Coastal Protection and Restoration Authority (CPRA) CIMS database)

criticism, particularly in relation to the heavy emphasis placed on using levees for hurricane protection. Indeed, levees had contributed to coastal erosion and then failed to perform adequately during Katrina due to improper designs and maintenance and because Congress had not appropriated the necessary funds so there were incomplete portions of the system. In response to the critiques that the CPRA received in regard to its initial draft release, the committee revised the Coastal Master Plan to rely less on levees for hurricane protection and pursued something closer to the multiple lines of defense strategy which envisioned a series of speed bumps from barrier islands to interior marsh restoration and to restorations of ridges and including levees.

Much of the 2007 Master Plan was visionary rather than a list of specific projects to pursue, and in that sense, the document resembled Coast 2050. There were some specific suggestions such as closing MRGO or building the Morganza-to-the-Gulf levee system (Louisiana Coastal Protection and Restoration Authority 2007). Overall, the CPRA's first Master Plan, which was accepted by the state legislature in March 2007, was a blueprint for the future (Schleifstein 2007). To implement more specific actions, the CPRA would release annual reports with more targeted suggestions (Louisiana Coastal Protection and Restoration Authority 2007). All ongoing projects – including ones conducted under the CIAP, the CWPPRA, and the Corps – needed to be consistent with the state's Master Plan (Louisiana Coastal Protection and Restoration Authority 2008).

Overall, reception of the 2007 Master Plan was mixed. Though the plan was not a radically innovative proposal, integrating restoration with hurricane protection was a new step. Further, there seemed to be an increased commitment to funding a plan that treated both activities as related after Hurricane Katrina. However, a review panel criticized the state over the "breakneck pace" at which the CPRA's Integrated Planning Team "attempted to craft solutions for a complex and all-important task." Other observers took issue with some of the proposals in the plan that were based on questionable scientific evidence. Technical reviews made clear that more complex modeling and scenario analyses were needed and in response the CPRA ramped up its scientific analysis and modeling efforts (Wiegman et al. 2018).

In preparation for the 2012 update to the Coastal Master Plan, the CPRA developed several new models that were linked to predict change in the Louisiana coastal system under two types of future management strategies: a future without the implementation of future protection and restoration projects and a future with implementation of individual projects (Peyronnin et al. 2013). This systems-based numerical modeling approach relied heavily on a decision support tool designed to provide an analytical and objective basis for comparing projects and developing alternative groups of projects for consideration in the final plan. Candidate projects were selected by mining earlier studies, reports, presentations, and plans to develop a final list of 397 candidate projects.

In 2012, the state released its first legislatively mandated update to the 2007 Master Plan, which included an assessment of the progress achieved in coastal restoration. Over the previous 5 years, the CPRA had administered projects related to building or improving 159 miles of levees, constructed 32 miles of barrier islands or berms, placed 150 in design or construction, and benefitted over 19,000 acres of wetlands (Louisiana Coastal Protection and Restoration Authority 2012). By 2014, 45 miles of barrier islands or berms had been built, and coastal restoration and protection programs had benefitted 26,241 acres of land. As of 2015, the state planned to monitor or maintain 230 projects, while overseeing the design and construction of 79 more (Louisiana Coastal Protection and Restoration Authority 2016). Overall, the CPRA reported that the rate of shoreline erosion was down significantly from its height in the 1970s. Despite the progress made, however, the state was still losing approximately 16 square miles of land per year, highlighting the importance of continued coastal restoration planning (Louisiana Coastal Protection and Restoration Authority 2012).

The 2017 update to the Master Plan was largely based on the same framework established in 2007 and reaffirmed in 2012. Coastal restoration projects will remain under the oversight of the CPRA, but the latest iteration of the Master Plan does contain an increased emphasis on nonstructural means of combatting storm-related flooding (Louisiana Coastal Protection and Restoration Authority 2017). Furthermore, officials appear to be less optimistic about the future of the coast in the 2017 update, citing increasing concerns about climate change. New Orleans and other low-lying areas in southern Louisiana are expected to become even more vulnerable to flooding and storm-related damages as sea levels rise in response to the warming planet (Marshall 2017). Worse, the funding problems that have undermined coastal restoration efforts since the 1970s have not been resolved. Louisiana has long been planning to use GOMESA funds to help pay for projects beginning after 2017. The state expected to get approximately $140 million in the first year but is currently slated to get half that amount. As a result, projects scheduled to begin in the 2019 fiscal year may have to be scaled back or placed on hold while officials search for additional funding (Schleifstein 1996b).

## 5.3 Shifting Costs and Benefits of Protection and Restoration: Coastal Planning as a Matter of Social Justice

The benefits of the Master Plan and other similar unified restoration plans are without a doubt broad and sweeping. These plans do, however, acknowledge that it is not possible to provide the same level of benefits to all coastal communities. Coastal management literature argues that while structural defenses can be justified in urban areas, they often fail to meet the cost-benefit test in thinly settled, rural locales (Colten et al. 2018). In coastal Louisiana, for example, much of the at-risk Native American populations reside in the small rural communities located along the land-water interface, as do many other minority communities who rely on subsistence fishing to supplement household resources. These communities, due to their proximity to the coast and their rural nature, make them especially vulnerable to natural hazards and risks (Dalbom et al. 2014). They also reside in locations where the construction of structural protection features is largely untenable. But it's not only geography that makes them vulnerable. Histories of displacement, segregation, and political disenfranchisement have made many Native communities economically under-resourced and comparatively less politically powerful than urban, white populations in south Louisiana.

When taken together, the historical and contemporary contexts that situate the negative outcomes for residents residing in small rural communities in the coastal zone are potentially magnified in areas that are highly dependent upon fisheries and other natural resources for their economic well-being. Changes in the distribution and abundance of species, for example, will likely have socioeconomic effects on fishers, hunters, and other harvesters who use the wetlands for commercial, subsistence, recreational guiding, and recreational activities. Fish and wildlife will likely adapt quickly, whereas it is harder and takes longer for resource harvesters to adapt (Peyronnin et al. 2017). This is a particular concern in coastal Louisiana, where projects focused on protecting the maximum number of residents over the long term are also projected to disrupt ecological conditions that sustain the natural resources that many coastal residents rely in the short term, creating a number of unique social justice concerns (Colten et al. 2018).

The shift from small-scale, localized projects to a science-driven, unified restoration plan has the potential to dramatically change the appearance of the coastal landscape, both natural and human (Table 5.1). Broadly speaking, numerical models are used to identify a suite of protection and restoration projects that will synergistically derive the greatest benefits for the greatest number of residents. In this

**Table 5.1** Restoration planning and policy outcomes

| Type of restoration planning | Louisiana examples | Policy outcomes | Advantages | Disadvantages |
|---|---|---|---|---|
| Small-scale projects | CIAP, CWPPRA (pre-1995) | Local areas benefit; short-term impacts | Individual communities benefit directly; multiple points of view considered | Multiple agencies involved in execution, which results in conflicting agendas; unable to stop net loss of wetlands over time |
| Large-scale projects | CWPPRA (post-1995), state of Louisiana, USACE | Broader areas benefit; short-term and long-term impacts | Individual communities and broader areas benefit directly; multiple points of view considered | Multiple agencies involved in execution, which results in conflicting agendas; unable to stop net loss of wetlands over time |
| Unified coastal restoration | Coast 2050, Coastal Master Plan for a Sustainable Coast | Mixture of area sizes benefit, but emphasis is on coastal ecosystem; short-term and long-term impacts | Streamlined administration; coastal ecosystem prioritized to slow overall losses | Viewpoints of individual communities have less influence and receive less attention; unable to stop net loss of wetlands in the near term |

approach, each numerical model derived for the analysis provides input to other models, produces outputs, and estimates how the landscape.

might change and how projects might perform on the landscape over time (Peyronnin et al. 2013). The idea that the final model outputs potentially identify that suite of projects that provide the greatest level of social benefit presents a powerful justification for comprehensive master planning efforts. An analysis of the 2012 Master Plan estimated that, if fully implemented, the planned risk reduction projects would provide heightened protection to over 86% of families and nearly 85% of poor families in Southeast Louisiana (Dalbom et al. 2014). This same study reveals that, by extending protection to the majority of the population residing in the developed areas of the coastal zone, the 2012 Master Plan will simultaneously reduce the anticipated level of risk for the urban African American, Asian, and Hispanic populations of the region.

However, there are social costs associated with the shift to a purely science-based approach to project selection. While all restoration and protection plans accept that change is inevitable across the coast, more recent science-driven plans are less constrained by the impacts of these changes on local populations (Reed 2009). As a result, some of the poorest and most geographically marginal coastal groups are often outside the purview of restoration and protection. This situation raises the question of how policy makers can fairly distribute the benefits and burdens of coastal restoration (Colten et al. 2018). As a result of a purely science-driven planning process, the impacts of protection and restoration projects on individual communities are devalued as building and maintaining land and reducing risk on a broad scale become the key decision drivers for selecting projects (Peyronnin et al. 2013). Indeed, while Louisiana's Master Plan is couched in terms of sustainability, it does not propose sustainability for all (Colten 2015). Because one of the primary goals of the Master Plan is to provide protection to the greatest number of individuals, many of the proposed projects will prioritize providing protection to urban residents and those residing in more densely populated areas. In short, as many coastal residents attest to, this approach to project selection runs the risk of sacrificing remote coastal areas home to already socio-spatially marginalized groups for the protection of environmentally viable urbanized coastal regions. This impacts a disproportionate number of small business and subsistence fisherfolks from Native American, African American, southeast Asian, and other minority groups as well as white residents with long histories of occupancy of the rural coastal areas.

Mapping onto existing geographies of racial and ethnic difference and economic inequality, the uneven distribution of risk and anticipated siting of protection projects raises the issue of social justice to the fore of contemporary coastal planning. An issue largely unexplored in Louisiana (Colten et al. 2018), social justice is comprised of three key elements, each of which can be impacted by coastal restoration programs in distinct ways: distributive justice, procedural justice, and contextual justice. The degree to which the outcomes of environmental projects address each of these elements can have a decisive impact on both the overall equity of the outcomes of the program and ultimately whether these efforts succeed or fail (Fischer et al. 2015).

### *5.3.1 Distributive Justice*

Scholars have noted that the last several decades of research into social justice have focused largely on one key dimension: distributive justice (McDermott et al. 2013; Schlosberg 2004). This dimension focuses on the allocation of material goods, including environmental quality, and generally conceives of social justice and distribution as equivalent concepts (Dobson 1998; Foster 1998; Pulido 2000) . Distributive justice, as it relates to coastal restoration and protection, focuses on the allocation among coastal residents of costs and benefits resulting from environmental policy, resource management decisions, and environmental modifications (McDermott et al. 2013). Advances in high-end computing, numerical modeling, and geographic information systems (GIS) have allowed coastal researchers to develop innovative analytical techniques to measure and forecast the impacts of environmental change on broad spatial and temporal scales. Through these techniques, the efficient distribution of social costs and benefits can be measured and used to assess the ability of environmental programs to maximize the social welfare that can be achieved under given biogeophysical and financial constraints.

Under the distributive justice framework, the costs and benefits of environmental adaptations may be unequally distributed among individuals for the sake of net social gain for the entire population (McDermott et al. 2013). Socioeconomically neutral coastal adaptation planning refers to the advancement of protection and restoration projects on the basis of scientific processes. This approach overlooks racial and economic inequality and the history of environmental inequity in both settlement and risk patterns. Climate change adaptation plans based upon socioeconomically neutral, physical science-driven numerical models can create winners and losers, potentially shifting the distribution of benefits or risks from one group to another (Lebel et al. 2009). The purely distributive focus of these models – greatest good for the greatest number of people – obscures the role that social structure and institutional context play in determining the patterns of distribution (Foster 1998). If such contextual issues go unrecognized, adaptation planning built upon science-driven numerical models may lead to restoration and climate change adaptation plans that benefit some populations while abandoning others (Hardy et al. 2017). Such measures may even exacerbate injustice, as when actions designed to maximize protection in urban areas or protect critical assets and infrastructure make some disadvantaged groups even more vulnerable than they were before (Lebel et al. 2009).

### 5.3.2 Procedural Justice

The limitations of a distributive justice framework for understanding environmental (in)justice are elaborated by scholars concerned with the ways existing and new social and economic inequalities are entrenched by practices for managing environmental hazards (Pulido 2000, 2015). By downplaying or ignoring the historical processes and causes that result in an inequitable distribution of risks and benefits across the coast, science-driven adaptation plans may inadvertently exacerbate existing inequities. The question then becomes whether promoting procedural justice by instituting inclusive, participatory processes within coastal restoration planning makes it possible to correct for any unfair distributional outcomes and potentially address causal origins (McDermott et al. 2013). The concept of procedural justice shifts the focus from the actual distribution of the costs and benefits of coastal restoration projects to the fairness of the process by which these costs and benefits are allocated and decisions are made (Clayton 2000). It involves recognition, inclusion, representation, and participation in the decision-making process by local residents and potentially impacted stakeholders (Ishiyama 2003; McDermott et al. 2013). Ultimately, reducing the risk of exposure to coastal hazards, both physical and economic, requires engagement with residents and stakeholder groups likely to be affected by policy actions and those who are especially vulnerable to risk. Engagement, in this sense, goes beyond legally mandated public comment protocols. Instead, aspirations to procedural justice would aim to give significant weight and representation to marginalized voices at all levels of the planning process and final decision-making.

### 5.3.3 *Contextual Justice*

To accurately assess the social impact or fairness of a project or program, it is necessary to identify not only the outcomes and processes of implementation but also the initial social conditions and origins of any existing environmental inequities (McDermott et al. 2013). Coastal planners need to understand current political processes and distributive outcomes within a historical context and address the fact that, in many cases, the playing field is already highly skewed against local communities due to a number of economic and social disadvantages (Larson and Ribot 2007). Such disadvantaged communities face a number of technical and bureaucratic hurdles that other communities may not face, often compounded by a lack of access to vital information and an inability to pay for needed technical expertise. For example, the lower a resident's income level, the less likely they are to be familiar with proposed restoration projects that could directly impact them and the more likely they are to think that the project will not change fisheries (Gramling et al. 2006). Without a clear understanding of the historical processes that have led to these disadvantages, coastal policy and implementation practices run the risk of exacerbating existing environmental inequities. An understanding of contextual justice, as it relates to coastal protection and restoration, takes into account those preexisting conditions that limit a community's access to decision-making procedures, resources, and benefits, effectively serving as a link between distributive and procedural justice (McDermott et al. 2013). Ultimately, to navigate these issues and effectively redress historical injustices while also promoting effective coastal planning, more and better knowledge is required about the development of those preexisting political, economic, and social conditions that limit people's capacity to engage in and benefit from the coastal planning process (Fischer et al. 2015). By incorporating aspects of contextual equity into the planning process, policy makers are more likely to identify uncover impacts that are harder to measure but are often crucial to local welfare (McDermott et al. 2013).

## 5.4 Public Participation in Coastal Planning

Within a procedural and contextual justice framework, those most at risk should be given opportunities to participate in reshaping and reducing risk to which they are to be exposed (Lebel et al. 2009). Echoing work in environmental justice (Checker 2011; Ishiyama 2003), such an approach must go beyond participation and token integration of marginalized voices to generate meaningful and politically efficacious modes of interaction in policy development that does eschew or co-opt the self-determination of less populated and economically marginal coastal areas.

The importance of public participation in the restoration planning process has been acknowledged by the state of Louisiana in the development of the Master Plan, which developed a set of four key outreach and engagement principles to ensure structured and transparent interactions with the public as well as key businesses and industries, federal agencies, nonprofits, academia, and fisheries interests. Key goals outlined for the state in both the 2012 and 2017 Master Plans include:

1. Stakeholders and citizens should be given opportunities to learn about and comment on the 2017 Master Plan tools and the processes that assist in creating the plan – not just the finished plan itself.
2. Comments and ideas should be received, reviewed, and incorporated while the 2017 is being developed, not after the fact.
3. Not every stakeholder or citizen preference will be included in the 2017 Master Plan. However, the state promises that each idea will receive a fair hearing and that questions will be answered promptly and with care.
4. The state has an obligation to provide a variety of ways for stakeholder and citizens to learn about and participate in the master planning process, including small group gatherings, web offerings, direct communication with local and state government, and public meetings (Speyrer and Gaharan 2017).

These goals highlight CPRA's desire to capture a wide swath of public feedback on the Master Plan. Further, they reflect that the state understands the persistent frustrations of citizens across the coast that they are engaged too late in the planning process and that their comments make little difference to what the state decides to do.

In developing the 2012 Master Plan, the CPRA attempted to respond to these staunch and persistent critiques through the development of numerous stakeholder groups and citizen outreach tactics. Stakeholder groups engaged with in the planning process included the following: (1) a Master Plan framework development team, residents from Louisiana representing federal, state, and local governments, NGOs, business and industry, academia, and coastal communities (this group was comprised of 33 members that met on an almost quarterly basis to review ongoing research and project selection processes for the Master Plan; (2) a fisheries focus group composed of approximately 15 members in the commercial fishing industry; (3) a group of about 10 members representing the oil and gas industry; and (4) a group of approximately 15 members representing navigation interests. These groups were variously consulted over the development of the 2012 Master Plan in order to assess the impacts of potential projects upon these industries. There were a handful of coastal citizens who participated in these groups, but the majority of participants were selected because of their professional and political affiliations.

The process for public engagement entailed numerous public meetings and official hearings for the 2012 Draft Master Plan. At the outset of the planning process, the CPRA held ten regional community meetings throughout the coast. Approximately 600 citizens participated these meetings, which were designed to gather local knowledge and identify public priorities and concerns. Additionally, the CPRA conducted a statewide telephone poll to elicit information from over 1000 additional residents. The results of the initial community meetings and polling indicated that, regardless of where they live, citizens were concerned about land loss, reducing flood risk, and the future of coastal fisheries (CPRA 2012). Input from these meetings was catalogued and posted to their public site alongside suggestions for citizens to become involved in the planning process. In total, community meetings, public forums, civic presentations, a community survey, and a telephone poll were used by state decision-makers to gather information on citizen preferences and ideas that could be incorporated into the decision-making process (Peyronnin et al. 2013).

Upon completion of the draft Master Plan in January of 2012, the CPRA hosted three additional public hearings to receive comments on the plan. The state received over 100 formal comments during these hearings and over 2200 additional comments received subsequently via email, website, and mail. After collecting comments, the CPRA had approximately 1 month to evaluate and address project-specific concerns before sending off a finalized version to the state legislature to vote on for approval. Comments related to both policy and implementation were also evaluated and catalogued to help guide the state as Master Plan projects and programs begin to be implemented in the future.

Accounting for how public comments become incorporated into and/or influenced the Master Plan was addressed in 2012 through specific tactics the CPRA used to test particular projects. As the Master Plan notes, projects were adjusted "based on local knowledge and stakeholder input where appropriate. The changes were principled responses to the feedback we received, grounded in science, and responsive to the needs of our coastal communities" (CPRA 2012, 112). The state noted that they considered all public comments, categorized them by major theme, and provided responses to each theme, specifically identifying the policy- and project-level adjustments to the final plan (CPRA 2012). Changes were reflected explicitly in several structural protection and flood risk reduction projects that were either added or adjusted in the final plan, based upon a combination of policy constraints, public input, and scientific models. For example, the CPRA used public comments to test preferences for and against large-scale river sediment diversions. Using data generated by seven integrated predictive models, nine decision criteria, and various project implementation constraints, the CPRA evaluated the presence and removal of several sizes of river diversions and evaluated what different public preferences for diversion sizes and locations would be.

Goals and approaches to the 2017 Master Plan were similar to 2012, but with several key changes. First, the state introduced a community focus group as one of the handful of advisory groups they met with regularly during the plan development. Community groups included leadership from local Native American tribes, community organizations serving Vietnamese fishermen, and organizations serving predominately African American communities in rural coastal areas. They met four times between April 2013 and October 2016, having anywhere from 4 to 15 members in attendance. Beyond CPRA presentations, participants in the focus group primarily discussed interest in attaining small grants for local community organizations to help with CPRA education and outreach, explicit concern for projected land loss in certain Native American communities along the coast, and how projections of future flood risk might impact low to moderate income populations on the coast (Speyrer and Gaharan 2017). Concern for expanding the geographic scope of community engagement was also a frequent topic of discussion between community focus group members and CPRA officials as was defining the scope and content of what nonstructural projects – such as home elevation, flood proofing, and relocation – would be. This latter point is particularly important among the community focus groups as most participants are residents of small coastal communities located outside the extensive levee and flood wall protection.

While the community focus group was not framed explicitly by the CPRA as an attempt to engage minority communities, in practice it was the most consistent and strategic engagement the state made with representatives from minority communities. In the context of coastal planning in Louisiana, there is no explicit representation or study of minority or economically marginalized communities within research that informs the Coastal Master Plan. Meetings with the community focus group reflect the ethos of socioeconomically neutral planning techniques that utilize a non-specified, generic notion of "community" as a stand-in for representing the experiences of marginalized communities without naming racial, ethnic, economic, or other forms of difference – let alone social justice – as a key motivation for the generation of the community focus group. For example, the needs and challenges inside bay subsistence and small-scale commercial fisherfolk face with impending coastal restoration projects and their changes to regional ecologies are distinct from those that local homeowners face: For one group, environmental changes for restoration mean potentially going out of business or taking on the financial burden of developing new fishing practices. For the other, those same environmental changes point to the possibilities of high flood risks for private property, a financial burden many might not be able to shoulder. Beyond these basic examples of difference within the generic category of "coastal communities," review of CPRA documents shows that it is unclear how and if this particular focus group, or any of the advisory councils CPRA engages, actively shapes the projects and decision-making frameworks that the state utilizes to develop coastal policy.

The CPRA also established several other focus groups, including landowners and parish floodplain managers. Inclusion of these groups reflects the state's gradual expansion of the stakeholder and resident types who they believe need to have a sustained engagement in the master planning process beyond limited public hearings. The state also began to publish materials in Vietnamese, Spanish, and French and developed a series of online flood risk and other informative tools in order to reach more diverse audiences around the coast. While communication techniques are crucial, they do not necessarily equate a more robust engagement with the coastal public or incorporation of social justice concerns into planning. They might, however, increase the likelihood that state representatives will develop a more consistent relationship to different groups and perhaps incorporate changes to Master Plan projects derived from coastal communities in concert with numerical models and scientific expertise.

As with master planning initiatives in 2012, the CPRA partnered with NGO groups to organize public meetings and series of open houses prior to the official public comment period that commenced in January 2017. In October and November 2016, the CPRA held community meetings in several coastal communities to solicit early feedback on draft lists of potential projects for the 2017 Master Plan in response to increasing public pressure to give individual citizens who are not members of select advisory or focus groups more opportunities to vet Master Plan ideas prior to the production of the draft Master Plan. According to CPRA, approximately 500 people attended 7 meetings held across the coast (Speyrer and Gaharan 2017). There is no information on the demographic or geographic composition of the crowds, and it is difficult to assess how, exactly, more meetings correlate to a plan that more effectively represents the diversity of values and interests associated with protection and planning projects.

Establishing new focus groups that include community organizations, landowners, parish floodplain managers, and commercial fishing interests and doubling engagement efforts (including expanding the range of linguistic outreach to access southeast Asian and Latino residents) reflects a recognition that the impacts of the Master Plan are geographically and economically diverse. Reluctance to name social, racial, or economic justice as a matter of concern for coastal planning, however, reflects that the state envisions coastal restoration as something that operates outside of explicitly racial, economic, geographic, and social disparities and histories. Recent efforts by the CPRA strive to achieve social justice through a scientific model-based distributive justice framework but often do little to address historical and ongoing power inequalities that circumscribe small coastal communities to disproportionately bear the burdens of environment risks. While public participation has been ramped up from 2012 to 2017, the extent of addressing social, economic, and spatial inequity in coastal planning remains vague.

## 5.5 More Meetings and Public Participation, More Justice?

Despite ongoing efforts by the state of Louisiana to actively engage with local residents and incorporate aspects of procedural justice into the restoration planning process, many community groups have condemned the process as exclusionary and undemocratic (Gotham 2016a). Additionally, many residents feel disenfranchised by what they perceive to be a repetitive and ambiguous public engagement process that often leaves them feeling fatigued, frustrated, and ignored by state policy makers and coastal planners (Carruthers et al. 2017). The stark contrast between the goals of the state's outreach and engagement plan and the experience of some coastal residents highlights an essential dilemma faced by CPRA and other coastal policy makers. Debates over coastal protection and restoration are not just about risk but represent struggles over access to resources and the power of residents to define and defend cultural forms (Gotham 2016b). The current focus on developing world-class, science-driven numerical models is perceived as coming at the expense of taking residents' concerns seriously. These same residents feel ignored or left to fend for themselves against the forces of nature and the economy because they are often the inherent "losers" of land loss *and* coastal planning. When used as a tool to substantiate the integrity and power of the state to make "decisions in the best interests of Louisiana's citizens," the coastal restoration planning process runs the risk of reinforcing a longer history of state and federal governments justifying their power over environmental management practices in the United States that frequently results in the political, economic, and geographic displacement of politically and economically disenfranchised groups (Kosek 2006; Spence 1999; Hardy et al. 2017). The question of whether or not the government intends to have disproportionate impacts on socially, politically, and geographically marginalized groups is difficult to answer and obscures the fact that supposedly objective decisions about where and how to protect the coast often struggle to move beyond limited notions of distributive justice to address broader social justice and equity issues.

To be sure, coastal restoration has not historically been designed to remediate or define ways to cultivate social justice and equitability when it comes to addressing Louisiana's coastal land loss crisis. Instead, planners and scientists have used numerical models to justify and legitimize the selection of specific risk reduction techniques to protect broad swaths of the coast, while residents often use perceptions of increased threat and a fundamental distrust of government at all levels as a justification for locally rejecting many of these techniques (Colten 2015; Gotham 2016b). While Louisiana's most recent iterations of the Coastal Master Plan boost extensive public engagement efforts operating in parallel with the systems-based scientific analysis of coastal projects that forms the backbone of the protection and restoration plan, the capacity to effectively integrate these streams is not readily apparent to many frontline coastal communities nor is it apparent to external scientific working groups (Wells et al. 2015). Coastal residents continue to struggle with bureaucratic processes related to how restoration projects are nominated, prioritized, and selected and understanding where restoration efforts and funds are spent (Carruthers et al. 2017). This highlights the fact that simply increasing the number of people touched by public engagement may reduce levels of procedural injustice inherent in the restoration planning process, but it is not a guarantee that social justice can be achieved for socially or economically marginalized groups. Instead, increasing levels of outreach and engagement often appears to recapitulate notions of distributive justice which, by and large, does not resolve the fact that there will be, as state officials are partial to saying, "winners and losers" in coastal restoration and protection planning. The decision-making process is still largely top-down and guided by scientific models that do not reflect or have the capacity to change the power dynamics inherent in the restoration planning process itself. While the state has significantly increased and documented the number of public meetings that have been held as part of the restoration planning process, a mechanism of accounting for input derived from these meetings has yet to be generated. It is therefore difficult to assess the extent to which holding more public meetings, or creating different interactive formats for public meetings, will result in any change in outcomes for groups who disproportionately bear the greatest risks from coastal hazards and land loss. In short, increased engagement is not a guarantee of risk reduction.

At a minimum, outreach and engagement attempts can build trust with citizens, trust on the part of citizens that state officials have their best interests in mind. As the introduction to the 2017 Master Plan succinctly captures, "our goal is to develop public confidence that CPRA is the primary technical authority on coastal protection and restoration for Louisiana and is making decisions in the best interests of Louisiana's citizens" (CPRA 2017, p.1). More outreach and engagement efforts aspire toward increased procedural justice and accountability to diverse coastal population and a genuine concern for coastal Louisiana's well-being. However, merely increasing levels of procedural justice does not necessarily result in increased social justice. If public participation drives the selection of certain specific projects that protect a small number of residents at the expense of a greater number of residents elsewhere on the coast, then it will have reduced the level of distributive justice, which is focused on maximizing net social gain for all residents of coastal Louisiana.

## 5.6 Mapping a Path Forward

Despite the power of incorporating local knowledge into the coastal planning process, to date it has been challenging to broadly implement due to difficulties in achieving scientifically rigorous, replicable, and widely accessible methods of data collection. In large part, projects that have taken such an approach have been wholly qualitative in nature, which, though valid, are still not as easily accepted across the sciences. However, with advances in geospatial technologies, a growing acceptance of mixed methods research, and awareness of the validity and importance of local knowledge, this situation is changing (Curtis et al. 2018). There is a growing literature on the potential of combining local knowledge systems with technical scientific knowledge to manage both ecosystems and resources, including the evaluation of climate change impacts and the management of fisheries, biodiversity, and landscape dynamics (Folke et al. 2005). The people who live and work in coastal communities are becoming recognized as repositories of valuable local knowledge of concentrated community risks that reduce capacity in preparedness, such as issues of safety, health, and education, as well as on the critical social infrastructure network that they would access in response and recovery. Community members also hold perceptions of risk that shape their preparedness and mitigation activities, such as which places in their community are dangerous and which are thought to be safe. Such local knowledge and environmental perceptions are often geographically explicit and are powerful influences on behavior (Curtis et al. 2018). It is essential that coastal planners account for these data to form a more complete evidence base in guiding the development of resilient coastal communities.

Several recent methodological advances that allow for the input of qualitative local knowledge into mathematical models have provided tangible ways to evaluate potential outcomes and shortcomings of ongoing and planned restoration and protection projects against projected results which can allow coastal planners to make adjustments that respond to the real-time needs of impacted communities. Methods such as local knowledge mapping, social return on investment, and competency groups have all been used in coastal Louisiana to collect, analyze, and map qualitative data with the goal of characterizing local community members' understanding of what ecological restoration has historically achieved, as well as a suite of potential short- and long-term outcomes of emerging ecological restoration projects identified by residents. Results from these approaches provide a new, geographically targeted, evidence base for planning strategies, especially those focused on coastal protection and restoration. These approaches are not designed to directly address issues of social injustice or change public policy. Rather, they present examples for state agencies and policy makers to follow as a means of anticipating, understanding, and attempting to alleviate unequal impacts before they occur, an important first step in addressing many of the social justice issues faced by coastal residents.

### *5.6.1 Local Knowledge Mapping*

Many science-driven planning processes, including Louisiana's Coastal Master Plan, rely upon quantitative, geospatial datasets as model inputs and to derive metrics as criteria for evaluating the effectiveness of protection and restoration projects. While these datasets are effective at locating any number of nonresidential, residential, and infrastructure assets at risk within an area, they are not able to specifically identify places that have social or cultural value to residents and communities. State planners recognize that protecting such places of value is vital to preserving the culture and identity of Louisiana's various coastal communities (Louisiana Coastal Protection and Restoration Authority 2012), yet modeling efforts have focused largely on the more tangible aspects of cultural heritage that can easily be captured by existing geospatial datasets, such as the presence of ethnic minority groups or historic properties. The overreliance on such datasets in the planning process, particularly when presented with no additional context, may result in any number of social justice outcomes. Local knowledge mapping is an approach that aims to encourage community member participation in sharing knowledge and perceptions of a given area and has been shown to provide an effective means of incorporating community and traditional ecological knowledge into a coastal protection and restoration framework. The incorporation of these data into the planning process would represent an important first step in ameliorating the impacts of past environmental inequities while reducing the risk of future disproportionate impacts on particular social or cultural groups. While local knowledge mapping typically involves having local stakeholders mark locations on paper maps, recent advances in mapping and in GPS-enabled technology that are low in cost, widely available, and accessible to the public have allowed researchers to directly gather geospatial data from local knowledge experts, which is particularly important when the pace and geographic scale of change is dynamic (Curtis et al. 2018). Qualitative data collected during local knowledge mapping exercises have been used to create a geospatially explicit baseline dataset allowing researchers to incorporate local knowledge into an assessment of ecological restoration projects. When incorporated into a GIS environment and assessed in combination with biophysical data, the resultant "Sci-TEK" data can potentially be used to refine the large conceptual footprints of restoration projects and aid in the identification of future restoration projects and identify associated areas of consensus and potential conflict between local stakeholders and policy makers (Bethel et al. 2011, 2014, 2015). The information gained in this way has also been used to determine the geographic specificity of local perceptions and develop community-informed prioritization tools that can be used to plan future ecological restoration projects (Barra 2017; Carruthers et al. 2017).

### *5.6.2 Social Return on Investment*

Ecological restoration and other activities that interact with environmental systems have typically relied on scientific analysis to predict the impacts of these projects and have operated on the assumption that good science could reveal and remedy potential problems (Colten and Hemmerling 2014). Because coastal planning is fundamentally a human activity, however, effective predictions of human impacts demand equal attention to the social, political, cultural, and economic systems in which environmental management takes place (Ludwig et al. 1993). Protection and restoration projects deliver variable costs and benefits to Louisiana coastal communities and the economies they depend upon, such as navigation and fisheries (Caffey et al. 2014). Residents impacted by these projects have recognized these variabilities, valuing some projects as vitally important and highly desirable, while questioning or opposing others (Colten 2014). Qualitative data analysis can successfully classify differences in the ways stakeholder groups potentially impacted by ecological restoration projects engage with the project sites and identify a suite of outcomes unique to each stakeholder group. Identifying these outcomes is integral to defining both the specific objectives and variables needed to develop a comprehensive assessment and monitoring framework.

In order to quantify locally specific social impacts and develop a framework amenable to measuring social change resulting from ecological restoration, qualitative data derived from focus groups, surveys, and one-on-one interviews with a selection of key stakeholders have been used to develop empirically grounded forecast and retrospective assessments of protection and restoration projects (Hemmerling and Barra 2017). Recent restoration work conducted in coastal Louisiana by the Restore the Earth Foundation used qualitative research to inform the calculation of economic, recreational, cultural, educational, and ecological values of ecological restoration projects on numerous stakeholder groups (Hemmerling et al. 2017a, b). Interviews, survey methods, and focus groups were centered around these discrete topics to develop a consistent analysis across groups and a framework for future research and monitoring. Conversations with participants were analyzed to determine which qualities or concerns were important to participants as well as how they weighted different social and environmental values derived from the restoration projects. The qualitative data derived through this process can provide new insight into the social impacts of restoration that cannot be gained through traditional scientific approaches and identify potential inequities in the distribution of costs and benefits. This knowledge can help to bound the uncertainty of a purely quantitative analysis and therefore makes it more useful in setting public policy and making cost-benefit decisions between different environmental interventions. The suite of methodologies used in this research can be translated into a longer-term monitoring program, tracking where and how different economically and geographically situated communities are unequally impacted by the changing material conditions that accompany restoration projects over time. Empirically derived information on residents' perceptions of the values – positive, negative, or otherwise – of restoration projects grounds anticipated social impacts in the material experiences of the residents themselves.

### *5.6.3 Competency Groups*

Despite recent efforts by public officials and scientists to actively engage with coastal residents and stakeholders, many of these residents still feel that their local knowledge is not ultimately accounted for in the coastal restoration planning process within their own communities (Carruthers et al. 2017). This is due in large part to the fact that science-based knowledge, including such expert devices as predictive models, risk indicators, monitoring instrumentation, environmental services calculations, and cost-benefit analyses, is still a priori granted priority over experience-based knowledge (Landström et al. 2011; Whatmore 2009). When this prevailing scientific expertise contradicts the direct experience and knowledge of coastal residents, knowledge controversies may develop, generating conflict and eroding public trust in both scientists and public officials. In coastal Louisiana, one such knowledge controversy has developed around the planned reintroduction of Mississippi River water and sediment into the Breton Sound Estuary in an attempt to mimic the natural functioning of the river delta (Barra 2016). Public opposition to this and other large-scale sediment diversion projects has developed around a number of perceived threats, including the over-freshening of coastal estuaries, displacement of fisheries, and assertion that nutrients in the river water will lead to wetland deterioration (Day et al. 2018). This location recently served as a pilot to investigate the utility of an innovative competency group approach to predictive modeling that utilizes a collaborative process to redistribute expertise between local residents and resource users, hydrological modelers, experts in numerical modeling, and members of an interdisciplinary project team. The scientists participating in the competency group were experts in numerical modeling who played key roles in the initial modeling of the sediment diversions, while many of the local residents were fishers, shrimpers, and oystermen who utilize the estuary on a daily basis. The competency group met on a regular basis over a 6-month period to define the scope and priorities for the creation of a new nature-based defense model (Fig. 5.4). The effort culminated with the co-development of a Delft3D flexible mesh model that incorporates local knowledge and input from the local community on preferred nature-based defenses and criteria for evaluating the effectiveness of the tool for cultivating

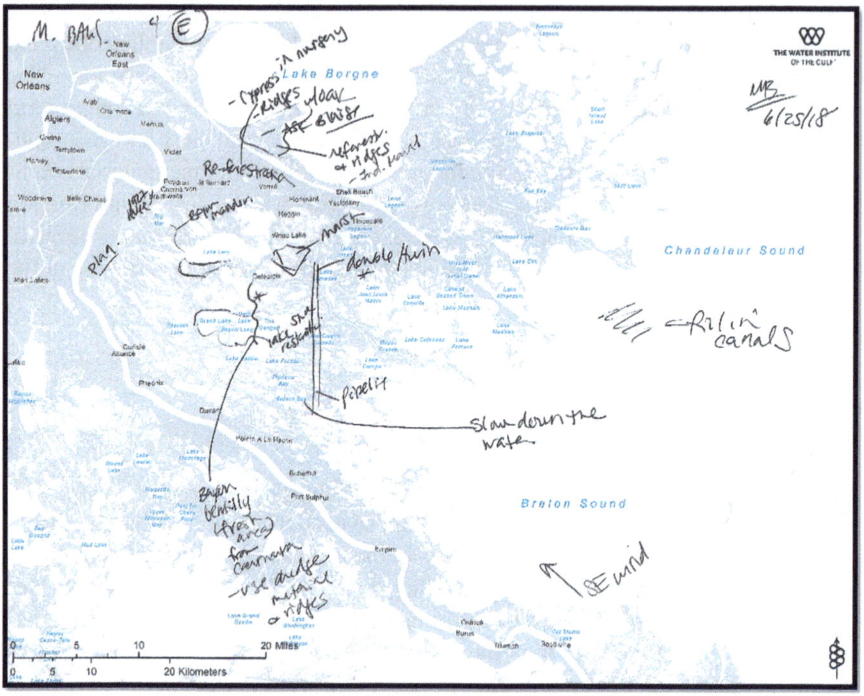


Fig. 5.4 Coastal protection and restoration projects designed and modeled by a competency group consisting of coastal residents and scientists. (Used with permission of The Water Institute of the Gulf)

coastal resilience in different geographic regions (Hemmerling et al. 2019). While the resultant model may bear a superficial resemblance to other scientifically derived models used in efforts like Louisiana's Coastal Master Plan, it is qualitatively different in that the model was co-designed to specifically address current and historical inequities identified by local resource users and residents. By bridging the information gap between local and technical knowledge experts, the competency group process provides a mechanism to bring issues of social justice to the foreground of the planning process.

## 5.7 Conclusion

Ultimately, for coastal protection and restoration to proceed in a socially just manner, the coastal planning process will need to strike an effective balance between science-driven processes and engagement with residents and stakeholder groups who are especially vulnerable to risk as well as those who are likely to be affected by policy actions. A central goal of restoration and protection planning should therefore be to create and sustain a process that is just, transparent, and accountable to those affected by its actions (Olsen et al. 2006). Many coastal residents feel that their local knowledge is not ultimately accounted for in the coastal restoration planning process, even within their own communities, and that new, meaningful, and actionable ways of accounting for and integrating community input into the management, planning, and decision-making process were seen as necessary to increase local support of restoration projects (Carruthers et al. 2017). It is not enough to simply introduce participation into a system that has historically been considered unfair or biased (Larson and Ribot 2007). To be both effective and sustainable, coastal management programs must be supported by the generation and incorporation of reliable knowledge that allows affected stakeholders and the project management teams to better understand and anticipate the consequences of different courses of action. This knowledge should be drawn from both the scientific community and from the observations and local knowledge of community members who reside and work in the systems of which they are a part. The participation of local knowledge experts in the planning process can provide insight into social, ethical, and political values that cannot be gained through scientific approaches alone and allows coastal planners to generate more alternatives, resulting in flexible actions and mutual benefits (Stringer et al. 2006; Zedler 2017). Such a participatory process should create opportunities for coastal planners and project managers, residents, and key stakeholders to assess project outcomes through every step of the process. To begin to ameliorate social justice issues, engagement needs to involve residents as full partners in the process. If their voices are heard but do not impact the process, then the process will fail to even begin to address deep-seated justice issues. By incorporating data derived from two-way dialogue with local knowledge experts into the coastal planning process, coastal managers will be able to more effectively adapt to local needs and changing circumstances, particularly when knowledge is transferred horizontally between stakeholder groups and vertically to higher institutional levels (Zedler 2017). It may be this institutional acceptance of the validity of local knowledge as an important data source, one on par with technical scientific knowledge, determines the ability of local residents to effectively influence the protection and restoration process. Ultimately, if the results of the engagement process are not used by coastal planners, then the engagement effort will be to no avail because it will fail to contribute to a better and more just coastal restoration.

## References

Alpert, B. (2005, July 25). Congress close to allocating money for coast – La. senators work to ensure $540 million budget isn't cut. *The Times-Picayune*.

Anderson, B. (1995a, September 21). Coastal restoration changes opposed. *The Advocate*,

Anderson, B. (1995b, September 22). Plan laid for large-scale coastal wetlands project. *The Advocate*.

Anderson, B. (1997, February 23). Coastal protection efforts insignificant. *The Advocate*.

Barra, M. (2016). Natural infrastructures: Sediment, science, and the future of Southeast Louisiana. (Blog). *Anthropology and Environment Society*. aesengagement.wordpress.com.
Barra, M. (2017). The new cartographers. *Louisiana Cultural Vistas*, Winter.
Bethel, M. B., Brien, L. F., Danielson, E. J., Laska, S. B., Troutman, J. P., Boshart, W. M., et al. (2011). Blending geospatial technology and traditional ecological knowledge to enhance restoration decision-support processes in Coastal Louisiana. *Journal of Coastal Research, 27*(3), 555–571.
Bethel, M. B., Brien, L. F., Esposito, M. M., Miller, C. T., Buras, H. S., Laska, S. B., et al. (2014). Sci-TEK: A GIS-based multidisciplinary method for incorporating traditional ecological knowledge into Louisiana's coastal restoration decision making processes. *Journal of Coastal Research, 30*(5), 1081–1099.
Bethel, M. B., Laska, S. B., Gremillion, M. M., Brien, L. F., Peterson, K. J., Philippe, R., et al. (2015). *Sci-TEK: Mapping traditional ecological knowledge of Barataria Basin with spatial and science-derived datasets*, Final Report for CPRA-UNO Interagency Agreements, Baton Rouge: Louisiana Coastal Protection and Restoration Authority.
Caffey, R. H., Wang, H., & Petrolia, D. R. (2014). Trajectory economics: Assessing the flow of ecosystem services from coastal restoration. *Ecological Economics, 100*, 74–84.
Carruthers, T. J., Hemmerling, S. A., Barra, M., Saxby, T. A., & Moss, L. (2017). *"This is your shield...this is your estuary": Building community resilience to a changing Louisiana coastline through restoration of key ecosystem components*, No. WISR-002-2017, Baton Rouge: The Water Institute of the Gulf.
Checker, M. (2011). Wiped out by the "greenwave": Environmental gentrification and the paradoxical politics of urban sustainability. *City & Society, 23*(2), 210–229.
Clayton, S. (2000). New ways of thinking about environmentalism: Models of justice in the environmental debate. *Journal of Social Issues, 56*(3), 459–474.
*Coastal Wetlands Planning, Protection and Restoration Act 1990*, Public Law 101–646, Title III CWPPRA § 3951.
Colten, C. E. (2014). *Scenario building workshops*. Baton Rouge: The Water Institute of the Gulf.
Colten, C. E. (2015). The place for humans in Louisiana coastal restoration. *Labor e Engenho, 9*(4), 6–18.
Colten, C. E. (2017). Environmental management in Coastal Louisiana: A historical review. *Journal of Coastal Research, 333*, 699–711.
Colten, C. E., & Hemmerling, S. A. (2014). *Social impact assessment methodology for diversions and other Louisiana Coastal master plan restoration and protection projects*. Baton Rouge: The Water Institute of the Gulf.
Colten, C. E., Simms, J. R. Z., Grismore, A. A., & Hemmerling, S. A. (2018). Social justice and mobility in coastal Louisiana, USA. *Regional Environmental Change, 18*(2), 371–383.
Committee on Homeland Security and Governmental Affairs. (2006). *Hurricane Katrina: A nation still unprepared*, No. S. Rep. 109–322, Washington, D.C.: Committee on Homeland Security and Governmental Affairs.
Curtis, J. W., Curtis, A., & Hemmerling, S. A. (2018). Revealing the invisible environments of risk and resiliency in vulnerable communities through geospatial techniques. In A. Barberopoulou (Ed.), *Tsunamis: Detection, risk assessment and crisis management*. Hauppauge: Nova Science Publishers.
Dalbom, C., Hemmerling, S. A., & Lewis, J. A. (2014). *Community resettlement prospects in Southeast Louisiana: A multidisciplinary exploration of legal, cultural, and demographic aspects of moving individuals and communities*, Issue Paper, New Orleans: Tulane Institute on Water Resources Law & Policy.
Day, J. W., Boesch, D. F., Clairain, E. J., Kemp, G. P., Laska, S. B., Mitsch, W. J., et al. (2007). Restoration of the Mississippi Delta: Lessons from hurricanes Katrina and Rita. *Science, 315*(5819), 1679–1684.

Day, J. W., Lane, R. R., D'Elia, C. F., Wiegman, A. R. H., Rutherford, J. S., Shaffer, G. P., et al. (2018). Large infrequently operated river diversions for Mississippi Delta restoration. In J. W. Day & J. A. Erdman (Eds.), *Mississippi delta restoration* (pp. 113–133). Cham: Springer International Publishing.

Dobson, A. (1998). *Justice and the environment: Conceptions of environmental sustainability and theories of distributive justice*. New York: Oxford University Press.

Dunne, M. (1998, September 14). Balance sought in Louisiana's Coastal Use Plan. *The Advocate*.

Fischer, J., Gardner, T. A., Bennett, E. M., Balvanera, P., Biggs, R., Carpenter, S., et al. (2015). Advancing sustainability through mainstreaming a social–ecological systems perspective. *Current Opinion in Environmental Sustainability, 14*, 144–149.

Folke, C., Hahn, T., Olsson, P., & Norberg, J. (2005). Adaptive governance of social ecological systems. *Annual Review of Environmental Resources, 30*, 441–473.

Foster, S. (1998). Justice from the ground up: Distributive inequities, grassroots resistance, and the transformative politics of the environmental justice movement. *California Law Review, 86*, 775–841.

Freudenburg, W. R., Gramling, R. B., Laska, S., & Erikson, K. (2009). *Catastrophe in the making: The engineering of Katrina and the disasters of tomorrow*. Washington, DC: Island Press.

Gagliano, S. (1994). *An environmental-economic blueprint for restoring the Louisiana Coastal Zone: The state plan*. Baton Rouge: Coastal Environments.

Gotham, K. F. (2016a). Antinomies of risk reduction: Climate change and the contradictions of coastal restoration. *Environmental Sociology, 2*(2), 208–219.

Gotham, K. F. (2016b). Coastal restoration as contested terrain: Climate change and the political economy of risk reduction in Louisiana. *Sociological Forum, 31*, 787–806.

Gramling, R., Laska, S., Woodell, G., Forsyth, C., Darlington, J. D., Green, B., et al. (2006). *Anticipating social effects of coastal restoration projects*. Baton Rouge: Louisiana Department of Natural Resources.

Groen, J. A., & Polivka, A. E. (2010). Going home after Hurricane Katrina: Determinants of return migration and changes in affected areas. *Demography, 47*(4), 821–844.

Hardy, R. D., Milligan, R. A., & Heynen, N. (2017). Racial coastal formation: The environmental injustice of colorblind adaptation planning for sea-level rise. *Geoforum, 87*, 62–72.

Hemmerling, S. A. (2017). *A Louisiana coastal atlas: Resources, economies, and demographics*. Baton Rouge: Louisiana State University Press.

Hemmerling, S. A., & Barra, M. (2017). Incorporating local knowledge into ecological restoration assessments – case studies in Louisiana. *SER News, 31*(3), 7.

Hemmerling, S. A., Carruthers, T. J., Hijuelos, A. C., Riley, S., & Bienn, H. C. (2016). *Trends in oil and gas infrastructure, ecosystem function, and socioeconomic wellbeing in coastal Louisiana*, No. WISR-001-2016, Baton Rouge: The Water Institute of the Gulf.

Hemmerling, S. A., Barra, M., & Bienn, H. C. (2017a). *Restore the Earth Foundation reforestation social return on investment report: Pointe-aux-Chenes Wildlife Management Area*. Baton Rouge: The Water Institute of the Gulf.

Hemmerling, S. A., Barra, M., & Bienn, H. C. (2017b). *Restore the Earth Foundation reforestation social return on investment report: Tensas River National Wildlife Refuge*. Baton Rouge: The Water Institute of the Gulf.

Hemmerling, S. A., Barra, M., Bienn, H. C., Baustian, M. M., Jung, H., Meselhe, E., et al. (2019). Elevating Local Knowledge through Participatory Modeling: Active Community Engagement in Restoration Planning in Coastal Louisiana. *Manuscript Submitted for Publication*.

Horst, J. (1997a, June 26). Study to focus on state's coastline. *Times-Picayune*.

Horst, J. (1997b, July 10). Envision the coast of 2050 at forum. *Times-Picayune*.

Ishiyama, N. (2003). Environmental justice and American Indian tribal sovereignty: Case study of a land–use conflict in skull Valley, Utah. *Antipode, 35*(1), 119–139.

Knabb, R., Rhome, J., & Brown, D. (2005). *Tropical cyclone report: Hurricane Katrina, August 23–30, 2005*. Miami: National Hurricane Center.

Kosek, J. (2006). *Understories: The political life of forests in northern New Mexico*. Durham: Duke University Press.

Landström, C., Whatmore, S. J., Lane, S. N., Odoni, N. A., Ward, N., & Bradley, S. (2011). Coproducing flood risk knowledge: Redistributing expertise in critical 'participatory modelling'. *Environment and Planning A: Economy and Space, 43*(7), 1617–1633.

Larson, A. M., & Ribot, J. C. (2007). The poverty of forestry policy: Double standards on an uneven playing field. *Sustainability Science, 2*(2), 189–204.

Lebel, L., Foran, T., Garden, P., Manuta, B. J., & Mai, C. (2009). Adaptation to climate change and social justice: Challenges for flood and disaster management in Thailand. In *Climate Change Adaptation in the Water Sector* (pp. 125–141). London: Earthscan.

Lopez, J. A. (2009). The multiple lines of defense strategy to sustain coastal Louisiana. *Journal of Coastal Research, SI*, 186–197.

Louisiana Act No. 8 2005, Pub. L. No. Senate Bill No. 71, House Bill No. 141. Baton Rouge.

Louisiana Advisory Commission on Coastal and Marine Resources. (1972). *Louisiana Government and the Coastal Zone*. Baton Rouge: Louisiana Advisory Commission on Coastal and Marine Resources.

Louisiana Advisory Commission on Coastal and Marine Resources. (1973a). *Louisiana wetlands prospectus conclusions, recommendations, and proposals of the Louisiana Advisory Commission on Coastal and Marine Resources*. Baton Rouge: Louisiana Advisory Commission on Coastal and Marine Resources.

Louisiana Advisory Commission on Coastal and Marine Resources. (1973b). *Wetlands '73: Toward coastal zone management in Louisiana*. Baton Rouge: Louisiana Advisory Commission on Coastal and Marine Resources.

Louisiana Coastal Protection and Restoration Authority (CPRA). (2007). *Integrated ecosystem restoration and hurricane protection: Louisiana's comprehensive master plan for a sustainable coast*. Baton Rouge: Louisiana Coastal Protection and Restoration Authority.

Louisiana Coastal Protection and Restoration Authority (CPRA). (2008). *Fiscal year 2009 annual plan: Ecosystem restoration and hurricane protection in Coastal Louisiana*. Baton Rouge: Louisiana Coastal Restoration and Protection Authority.

Louisiana Coastal Protection and Restoration Authority (CPRA). (2012). *Louisiana's comprehensive master plan for a sustainable coast*. Baton Rouge: Louisiana Coastal Protection and Restoration Authority.

Louisiana Coastal Protection and Restoration Authority (CPRA). (2016). *Fiscal year 2016 annual plan: Ecosystem restoration and hurricane protection in Coastal Louisiana*. Baton Rouge: Louisiana Coastal Restoration and Protection Authority.

Louisiana Coastal Protection and Restoration Authority (CPRA). (2017). *Louisiana's comprehensive master plan for a sustainable coast*. Baton Rouge: Louisiana Coastal Protection and Restoration Authority.

Louisiana Coastal Wetlands Conservation and Restoration Task Force. (1993). *Louisiana Coastal Wetlands restoration plan: Main report and environmental impact statement*. Baton Rouge: Louisiana Coastal Wetlands Conservation and Restoration Task Force.

Louisiana Coastal Wetlands Conservation and Restoration Task Force. (1996). Funding shifts to large-scale projects. *WaterMarks*, (Spring), 1–2.

Louisiana Coastal Wetlands Conservation and Restoration Task Force. (1999). One vision, one voice: Coastal Louisianans support coast 2050. *WaterMarks*, (Spring), 3–4.

Louisiana Coastal Wetlands Conservation and Restoration Task Force and Louisiana Wetlands Conservation and Restoration Authority. (1998). *Coast 2050: Toward a sustainable Louisiana coast*. Baton Rouge: Louisiana Department of Natural Resources.

Louisiana Department of Natural Resources. (1980). *Louisiana Coastal Resources Program: Final environmental impact statement*. Baton Rouge: Louisiana Department of Natural Resources.

Louisiana Governor's Office of Homeland Security and Emergency Preparedness. (2015). *Katrina/Rita: Building a smarter + safer + stronger + more resilient Louisiana*. Baton Rouge: Louisiana Governor's Office of Homeland Security and Emergency Preparedness.

Ludwig, D., Hilborn, R., & Walters, C. (1993). Uncertainty, resource exploitation, and conservation: Lessons from history. *Science, 260*, 17.

Managing Our Coastal Resources. (1980, October 1). *New Orleans Times-Picayune*..

Marshall, B. (2017, April 14). Unless emissions drop, much of Coastal Louisiana will be swamped. *The Times-Picayune*.

McDermott, M., Mahanty, S., & Schreckenberg, K. (2013). Examining equity: A multidimensional framework for assessing equity in payments for ecosystem services. *Environmental Science & Policy, 33*, 416–427.

McKinney, J. (1989, August 4). Senate kills La. wetlands amendment. *The Advocate*.

McMahon, B. (1989, October 8). Voters approve 5 amendments. *The Advocate*.

Olsen, S. B., Padma, T. V., & Richter, B. D. (2006). *Managing freshwater inflows to estuaries: A methods guide*. Narragansett: Coastal Resources Center.

Peyronnin, N., Green, M., Richards, C. P., Owens, A., Reed, D., Chamberlain, J., et al. (2013). Louisiana's 2012 coastal master plan: Overview of a science-based and publicly informed decision-making process. *Journal of Coastal Research, 67*(sp1), 1–15.

Peyronnin, N., Caffey, R., Cowan, J., Justic, D., Kolker, A., Laska, S., et al. (2017). Optimizing sediment diversion operations: Working Group recommendations for integrating complex ecological and social landscape interactions. *Water, 9*(6), 368.

Pulido, L. (2000). Rethinking environmental racism: White privilege and urban development in Southern California. *Annals of the Association of American Geographers, 90*(1), 12–40.

Pulido, L. (2015). Geographies of race and ethnicity 1: White supremacy vs white privilege in environmental racism research. *Progress in Human Geography, 39*(6), 809–817.

Radtke Russel, P. (2007, April 17). State will receive millions in oil and gas revenues –hurricanes reduce amount allocated for coastal restoration. *The Times-Picayune*.

Reed, D. J. (2009). Planning for the future of the Pontchartrain Coast. *Journal of Coastal Research, 10054*, 198–205.

Schleifstein, M. (1996a, March 26). Early warning went unheeded. *The Times Picayune*,

Schleifstein, M. (1996b, March 26). Sinking treasure. *The Times Picayune*.

Schleifstein, M. (2006, November 9). State maps plan for coastal projects – billions of dollars of levees, dikes urged. *The Times-Picayune*.

Schleifstein, M. (2007, October 14). Harnessing the river – environmental groups offer their own vision to protect Louisiana. *The Times-Picayune*.

Schleifstein, M. (2017, April 19). Louisiana's 2017 Coastal Master Plan unanimously approved by state authority. *The Times-Picayune*.

Schlosberg, D. (2004). Reconceiving environmental justice: Global movements and political theories. *Environmental Politics, 13*(3), 517–540.

Schon, S. (1998, September 4). Coast 2050 Plan nears completion. *Daily Star*.

Sentell, W. (2006, February 17). 'Levee plan on course to pass – House sends plan to friendly Senate. *The Advocate*.

Shields, F. D. (2006). Fate of lower Mississippi River habitats associated with river training dikes. *Aquatic Conservation: Marine and Freshwater Ecosystems, 5*(2), 97–108.

Spence, M. D. (1999). *Dispossessing the wilderness: Indian removal and the making of the national parks*. New York: Oxford University Press.

Speyrer, N., & Gaharan, A. (2017). *2017 coastal master plan appendix G: Outreach and engagement*. Baton Rouge: Coastal Protection and Restoration Authority.

Stringer, L. C., Dougill, A. J., Fraser, E., Hubacek, K., Prell, C., & Reed, M. S. (2006). Unpacking "participation" in the adaptive management of social–ecological systems: A critical review. *Ecology and Society, 11*(2), 39.

Walsh, B. (2006, December 10). State takes long road to share oil revenue – Louisiana rejected Truman's 1949 offer. *The Times-Picayune*.

Weinstein, M. P., Baird, R. C., Conover, D. O., Gross, M., Keulartz, J., Loomis, D. K., et al. (2007). Managing coastal resources in the 21st century. *Frontiers in Ecology and the Environment, 5*(1), 43–48.

Wells, J. T., Battaglia, L., Berke, P., Boyd, J., Deegan, L. A., Espey Jr, W., et al. (2015). *Expert panel on diversion planning and implementation: Report 5,* No. 5, Baton Rouge: The Water Institute of the Gulf. Funded by the Coastal Protection and Restoration Authority.

Whatmore, S. J. (2009). Mapping knowledge controversies: Science, democracy and the redistribution of expertise. *Progress in Human Geography, 33*(5), 587–598.

Wiegman, A. R. H., Rutherford, J. S., & Day, J. W. (2018). The costs and sustainability of ongoing efforts to restore and protect Louisiana's coast. In J. W. Day & J. A. Erdman (Eds.), *Mississippi delta restoration* (pp. 93–111). Cham: Springer International Publishing.

Zedler, J. B. (2017). What's new in adaptive management and restoration of coasts and estuaries? *Estuaries and Coasts, 40*(1), 1–21.

# Part III
# Relocation and Resettlement: An Extreme Adjustment

# Chapter 6
# Community Resettlement in Louisiana: Learning from Histories of Horror and Hope

**Nathan Jessee**

## 6.1 Introduction

It was a warm evening in early May of 2016. I made my way up two flights of stairs to Victor's wooden wraparound porch. His family's house is a comfortable prefab that looms 12 feet above the banks of Bayou Pointe-au-Chien on the border of Lafourche and Terrebonne parishes in Southeast Louisiana. It was built in 2008 after Hurricane Gustav blew the roof off their previous one—which itself had sustained extensive flood damage and needed to be raised after Hurricane Lili in 2002. I met Victor a little over a year earlier when he participated in a series of film screenings and panel discussions I co-organized in the Northeast United States. The events were coordinated to raise awareness about the recurrent disasters affecting his tribe, the Isle de Jean Charles Biloxi-Chitimacha-Choctaw Indians of Louisiana (IDJC), and their plans for the future. Victor and the rest of the IDJC Tribe trace their heritage to Choctaw, Biloxi, and Chitimacha ancestors who, by the early 1840s, had escaped Indian Removal-era violence and resettled on a ridge of land 90 miles southwest of New Orleans called the Isle de Jean Charles, referred to locally as "the Island."

---

"This is not the first time we have had to resettle. Our ancestors were displaced by treaties and Indian Removal. My papa's generation was displaced from the Island. We're already a displaced Tribe. That's why we've got to get it right."
Tribal member, Isle de Jean Charles Biloxi-Chitimacha-Choctaw Tribe
"The question of migration and climate change is not a contingent problem to be solved (or that can be solved) by some technocratic protocol—but rather a metaphor carving out space to pose, contest and struggle for the highly political questions about the climate, mobility, economy, and the society we want."
Giovanni Bettini (2017, p. 90)

---

N. Jessee (✉)
Department of Anthropology, Temple University, Philadelphia, PA, USA
e-mail: nathan.jessee@temple.edu

 
S. Laska (ed.), *Louisiana's Response to Extreme Weather*,
Extreme Weather and Society, https://doi.org/10.1007/978-3-030-27205-0_6

For generations, the people of Isle de Jean Charles fished, hunted, trapped, tended to livestock, and grew their food. Today, adults longingly remember shrimping and crabbing with their families in the bayou that ran the length of the Island—a sanctuary in an area once considered "uninhabitable swampland" by government officials. The Isle de Jean Charles Biloxi-Chitimacha-Choctaw Indians are one of the many indigenous nations who inhabit the land currently occupied by the state of Louisiana. They are one of 15 tribes recognized by the state government, and though they have yet to be formally recognized by the US Bureau of Indian Affairs, the Isle de Jean Charles Biloxi-Chitimacha-Choctaw Tribe has participated in programs administered by multiple federal agencies including the Environmental Protection Agency, Department of Agriculture's Natural Resources Conservation Service, and Department of Housing and Urban Development. As someone who was raised and lived as a young adult on the Island, Victor maintains intimate knowledge of how life and landscape have changed there. And after having moved about 6 miles "up the bayou" to nearby Pointe-aux-Chenes in large part due to the recurrent flooding of the one road that connects the Island to the rest of Louisiana, he knows what it means to leave as a result of changing environmental conditions. Additionally, as the son of a former Chief and someone who has been active within their tribal-driven plans to resettle inland, Victor is able to speak to the long history of organizing to bring resources and support to the Island Tribe.

Upon entering his house, I was greeted by a number of familiar faces sitting around the supper table for an informal meeting: Victor and his family, Tribal Chief Albert Naquin, two teenagers who have represented the Tribe at a number of conferences and in media coverage of land loss, and their father. Also sitting at the table were Dr. Shirley Laska and Dr. Kristina Peterson, co-founders of the Lowlander Center—a nonprofit advocacy and education center who in 2010 established a partnership with the Isle de Jean Charles Biloxi-Chitimacha-Choctaw Tribal Council to support their plans to reunite their displaced people and rejuvenate their traditional ways of life together on higher ground. The Tribal leadership and Lowlander Center were finalizing plans for an upcoming trip to Washington D.C. where Chief Naquin was invited to share his experiences of coastal land loss and resettlement planning as part of a congressional forum entitled "Confronting a Rising Tide: The Climate Refugee Crisis." At first the conversation was mostly light-hearted and speckled with jokes and laughter, meandering from the teenagers' college plans to food and family updates. As the discussion began to focus on the upcoming trip, the tribal members in attendance began to discuss their uncertainty as to the future of the resettlement, despite securing recent financial support through the state of Louisiana's application to the National Disaster Resilience Competition (NDRC).

"It's hard to know what to say to people, because we don't hear anything," Chief Albert explained. "We haven't really heard what the state wants to do since they got the money over three months ago. We don't know what's going to happen. Maybe they want to make our tribal resettlement into just another subdivision? We don't know." Responding to the concern, Dr. Laska asked, "What do you think is most important for you to convey on the climate migration panel?" Chief Albert reflected, "You know, treaties are made to be broken," he said with a pause. "Well, they aren't supposed to be broken, but they always are. This is going to be the same old forced relocation again. The white people pushed the Indians out. We came down here, and some would say we were cowards. Others would say we were smart. Then they had white people drilling up the oil and so now they're pushing us back. We hope this is not another treaty made, another treaty broken." Everyone listened intently, and after a moment, Victor responded. "Let's tell them." He went on to suggest that on the upcoming panel, they should be direct in explaining how their uncertainty as to how the state planners will approach their resettlement builds upon a long history of being excluded from the decisions that affect their peoples' lives. "They need to hear what's been going on," he said.

This moment was not the first or the last time that I heard someone from the Island reference historical violence and the colonial history of the United States while explaining land loss or their efforts to adapt. A number of Native American tribes and communities of color who have been pushed to what is now the edges of Louisiana's coastal zone—in large part due to histories of violence and forced displacement—are grappling with land loss, extreme weather, and the various institutional responses to those hazards as part of their long-standing efforts toward collective survival and justice. Contemporary experiences of environmental catastrophe and approaches to governing them are not experienced within a bubble. Rather, they are encumbered by memories and legacies of historical injustices. Meanwhile, state-level policy-oriented conversations about resilience planning, and community resettlement in particular, have emerged during a moment of stark realization regarding the state's capacity for coastal ecological restoration, increasing extreme weather, and future flood risks. Louisiana's 2017 Coastal Master Plan identifies a number of locations where coastal erosion, subsidence, and recurrent flooding due to extreme weather threatens traditional ways of life and future human habitability (Clipp et al. 2017). In response, scientists and planners have advanced ambitious plans to restore what is possible of the wetlands while beginning to reimagine development along the coast. Louisiana's 2017 Coastal Master Plan, however, makes clear that conservation and restoration will not always be possible, and while restoration can be expected in some areas, others will sink or wash away (see also Jankowski et al. 2017). Human and nonhuman residents of the region must continue to adapt to a shrinking coast and increasingly extreme weather. For many, this means navigating exceedingly complex policy worlds and relocating out of harm's way.

This chapter explores one of the most pressing challenges that risk reduction professionals, scholars, policy-makers, and Louisiana residents face during the implementation of community resettlement planning activities in Louisiana: the need to reckon with, on the one hand, the increasing risks to flooding and extreme weather and, on the other hand, the experiences and initiatives advanced by those whose vulnerability to these risks is connected to histories of forced displacement, dispossession of land and resources, and social marginalization. In the following pages, I describe some tensions that have emerged as long-standing tribal-driven resettlement efforts have been incorporated into state-level "coastal retreat" planning utilizing existing federal funding sources. I first describe some background to local policy conversations focused on resettlement as a strategy for adapting to environmental change in coastal Louisiana. Through the Louisiana Strategic Adaptations for Future Environments (LA SAFE) and state support of the Isle de Jean Charles resettlement, Louisiana's government has begun rethinking floodplain and coastal development in anticipation of increasing future flood risk and exposure to extreme weather. In their work, however, there remains a risk of disconnecting current and future exposure to coastal hazards with the development practices and legacies that have produced vulnerability unevenly among particular groups of people, such as indigenous peoples and coastal communities of color. I refer to such disconnections as *ahistorical adaptation*. Then, drawing on my work following the resettlement efforts of the Isle de Jean Charles Biloxi-Chitimacha-Choctaw Tribal leaders, I suggest a number of histories salient to adaptation and resettlement policy in Louisiana. I also consider the impacts of avoiding local histories of injustice and conclude with some ideas for honoring experiences and initiatives of local communities and tribes.

It is important to disclaim that my goal is not to dismiss buyout programs, community resettlement, or efforts to critically rethink floodplain and coastal development as a whole. These are important pathways for reducing exposure to hazardous environmental conditions. The multiple severe hurricanes to make landfall on the Atlantic and Gulf of Mexico coasts over the last 3 years make these tools more important and urgent than ever before, and investing predominantly on structural flood protection and environmental restoration without adequate planning for the communities in coastal Louisiana at this point would be foolish. Rather, I hope this chapter encourages reflection on a particular problem, the ahistorical framing of resilience, adaptation programs, and disaster recovery policy. This chapter also provides additional support for the growing demand for climate change and environmental adaptation policies that protect the rights of indigenous peoples and provide adequate resources to local and tribal-driven plans for resettlement.

## 6.2 Context of Recent State-Level Resettlement Planning

Since 1932, Louisiana has lost over 1866 square miles of coastal wetlands—an area nearly the size of Delaware (Couvillion et al. 2017). Floodplain development and industrialization have exacerbated subsidence and erosion throughout the state's coastal zone. An expansive web of oil and gas pipeline and shipping canals crisscross through the coastal wetlands allowing saltwater to seep into freshwater marshes and creating ecologically disruptive sediment deposits, killing flora and breaking up the soil (Turner and McClenachan 2018). Meanwhile, flood protection levees and river control structures have prevented the Mississippi and Atchafalaya Rivers from replenishing the sinking delta with new sediments (Barry 1997). The loss of land and marsh has increased Louisiana communities' exposure to hurricanes, which then erode more of the marsh as well. On top of these factors, global climate change poses an increasing risk of more extreme weather and sea level rise in the region (Carter et al. 2018).

As the effects of climate change become more visible, there is an emerging debate about the number of people who risk displacement as a result of climate-related hazards (Bronen et al. 2018; Wilson and Fischetti 2010). Displacement from coastal areas has long been expected to increase due to the impacts of climate change. However, projections have recently become more dire. Rigaud et al. (2018) estimate as many as 143 million people in Sub-Saharan Africa, South Asia, and Latin America could be displaced by 2050 due to environmental changes associated with climate change. Meanwhile on the US coasts, Crowell et al. (2010) found that 8.6 million people, about 3 percent of the US population, inhabit the 100-year flood zone, and Hauer et al. (2016) estimated that as many as 13.1 million people living in coastal counties could be at risk of displacement due to sea level rise by the next century. Recent news coverage of extreme weather also speculates the massive displacements that may be anticipated due to rising seas and extreme weather (e.g., Cusik and Aton 2017; Gohd 2018). In the aftermath of every flood, fire, or hurricane, journalists and editors dub the people facing these challenges as the next potential *climate refugees*—a term without legal meaning that can render the coordinated responses of communities and their collective agency invisible by implying individuals in need of saving.

A number of migration scenarios, however, already occur after, during, and in anticipation of environmental disasters—some of which empower, and are even led by, local efforts, while others marginalize and disempower those resettling. Scenarios exist for a variety of social units from individuals, to families, to neighborhoods to communities of different sizes and with varying degrees of planning. After Hurricane Katrina, for example, many relocations that were unplanned before the floods turned into more permanent resettlements as families relied on distant relatives and social ties around the country for support (Weber and Peek 2012). Residents of St. Bernard Parish, for example, moved as households to nearby areas on the north shore of Lake Pontchartrain where family members, friends, and former neighbors already relocated (Lasley 2012). Liz Koslov's work after Hurricane Sandy documented another migration pattern during which activists organized buyout groups in Staten Island, New York, to advocate for their dispersal (Koslov 2016). The Allenville, Arizona, resettlement in the early 1980s (Perry and Lindell 1997) and the relocation of Pune, India (Cronin and Guthrie 2011), were advanced by extensive community-oriented activism and with the aim of keeping people together. There are also more than a dozen other resettlements currently being planned by indigenous communities and tribal nations of North America (Keene 2017). International frameworks for planned resettlement and the sensational media representations of so-called climate refugees rarely unpack the tensions between community-based or tribal-driven resettlements and government-led planning processes. Moreover, the divergent migration scenarios mentioned above exist simultaneously and within overlapping geographies in coastal Louisiana—a dynamic that deserves further scholarly attention.[1]

There is currently no singular national agency or policy framework that guides community resettlement despite growing calls for one (Bronen 2011; Maldonado et al. 2013). The Federal Emergency Management Agency (FEMA) offers the primary form of resettlement support available pre- and post-disaster, though this support is only available for the relocation of individual households, not the entire communities (Bronen 2011; Marino 2012). The Department of Housing and Urban Development's Community Block Grant-Disaster Recovery can similarly be used to fund buyout programs. The US Army Corps of Engineers has coordinated a number of group resettlements in response to riverine flooding and the impacts of public works. These efforts have resulted in varying outcomes for the people resettling (Perry and Lindell 1997; LDOA 2015a).

Since Hurricane Sandy in 2012, design competitions sponsored by philanthropic-public partnerships have been highly publicized for potentially offering new opportunities for communities working toward resettlement as a form of environmental adaptation. The Rockefeller Foundation and the US Department of Housing and Urban Development (HUD) partnered to sponsor the Rebuild by Design competition. This effort has been celebrated for spurring innovation and prioritizing public input throughout the design process (Collier et al. 2016). Based on perceived successes of the Rebuild by Design process, the Rockefeller Foundation and HUD administered the National Disaster Resilience Competition, which offered cities and states the opportunity to compete for part of approximately $1 billion Community Development Block Grant Disaster Recovery funds left over from the post-Hurricane Sandy expenditure. Despite the innovation engendered by these competitions, however, HUD and local jurisdictions who implement funded projects or programs have guidelines, planning conventions, and regulatory frameworks that constrain such innovative designs. As this chapter describes below, and Laska explains in the Introduction to this volume, Chap. 1, moving from the old regulations to ones that embrace adaptation is a clunky and uneven process that involves a number of legal contradictions and potential harm for communities or tribes who have invested so much throughout their lives and potentially during the program design processes. Additionally, more research is needed to investigate the various dimensions of post-design implementation processes (for some initial critiques of Rebuild by Design, see Dawson 2017 and Flemming 2019).

Louisiana's state government is only beginning to meaningfully devote attention and resources to resettlement as an approach to environmental adaptation. Though the 2017 Coastal Master Plan incorporates nonstructural risk reduction strategies, robust investment and planning for social and cultural preservation and adaptation on our shrinking coast is much needed. The plan identifies 11 locations where, within 50 years, "flooding will be high enough to make daily life next to impossible, even without future hurricane damage" under a medium scenario of sea level rise

[1] I am indebted to Dr. Shirley Laska for pointing out the importance of and lack of scholarship addressing this dynamic.

without future restoration and mitigation efforts. The identified locations from east to west are Delacroix, St. Bernard Parish; Venice, Plaquemines Parish; Grand Isle, Lafitte/Crown Point/Barataria, Jefferson Parish; Paradis, St. Charles Parish; Kraemer and Leeville, Lafourche Parish; and Cocodrie, Dulac, Isle de Jean Charles, and Lower Pointe-aux-Chenes, Terrebonne Parish (Clipp et al. 2017). Additionally, 2400 structures are targeted for potential voluntary acquisition due to their being in locations where the flood depths are forecasted to exceed 14 feet. The Coastal Protection and Restoration Authority (CPRA)'s work has received criticism, however, for not articulating a detailed buyout planning program or notifying those who live in the areas that would be targeted if one existed, and at the time of this writing, the state has yet to secure or invest the estimated $1.2 billion necessary for the acquisitions (Wendland 2018). Policy-makers here, like elsewhere around the country, have been reluctant to embrace coastal relocation for fear of upsetting their coastal constituents as well as the administrative burdens that a relocation program would entail (Manning-Broome et al. 2015). Additionally, many communities and tribes throughout coastal Louisiana themselves do not plan on relocating and instead envision continuing to adapt in-place. Though the writing in the plan indicates people will need to move, there is no robust plan for what that process actually looks like, especially for those who plan on resettling while enhancing communal social structures or building upon traditional ways of life together in a new location.

In 2014 Louisiana's Office of Community Development began applying for funds available through the National Disaster Resilience Competition (NDRC), an initiative that has been viewed by some as an essential step in developing a more robust land-use policy to supplement the shortcomings of the Coastal Master Plan (LDOA 2015a). Among the primary goals of the competition was "to create multiple examples of local disaster recovery planning that applies science-based and forward-looking risk analysis to address recovery, resilience, and revitalization needs" (HUD 2015). The state's NDRC funding is allocated for two programs: the Isle de Jean Charles resettlement—discussed at length below—and the Louisiana Strategic Adaptations for Future Environments (LA SAFE) program which advances resilience planning in six parishes affected by recent tropical storms and hurricanes. Louisiana's LA SAFE policy framework draft refers to areas that will experience +14 feet of flooding during a 100-year flood scenario within the next 50 years as "resettlement zones," while those that will experience between 3 and 14 feet of flooding are seen as "retrofit zones," and locations that can expect less than 3 feet of water "reshaping zones," in which development and growth will be encouraged (LDOA 2017a).

With regard to resettlement, the state "envisions a systems-based approach to community-led planning and group migration" and advances nine principles for community resettlement (LDOA 2017a, p. 14–15). The principles foreground the need for community resettlements to (1) be community-driven and voluntary; (2) be responsive to future risks and opportunities; (3) build social networks; (4) "where prudent, appropriate, and desirable, approaches should envision scenarios by which resettled communities retain access to abandoned lands for cultural, social, or economic reasons"; (5) reduce current and future risk; (6) migrations should stay within a single jurisdiction; (7) "embody worldwide best practices in water management, energy conservation, wetlands restoration and habitat preservation"; (8) approach development holistically to include "cultural, social, and economic growth opportunities and techniques"; and (9) consist of a "total residential abandonment of original community" (LDOA 2017a, p.15).

These principles represent a strong foundation, though there is much room for improvement in the specificity of the language used and there remain major questions as to their implementation. First, we cannot rely on bureaucratic good will and discretion. Rather, there must be legally binding policies. Second, what does risk actually entail? According to the Louisiana Budget Project, Louisiana is currently tied for second highest poverty levels in the nation, with nearly 20% of the population living in poverty, and fourth in the nation in racialized income inequality (Louisiana Budget Project 2018). Racialized and classed health disparities also persist throughout the state (Macklin 2009). The unemployment rate is higher than the national average (U.S. Bureau of Labor Statistics 2018), and many who are employed work in contingent, part-time, or low-paying positions that do not provide a living wage. Louisiana renters have long faced an affordable housing crisis, and the state was recently ranked fifth highest rent-stressed state in the country (Louisiana Housing Corporation 2019). (See Chap. 9 of this volume by Andreanecia Morris and Lucas Diaz for expanded discussion of housing risks.)

The goal of "reducing risk" cannot be isolated from broader political, economic, and social conditions. These conditions not only stand as risks themselves but exacerbate capacity to respond to extreme weather events. Regarding the fourth principle—retaining access to original land—who defines "prudent, appropriate, and desirable" and what does "access" actually mean? What other types of land tenure must be honored, including collective land ownership or continued ownership of original lands for those whose ways of life and identities are so deeply rooted in place? Finally, within these principles, there is no commitment to the protection of human rights and no reparative, restorative, or regenerative measures for the historical social production of risks and vulnerabilities that have led some people to need to resettle in the first place. The second principle even seems to discourage consideration of historical atrocities and their legacies by explicitly orienting state community resettlement activities around *future* opportunities and risks.

## 6.3 Ahistorical Adaptation

The above context for policy-oriented work on community resettlement demonstrates how such efforts are situated within a future-facing discourse of disaster risk reduction that contributes to the technocratic politics of climate change adaptation and resilience (see Swyngedouw 2011). Due to the severity of existing and expected environmental hazards, investing in mitigation and adaptation is necessary. There is also an urgent need to alter coastal and floodplain development practices and address a lack of adequate land-use planning and policy at local, state, and federal levels. When taken together, however, future-oriented discourses may reduce the complex community-oriented and tribe-driven adaptations, which are responsive to social and environmental experiences, to a technocratic process of encouraging people to move solely in relation to future flood risk as outlined in the Coastal Master Plan. When social complexities are acknowledged by state adaptation and resilience planners, they are often devoid of historical context. For example, Louisiana's policy framework draft discussed above describes their approach to resettlement as "a small-scale, targeted strategy for culturally-sensitive at-risk communities and special needs groups, including the disabled, the elderly, disaffected minority groups and very low-income populations" (LDOA 2017a). The framework does not, however, point to any processes or conditions that have led some groups to become so-called culturally sensitive or at-risk. Though the document briefly cites the atrocities of previous federal and state relocations as part of the bad "track record" of forced relocation (ibid. pp.14), it offers no examination as to why so-called minority groups and very low-income populations may have become "disaffected" in the first place.

Recent scholarship has pointed to some of the ways in which practitioners and policy-makers avoid fraught social and political histories that produce environmental risks. Katrina Kuh (2016) examines how government-sponsored adaptation strategies and outreach must sometimes separate pro-adaptation behaviors from the causes of climate change in order to engage climate change deniers, a process she dubbed *agnostic adaptation*. Drawing on information provided to farmers by the United States Department of Agriculture's Farm Service Agency Climate Change Adaptation Strategy, Kuh describes the ways in which adaptation tools are promoted by deemphasizing the human factors that contribute to climate change by only briefly referring to anthropogenic causes of global warming while devoting more elaborate descriptions to geophysical explanations of warming. Kuh advances important research questions as to the efficacy of agnostic adaptation and how it might create barriers to climate justice. Meanwhile Hardy et al. (2017) coined the term *colorblind adaptation* as a way of understanding, "vulnerability, mitigation, and adaptation planning projects that altogether overlook racial inequality—or worse dismiss its systemic causes and explain away racial inequality by attributing racial disparities to non-racial causes." Their work demonstrates some of the ways that legacies of forced migration, chattel slavery, exclusionary employment and housing policies, and local politics affect risk and adaptation possibilities for African American residents in coastal Georgia.

With the spectacular media narratives of impending massive climate displacements and the future orientation of resettlement planning as a strategy for risk reduction or climate resilience, it is worth restating what social scientists have been yelling from the rooftops for four decades: Vulnerability to natural hazards, disasters, and displacement cannot be disentangled from harmful historical social and political-economic processes (O'Keefe et al. 1976; Oliver-Smith and Hoffman 1999; Wisner et al. 2004). Anthropologists and other social scientists have demonstrated many of the ways that coastal governance, disaster policies, and environmental adaptation are fraught with legacies of injustice and human rights violations (Burkett et al. 2017; Hardy et al. 2017; Jerolleman 2019; Maldonado et al. 2013; Marino 2015; Whyte 2016). In an effort to address these legacies, scholars, practitioners, advocates, and resettling communities have urged government officials to guide any framework for planned resettlement by an approach that affirms and protects the human rights of those resettling (Bronen 2011; Ferris 2012; Maldonado et al. 2013).

The above scholarship demonstrates the unevenness of vulnerability, whereby people with certain social characteristics or belonging to marginalized groups are made more sensitive to the effects of hazards due to the broader inequitable political economy. Hardy et al. (2018) argue that understanding local histories is critical for recognizing (1) the ways that exposure to hazards has unfolded, (2) the uneven sensitivities of different groups within a region, (3) the cultivation of adaptive capacity, and (4) rationales for particular kinds of local adaptation. Historical perspective can also reveal barriers or constraints to future possibilities (ibid.). According to the authors, by ignoring local histories, "we risk missing many factors that contribute to vulnerability and suboptimal adaptation measures" (Hardy et al. 2018, p. 10). Ignoring these histories also implicitly blames those who must deal with the immediate impacts of unsustainable aspects of our civilization for the risks they face (Tierney 2014). Inspired by the important critiques above and the questions raised by Kuh as to the efficacy of agnostic adaptation, in this chapter I hope to contribute to a conversation about how the institutionalization of environmental adaptation as an ahistorical discourse actively produces new kinds of vulnerability while aggravating existing experiences of injustice.

## 6.4 Historicizing Isle de Jean Charles Resettlement Planning

In the following sections, I reflect on my work following the Isle de Jean Charles Biloxi-Chitimacha-Choctaw Tribal leaders' resettlement efforts to consider how particular historical processes are relevant to current community resettlement planning in Louisiana. Not all of the potentially important local, regional, national, and international historical threads that could help one understand adaptation and resettlement are reflected in this chapter. However, I have found that the following histories are especially salient within the particular context in which I did my research: (1) legacies of environmental injustice, (2) recent experiences of displacement, (3) US settler colonialism, and (4) the Tribal-driven planning process that immediately preceded and directly contributed to Louisiana's successful application to the National Disaster Resilience Competition.

### 6.4.1 *Extraction, Exclusion, and Injustice*

There is no way to really understand the Isle de Jean Charles resettlement effort without reckoning with the tragic social and environmental impacts of extractive industries and regional development on and around the Island. Over the last 75 years, regional oil and natural gas infrastructure and development has transformed the place (Austin 2006). Companies like Apache, BP, Texaco, Chevron, and Shell have devastated the surrounding landscape with pipeline canals and oil and gas infrastructure (Maldonado 2018). (See Fig. 6.1 for a map of oil and gas infrastructure density throughout the region.) The Island itself sits between multiple oil and gas fields and is surrounded by pipeline canals, and those from the Island point to a long history of land grabs by land and oil companies preceding the current environmental crisis (Maldonado 2018). Since 1955, 98% of the Island's landmass has sunk or washed away (CPRA 2017). Over 22,000 acres of land and surrounding marsh have been lost (LDOA 2015a). According to the summary of a satellite imaging study conducted by the Earth Resources Observation and Science Center and United States Geological Survey, "Once 5 miles wide and filled with lush cypress groves and cow pastures, barely a half square mile of the Island remains above water" (USGS 2017). A former state senator explained that the Island road (built in 1953) sits on "some of the most unstable land in all of coastal Louisiana" (Reggie Dupre, personal communication, November 2017). According to Tribal leaders, the Island flooded from six major storms and multiple less severe storms since 2005. Maldonado et al. (2015) and other critics have referred to the region as a "sacrificial zone of resource extraction" due to the exploitation and subsequent disaster produced by industrial development.

Flood mitigation projects have also had devastating social and environmental consequences, and the Tribe's resettlement efforts are actually rooted in impacts of

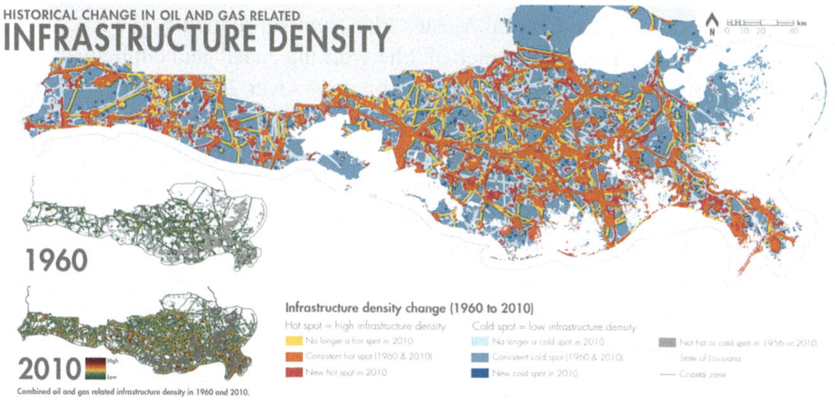


**Fig. 6.1** Maps showing historical change in oil- and gas-related infrastructure density in southern Louisiana between 1960 and 2010 as presented in a 2016 synthesis report by *The Water Institute of the Gulf*

one effort to address coastal hazards regionally. In 1998, the US Army Corps of Engineers rerouted designs for a 98-mile levee, the Morganza to the Gulf Risk Reduction System. The new route excluded the Island and a number of other communities down the bayou further materializing and intensifying their vulnerability (Maldonado 2018). The realignment increased the risk of flooding for those who remained outside of its protection (USACE 2013) and, along with the severe 2002 storms, led the Tribal Council to begin planning their resettlement to reunite their Tribe and ensuring the survival and growth of their culture together (Maldonado 2018). The US Army Corps offered to relocate the Island residents but backed out after demanding a 100% consensus in favor of relocation among Island residents (Simon 2008)—a naive ideal given the complexities involved in any relocation decision-making process. Without social scientists in Corps staffs (other than economists and archaeologists), the Corps could not have appreciated the unrealistic nature of the demand.[2] Then in 2008, Tribal leaders appealed to Terrebonne Parish for CDBG funds to support a resettlement that would reunite their already displaced tribal citizens and support their cultural and social revitalization, but the effort fell apart when confronted with resistance from the white residents adjacent to where they were hoping to move (Maldonado 2018). Tribal leaders continued to organize both for resettlement and sustaining the Island despite these disappointing setbacks from potential government partners.

The above histories of injustice are often obscured within state restoration and resilience planning. Colten (2017) compared how the causes of wetland loss are discussed in a number of reports released since 1990. He found that over time these reports have reframed oil and gas companies from culpable damage to the wetlands to those also in need of support and protection from environmental change. According to Colten, more recent reports emphasize the role of subsidence while refusing to acknowledge the impact of oil and gas pipeline and navigation canals in destroying the marsh (Colten 2017, p. 706). The Coastal Protection and Restoration Authority has continued to obscure causes in their efforts to build broad public support for restoration. The Coastal Master Plan, and subsequently the LA SAFE framework, emphasizes the importance of "the working coast" and euphemistically provides support for the interests of industrial actors. Over 20% of the nation's oil and gas, and 90% of the offshore oil, comes from or passes through the region for processing and refinement, and many of those made most vulnerable by oil and gas industries rely on them for their livelihoods (Clipp et al. 2017). This bind may discourage critique and makes the movement for a just transition from fossil fuels particularly relevant to coastal Louisiana.

Adaptation strategies that avoid addressing these histories of injustice while highlighting the "working coast" dynamic accommodate oil and gas industries by not holding them accountable for the destruction they have caused (Randolph 2018; Turner and McClenachan 2018). They also potentially stifle support for a much-needed just transition from fossil fuels in a key place of extraction, production, and transport. Moreover, these frames reproduce the notion that federal investment in

[2] Personal communication with Shirley Laska

resilience and resettlement planning may constitute a "handout" by neglecting the price that the Isle de Jean Charles Biloxi-Chitimacha-Choctaw Tribe and other coastal communities who have been made vulnerable by extractive industries and flood protection measures have already paid (Jessee 2016). They also obscure rationales and thus potential support for many of the social dimensions of the resettlement, which, for the Isle de Jean Charles Biloxi-Chitimacha-Choctaw Tribe, have included reuniting those who are already displaced or have moved from the Island, building energy and food independence, and bolstering the Tribe's capacity to assert their rights of self-determination and sovereignty (LDOA 2015a).

### *6.4.2 Ongoing Displacement and the Complex Notion of Community*

Another challenge for resettlement as a form of environmental adaptation involves addressing recent histories of displacement and complexities surrounding the notion of community. Displacement has dramatically changed life on the Island, and decades of outward migration have transformed the composition and social dynamics of the Isle de Jean Charles community. There were 78 houses and approximately 325 people living on Isle de Jean Charles in 2002, and 10 years later only about 25 houses and 70 people remained there (Maldonado 2014). Over 75 percent of the Island community have been displaced over this multigenerational environmental catastrophe of land loss. Many tribal members moved from the Island because regular flooding on the road kept them from off-Island jobs, while others recounted moving due to a decline in regional industries like shrimping and trapping in the 1980s. Though some Isle de Jean Charles Biloxi-Chitimacha-Choctaw Tribal Council and elders still live on the Island, much of the tribal leadership now lives just a few miles away in Pointe-aux-Chenes and Houma. Displaced Tribal members however maintain important social connections with their family and former neighbors who remain on the Island. The Island remains a cohesive aspect of displaced Tribal members' identities and social memory, and the expansive networks of former and current Island residents remain an essential aspect of resilience (Simms 2016).

Many self-identified communities and tribes in Louisiana who may consider resettling are likely already experiencing this kind of displacement (Colten et al. 2018). Numerous locations along the state's coast have seen extensive outward migration over the last 20 years: Dulac has lost 29%, Golden Meadow 3%, Lafitte 31%, and Buras-Triumph 67% of their populations between the years 2000 and 2010 (see Fig. 6.2). Coastal storms most often diminish local populations in "waves" after strong hurricanes rather than one event causing the abandonment of a location in one storm (Laska, personal communication, April 16, 2018). Dr. Shirley Laska likens it to the rate of coastal land loss which occurs in "waves" during powerful storms, not a little each year, but rather surging. Relocation decisions are often

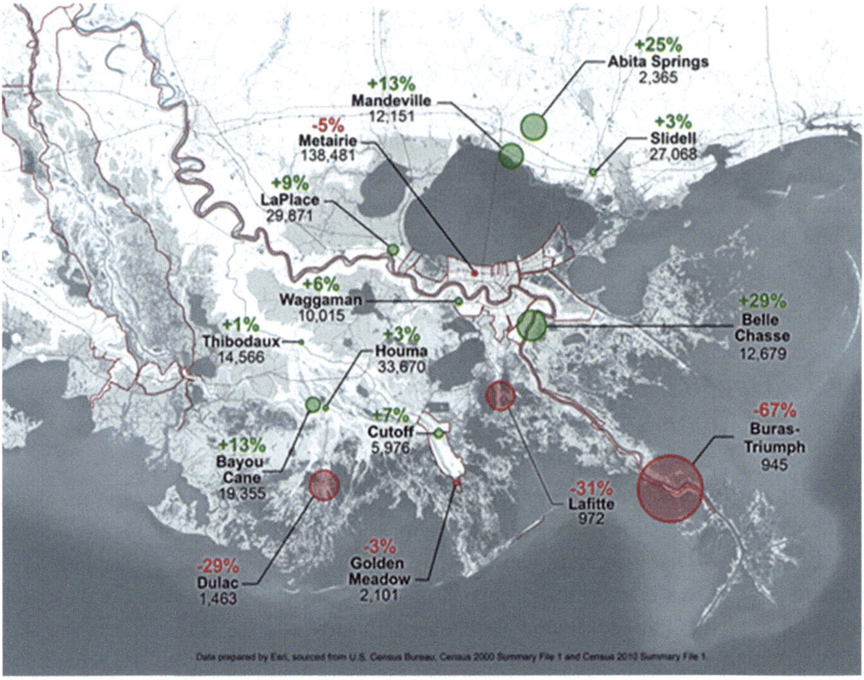


**Fig. 6.2** Map showing regional population shifts in southeast Louisiana between 2000 and 2010. (Image source: LA SAFE)

linked to the accessibility of schools or work, as younger people often leave, while elders have remained in place (Colten et al. 2018). Additionally, long-time residents often cannot afford to elevate their flood-impacted homes, continuing their vulnerability and preventing them from affording the flood insurance. According to Hobor et al. (2014), Louisiana residents who work in coastal industries increasingly travel from residences inland to their coastal job sites. Those who work on the coast, such as fishers, must assess how far inland they might live for safety while commuting in reverse for work (Laska et al. 2015). Meanwhile, since the 1970s, coastal areas like Cocodrie and Isle de Jean Charles have also seen an influx of wealthy white "weekend warriors" who can afford to maintain camps along the coast and enjoy the landscape for recreational activities—a dynamic that stirs concerns of coastal gentrification among the historic residents (Peterson and Maldonado 2016; Solet 2006).[3] Additionally, research into development-forced displacement and resettlement has revealed the risk of further severing social ties mediated by relationships to place, a process described by Michael Cernea (1997)

[3] For readers unfamiliar with "camps" in Louisiana, they are often more like lavish fishing vacation homes with docks and modern amenities than, say, a tent or a yurt. There is an incentive for parishes to encourage camp development for more tax dollars than low-lying homes, which are often valued at much less and have homestead tax exemptions.

as *social disarticulation*. Displaced people who experience disrupted networks, shared rituals, religious institutions, kinship structures, and social cohesion experience consistently worse outcomes during resettlement processes (Oliver-Smith 2009). In coastal Louisiana social disarticulation caused by displacement may already be underway and indeed has been a major rationale for Isle de Jean Charles Tribal leaders to use resettlement to reunite their Tribe.

"Community" is a slippery and contested concept in any context, and many scholars have debated its utility and application (Titz et al. 2018; Williams 2002). Brett Williams (2002) reviews the myriad of ways community can be produced through shared experiences, senses of place, cultural practices, memories, and relationships but describes how bonds of community are constantly negotiated, contested, and adjusted. State agencies and international frameworks, however, often advance narrow notions of risk and vulnerability that emphasize geographic exposure to hazards over social risks (Faas 2016). These frameworks also inscribe notions of community as bounded geographic units (e.g., Clipp et al. 2017; LDOA 2019a). Media coverage of land loss that relies on particular visual tropes, like aerial shots of the Island, may also reinforce such notions of vulnerability and community, obscuring broader political-economic causes of vulnerability and geographically expansive yet intimate social networks of care critical for resilience in the region. After reviewing the use of "community" throughout development and disaster risk reduction, Titz et al. (2018) conclude that the term should be abandoned in such contexts because of the harm that can be caused by its vagueness. They argue that more specific identifiers should be used. The ambiguities of "community" are part of a set of conditions that has potentially enabled state planners to reduce their commitments to and partnership with the Isle de Jean Charles Biloxi-Chitimacha-Choctaw Tribe after the award money for the resettlement was garnered. Divergent notions of "community" as location-specific versus based on tribal identity have troubled the Isle de Jean Charles resettlement process (Jessee Forthcoming). While the Office of Community Development considered geographic exposure to hazards at length in their application for NDRC funds, they also importantly operationalized the Isle de Jean Charles Biloxi-Chitimacha-Choctaw Tribe as the community during application phases, building on the long-standing tribal resettlement efforts that had preceded the competition.

Community resettlement planning efforts need to take careful steps toward the rearticulation of potentially fragmented or strained already displaced self-identified communities. Investing in the enhancement of communal structures and honoring collective identities and their histories early and throughout resettlement processes may provide a pathway to addressing social disarticulation. After Hurricane Katrina, those who relocated *clustered* into family and neighborhood groups in their new locations (Lasley 2012). The same is true among the tribes of lower Terrebonne Parish (Maldonado 2018). Clustering might be instrumental in the long-term well-being of coastal peoples as they move inland to a somewhat foreign culture. The historic migrations of Europeans to US cities over the last two centuries often also led to clustering of the new migrants in urban enclaves, which informally supported social networks, economic needs during transition, and personal preferences (Logan et al. 2002). Over the first 3 years of the state's administration of federal support for the Isle de Jean Charles resettlement, there has been much more attention devoted to outreach to individual households on the Island, residential options, and the overall site design, rather than specific planning for the continuation of tribal and communal institutions and organizations from the Island (Jessee Forthcoming). The tendency to focus on individual households at the expense of complex social realities of community is documented in the administration of other community resettlement processes as well (Wilmsen and Webber 2015). According to Cernea (2000), one way to mitigate social disarticulation, which I think might also support clustering and nurture social networks that may have grown distant throughout the years of displacement, is to establish common property and communal facilities and programing at new locations early on in the process and to engage with community-oriented or tribe-based organizations rather than having the focus be so oriented around individuals throughout the planning processes.

Despite the trickiness of defining community, social scientists have also stressed the importance of existing community-based institutions, organizations, leadership, and expertise. Meaningful partnerships with community leadership have been essential during post-disaster recovery (Nelson et al. 2007; Laska et al. 2010). Indeed, the successes attributed to well-known community resettlements in North Bonneville, Washington, and in Valmeyer, Illinois, can be attributed to the ability of existing leadership to maintain continuity throughout the planning processes (Comstock and Fox 1993). Perry and Lindell (1997) also found that the ongoing efforts of a nonprofit, the Allenville Citizens for Progress, representing the unincorporated people of Allenville, Arizona, were essential in the successes of their relocation from the Gila River floodplain. Existing leadership, like that of the Isle de Jean Charles Biloxi-Chitimacha-Choctaw Tribal Council, is invaluable, and often demonstrate a history of influence and resource provision, maintain shared memories and knowledge about local worldviews and higher incomes, and have access to social networks for further support into the future. Moreover, local community-based or tribal leadership is likely more committed to the future well-being of community members than government planners whose investment is typically limited to a grant, program, or employment timeline.

### 6.4.3 Reshaping Louisiana's Coastal Frontier: From Doctrine of Discovery to Climate Catastrophe

Contemporary environmental adaptation, US settler colonization, and struggles for indigenous self-determination are deeply intertwined. First of all, as Chief Naquin explained in Victor's kitchen, the Isle de Jean Charles Biloxi-Chitimacha-Choctaw Tribe descends from ancestors who survived Indian Removal Act-era violence by resettling down the bayou, away from and invisible to the nearby settler society. In addition to indigenous peoples, Vietnamese families, who resettled in the Mississippi

Delta through the 1970s through wartime refugee programs; Cajuns, whose ancestors were displaced from Canada; and African American communities who trace their histories to those who escaped the atrocities of chattel slavery also live in potential resettlement zones of coastal Louisiana (Dalbom et al. 2014; Laska et al. 2015). Legacies of forced displacements have long-term social, economic, and health consequences, as researchers have demonstrated by examining the longitudinal and intergenerational effects of Indian relocation (Walls and Whitebeck 2012), Japanese internment (Nagata et al. 2015), the holocaust (Matz et al. 2015), and urban renewal (Fullilove 2005). The histories of colonialism, forced migration, slavery, Jim Crow segregation, and other forms of racialized dispossession continue to shape where people live in relation to coastal risks, how sensitive communities are to hazards, and the extent to which oppressed peoples have the resources to adapt.

Additionally, the archaeological record indicates a long history of regular indigenous migrations that involved the movements of small kinship units and much larger political groups between coastal areas, along bayous, and further inland for many reasons including as a response to regular flood events (Sassaman and Anderson 2004; McCintire 1954). Recent anthropological and social science research has examined this longer history of migration as an adaptation strategy among indigenous peoples, the impact of settler colonial formations, and the need to recognize the wisdom passed on intergenerationally among indigenous peoples (see Marino 2012; Whyte 2016; Wildcat 2009). Elizabeth Marino (2012) draws on ethnographic work alongside the Inupiat people of Shishmaref, Alaska, documenting how colonial infrastructures like roads, post offices, and schools led to sedentarization of previously mobile communities and created risk by developing increasing reliance on now failing infrastructure. Kyle Whyte (2016) expands upon Marino's work describing how 500 years of colonial treaties, laws, and institutions have threatened indigenous adaptation by imposing regimes of containment and the erasure of indigenous social institutions. These scholars demonstrate both that colonial development has produced conditions that have exposed indigenous communities to climate change-related hazards and undermined traditional effective modes of adaptation. They also describe the harm that is produced when colonial models of communal life, which align more with individualism and capitalist social relations, are imposed and replace indigenous modes of social organization. Contemporary initiatives to adapt as a whole tribal community and through traditional forms of organizing, like the Isle de Jean Charles Biloxi-Chitimacha-Choctaw resettlement, extend and build upon the survival and adaptation to social change throughout histories of violence, forced displacement, and physical and social marginalization.

The historic and ongoing dispossession of lands from indigenous peoples provides necessary historical context to current issues of land tenure and migration in settler colonial states like the United States (Tuck and Yang 2012; Wolfe 2006). US claims to sovereignty are rooted in the Doctrine of Discovery—a set of principles established among European monarchies in the mid-fifteenth century that attaches land rights to the "act of discovery." As recently as 2005, the Doctrine of Discovery was referenced by the US Supreme Court in a challenge to the land rights of the Oneida Nation (Dunbar-Ortiz 2014). A fundamental history of the United States has been one of treaties, legal actions, and genocidal violence by European and White American settlers aimed at the dispossession of land from indigenous peoples and establishing the conditions for white supremacy, capitalist, and imperial expansion in the United States and around the globe (Ibid.). This history has created apocalyptic conditions for the indigenous nations who previously and still inhabit this land (Estes 2019; Whyte 2016). France, Spain, and the United States have all established colonial governments in the region of Isle de Jean Charles, and colonial infrastructure, private property regimes, and land grabbing by developers and oil and gas companies established key conditions for the current ecological crisis and the tribal-driven efforts to resettle.

Restoring capacity for seasonal habitation, communal gathering, subsistence, and the reinvigoration of traditions are thus potentially essential resettlement planning activities. Such considerations and the retaining of Island land have long been prioritized in the discussions among the Isle de Jean Charles Biloxi-Chitimacha-Choctaw Tribal leaders and their partners (Maldonado et al. 2015; Peterson and Maldonado 2016). Retaining previously held territory is seen as critical for the cultural survival of many indigenous communities and nations like the Isle de Jean Charles Biloxi-Chitimacha-Choctaw Tribe (Burkett et al. 2017; Marino 2015). The fundamental logic behind CDBG and FEMA buyout programs, however, is that a place has been identified as geographically, physically at risk to recurrent damage and therefore should not sustain future habitation. FEMA and HUD's general approach to buyouts for hazard mitigation in floodplains aims at returning residential areas exposed to repeated disasters to open space (see 44 C.F.R. § 79.6 2018 and Notice of National Disaster Resilience Competition Grant Requirements 81 Fed. Reg. 109, June 7, 2016). Typically, this involves demolishing structures and residents giving up their property in exchange for the buyout funds. The Isle de Jean Charles Biloxi-Chitimacha-Choctaw Tribal Council's adaptation goals have included the continued ownership of family parcels and protection of the Island while also investing in their collective future elsewhere therefore clash with federal agency expectations and regulations. After pressure from Tribal leaders and Island residents, state officials in charge of administering NDRC funds for the Isle de Jean Charles resettlement have taken some steps to ensure the continued "access" to the Island "for ceremonial, cultural, historic and recreational purposes" (LDOA 2019a, pp.7). At the time of this writing, however, there remain important questions regarding the future of Island properties, and tribal leaders have concerns as to restrictions on that access and potential continued ownership.

There are a broad set of questions engendered by the Isle de Jean Charles resettlement and the historical dispossession of indigenous lands. In terms of immediate planning, what may motivate those who administer federal grants to advocate for amendments that embrace adaptation innovations when they depart from established regulations, and how might tribal and community leaders inform that process? More importantly, what will it take to institutionalize the legal and governance innovations needed to implement plans developed by communities and tribes? How will policy-makers ensure that adaptation programs will not perpetuate the dispossession of tribal lands or further fragment traditional relationships to place? How can adaptation planning support and extend indigenous peoples' traditional forms of land tenure? And how will such programs provide reparations for the historical forced migrations and dispossession of indigenous peoples' lands sponsored by the US and local governments and private sector?

### *6.4.4 Between Recognition and Retreat*

Political recognition of tribal sovereignty has become important to many indigenous nations as they adapt to coastal land loss (Katz 2003; Ferguson-Bohnee 2018; Sneath 2018). The policy worlds of tribal recognition are complex and contested, and many indigenous leaders and scholars have criticized both the intentions and impacts of the federal recognition process in the United States. One set of criticisms points to the petition process administered by the US Bureau of Indian Affairs. The agency demands that petitions for acknowledgment rely on written historical evidence despite the oral traditions of many indigenous nations and the misrepresentations that proliferate among colonial record-keeping and academic analyses (Miller 2004). Additionally, the petitions must present narratives that demonstrate community, political influence, and other specific criteria in a way that satisfies what the US government considers legitimate indigenous identities, histories, and forms of social organization. The criteria for petitions for federal recognition are often seen as rooted in racist colonial ideologies rather than the experiences, expressions, notions of belonging, and social institutions maintained traditionally by diverse groups of indigenous peoples (Barker 2011). Additionally, the petitions are judged by ambiguous standards that seem to shift depending on who is applying for recognition (Miller 2004). There is also a set of criticisms that point to the many examples of federal recognition processes dividing indigenous groups over limited resources (Miller 2004). Moreover, once a Tribe has submitted a petition for recognition, they must wait sometimes years for a response, and crafting a compelling submission requires enormous investment of time and money for legal and archival research that can influence, overwhelm, and undercut indigenous leaders and indigenous social efforts (Den Ouden and O'Brien 2013). In 1992, anthropologist William Starna referred to the federal recognition process as "administrative genocide."

Despite these critiques, federal recognition may also provide needed resources to Louisiana's coastal and bayou tribes. Federal recognition might enable local tribes to protect what remains of their eroding and subsiding homelands (Katz 2003; Rivard 2015). It would also open up the possibility for further partnerships among the tribes and government agencies to implement adaptation and resilience planning or for disaster assistance. A number of regulations describe federal recognition as part of the eligibility for various kinds of disaster recovery and other kinds of public support (see 44 CFR § 201.2, Emergency Management and Assistance, and 2 CFR

§ 200.54, Grants and Agreements). Isle de Jean Charles Biloxi-Chitimacha-Choctaw tribal leaders have been pursuing federal recognition while also planning their resettlement, devoting an enormous amount of time and resources to both. For example, while Tribal leaders were grappling with the various components of the state's planning process, they continued archival research and confirmed their rolls—a core requisite of the federal recognition petition process that also provided proof to the state that the Island residents were overwhelmingly members of the Tribe. The urgency and logistic demands of both resettlement and recognition weigh heavy on Tribal leaders, leading to a dilemma as to where to devote energy at any given moment.

Some state governments, including Louisiana, have also formally recognized the existence and rights of indigenous nations, even when they are not federally recognized tribes. The processes by which recognition is administered and the rights ensured by so-called state recognition vary among those states (Koenig and Stein 2013). The Isle de Jean Charles Biloxi-Chitimacha-Choctaw Tribe is one of 11 non-federally recognized tribes currently recognized by the state government (LA Indian Affairs 2018), having first been acknowledged by the state legislature in a 2004 senate resolution passed to provide the Tribe access to education and healthcare opportunities (Sen.Con.Res. 105 2004). They are also one of the 15 indigenous nations recently invited to participate in the Louisiana Native American Commission by recent state legislation (LA ACT 102 HB 660 2018).

The state of Louisiana, however, has been inconsistent in their approach to recognizing the rights of indigenous peoples. Koenig and Stein (2013, p. 133) observed that state recognition in Louisiana was described by the former director of Louisiana's Office of Indian Affairs as establishing a "government-to-government relationship" between the state government and the tribes that lived within the state's borders. At some point after their interview, during Bobby Jindal's time as governor, the Office of Indian Affairs became inactive. In 2017 and 2018, the Edwards administration and state legislature took some initial steps to further develop and clarify the meaning of state recognition (e.g., LA ACT 102 HB 660 2018). During a recent interview, however, one Louisiana state planner working on the Isle de Jean Charles resettlement pointed to the limits of state power in recognizing tribal sovereignty and rights to self-determination. He explained that only if the Isle de Jean Charles Biloxi-Chitimacha-Choctaw Tribe were federally recognized, "We would have government to government relations. But we cannot approach it that way. It is not legal." This reported inability of state government, however, runs counter to interpretations of the US Constitution and Indian law by government officials in at least one other state (see Lindemuth 2017) and forecloses an exploration of how state planners and policy-makers can honor the inherent sovereignty of non-federally recognized tribes. Legal scholars have argued that state governments have considerable flexibility as to how they approach and honor the rights of state-recognized tribes (Cohen 2005). For example, some states maintain special programs to serve non-federally recognized tribes, and one has even established a legal process for devoting land to non-federally recognized tribes (Cohen 2005).

The state's inconsistency with regard to tribal recognition has haunted the Isle de Jean Charles resettlement process. During a recent state-held resettlement planning meeting, I asked an Office of Community Development planner about the minimal references to the indigenous heritage of the Island in recent resettlement program guidelines and designs. As a counterpoint, he pointed me to a recently produced document, which, according to him, appreciated this important history (LDOA 2019a). The document, however, presents a very narrow historical overview in three brief bullet points that begin in 1979, neglecting a much longer documented history of indigenous habitation of the Island and any celebration of indigenous adaptation and life ways. Moreover, according to tribal leaders, they were not consulted in crafting this representation of their history. The final bullet point in the document—which, again, was conveyed to me as evidence that the Office of Community Development appreciates the indigenous heritage of the Island—explicitly calls into question the state's recognition of the Isle de Jean Charles Biloxi-Chitimacha-Choctaw Tribe by pointing out the specific aims and scope of the 2004 state senate resolution to ensure their access to educational and healthcare opportunities, as if to notify readers that state recognition is irrelevant to the Tribe's resettlement efforts. The same 2004 senate resolution, however, also articulates a broader state policy "to provide for recognition of Indian tribes within its borders, to support their tribal aspirations, to preserve their cultural heritage and improve their economic condition and to assist them in the achievement of their just rights" (Sen.Con.Res. 105 2004). Why did the Louisiana Office of Community Development emphasize the narrow focus of one aspect of the legal act in their resettlement background and overview?

Tensions over the role of state and federal recognition reflect the difficult position in which tribal leaders have been placed. Disparate "opportunities" within complex and contradictory policy worlds constrain actual possibilities for indigenous planning and resilience. The Isle de Jean Charles Biloxi-Chitimacha-Choctaw Tribal Council is struggling for self-determination while caught between notions of community, between a state agency and the ambiguities of state and federal policy, between state and federal government approaches to recognizing indigenous sovereignty, and between the politics of tribal recognition and the liberal and colonial politics of so-called "managed retreat." Ahistorical adaptation enables the continuation of administrative barriers and prevents indigenous peoples' in the United States from realizing their rights to self-determination, historic lands, and cultural preservation in times of climate change. In order to provide a meaningful opportunity for coastal and bayou tribes in Louisiana to protect their heritage and land and ensure their cultural survival for generations to come, state and federal governments must quickly and fully recognize the sovereignty of each of the tribes, who are seeking formal recognition. The United States must also become a full signatory and legally adopt the United Nations Declaration of Rights of Indigenous Peoples and protect the rights described therein, which include the right to self-determination, right to land, right to protect historic sites, and the right to receive redress from experiences of land dispossession, forced assimilation, and the deprivation of distinct cultural values or ethnic identities (UN General Assembly 2007).

### 6.4.5 *Reframing Resettlement*

The most urgent form of ahistorical adaptation, however, for Isle de Jean Charles Biloxi-Chitimacha-Choctaw Tribal leaders has emerged as government planners and their contractors have not adequately honored the existing resettlement plans driven by tribal leaders long before resources arrived. Isle de Jean Charles Biloxi-Chitimacha-Choctaw Tribal leaders had been actively working on resettling inland and reuniting their displaced Tribe for over a decade prior to the National Disaster Resilience Competition. The Tribe's pre-NDRC resettlement efforts are documented extensively by journalists, academics, and filmmakers (Jessee 2015; Maldonado et al. 2015; Simon 2008). Despite multiple setbacks in the previous efforts the US Army Corps and Terrebonne Parish described in a previous section of this chapter, the Tribal Council has continued to search for resources and pathways to reunite their people and ensure their future inland—an effort described by Tribal Executive Secretary, Chantel Comardelle, during an interview as "creating a living and active bridge from our ancestral land to the new Isle de Jean Charles Tribal community."

Secretary Comardelle had lived on the Island until she was 4 years old. Her mother and father—the current Deputy Chief of the Tribe—left in the mid-1980s after Hurricane Juan. She had developed chronic respiratory problems from mold that proliferated from repeated flooding. Secretary Comardelle has a full-time job outside of her unpaid role supporting the Tribe and regularly works all hours of the night on resettlement planning, federal recognition, and other tribal-driven initiatives. She often participates in planning conference calls while at her children's various sports practices and gymnastics meets. In addition to all of this, in 2017 Chantel completed a Museum Studies Certificate Program from the Institute of American Indian Arts. Her aim has been to learn curatorial skills that will help future generations absorb their heritage and maintain relationships to their ancestral Island, even if the land mass erodes away. Secretary Comardelle is someone I often think of when I encounter the term "climate refugees" attributed to her Tribe. The term fails to capture the amount of time, work, planning, and passion that she and the Tribal Council have put into their resettlement over the years. It dehumanizes the Island people by representing them only in terms of vulnerability and environmental risk, rather than as whole human subjects with multiple dimensions, experiences, relationships, aspirations, fears, and dimensions of their identities. Nor does it capture the reality that the effort is with the entire tribal social organization and cultural survival in mind, not individuals fleeing one particular hazard. According to Secretary Comardelle, "For us, resettlement is an act of cultural survival."

In 2010, Tribal leaders began working with long-time collaborators at the Lowlander Center—a nonprofit organization that uses participatory action research to support the efforts of a number of bayou communities and tribes. Together, they assembled a team of academics, architects, indigenous leaders, and other professionals who could provide resources, support, collaboration, and fellowship as the Tribe pursued their resettlement. The Tribe, Lowlander Center, and other

**Fig. 6.3** A conceptual plan for the relocated Isle de Jean Charles community includes about 100 homesites and a village center. (Image courtesy of Evans + Lighter Landscape Architecture)

collaborators planned for key tribal institutions like a museum, tribal cultural and community center, health facilities, and gardens. Their planning efforts embraced renewable energy, mitigation standards for infrastructure and design, tribal governance capacity building, and tribal-driven economic development principles (see Fig. 6.3 for one pre-NDRC rendering produced by Evans & Lighter Architecture in collaboration with tribal leaders and their resettlement partners). When I began working with Tribal leaders and their resettlement partners in late 2014, they were in the midst of conducting extensive outreach to build support for their tribal resettlement plans and synthesizing existing literature in sustainable development, resettlement, and resilience to use as a baseline for critically analyzing their own planning process (Jessee 2015). The tribal-driven resettlement work has been celebrated internationally as the Tribal leadership has garnered awards like the Mississippi-Alabama Sea Grant Consortium's "Spirit of the Community" Award, the US Environmental Protection Agency's "Guardian of the Gulf Award for Environmental Justice," and the Rising Voices "Bob Gough Climate Justice in Action Award." Additionally, the Lowlander Center was one of five awardees of the 2018 Climigration Network Award for their approach to community relocation.

After years of struggling to secure financial resettlement support, in 2015 the Isle de Jean Charles Biloxi-Chitimacha-Choctaw Tribe and the Lowlander Center partnered with Louisiana's Office of Community Development (OCD). The Tribal Council, Lowlander Center, and their network of scholars and architects attended public meetings and multiple planning sessions with the state agency, contributing extensive time and labor to the State's process of crafting a successful submission to the National Disaster Resilience Competition—an effort that garnered $48.3 million in federal funds to support their tribal resettlement (LDOA 2015a). The application importantly spoke to the social as well as environmental concerns of tribal leaders by assuring that "All factors of design and process will help to support and enhance tribal identity, sovereignty, and dignity" (LDOA 2015b, pp. 107). Once the award was granted, the Isle de Jean Charles Biloxi-Chitimacha-Choctaw Tribe and Lowlander Center each received partnership letters from Louisiana's Office of Community Development (OCD) that thanked them, acknowledged their contributions to crafting a winning application, and ensured a continued partnership.

It appears as though state planners and US Department of Housing and Urban Development officials, however, did not thoroughly understand the social and historical complexities of the Island or the contested space that had emerged from decades of land grabs, displacement, gentrification, and the social demands on multiple tribes who inhabit the region. Soon after the NDRC award was announced, some of the complex realities of the Island as a social space became evident to state officials by way of a letter to the Governor from the Chief of the United Houma Nation—another indigenous nation from the region—expressing their concern over being excluded from the State's plans for using federal resilience funds despite their also having ties to the Island (Batte 2016).[4] The reason the United Houma Nation was not included in Louisiana's initial NDRC application remains unclear, though Crepelle (2019, pp. 27–28) suggests it might be the result of the United Houma Nation having not sent a representative to state-led meetings during the application process. State documentation portrays at least one person from the Tribe in attendance at a phase II workshop (LDOA 2015c, p. 350). However, the documentation does not capture the rationale behind the state's phase II outreach process or the content of specific conversations during that period. This confusion points to how critical it is for state agencies to more fully appreciate local histories and complex social relations before developing and committing to programs, something that the timelines of design competition application processes and the pace of grant cycles discourages.

Louisiana's Office of Community Development began redefining the character, scope, and beneficiaries of the resettlement after the letter from the United Houma Nation. What was in the application framed as a tribal resettlement that included current and former Island residents (most of whom also live in low-lying areas in the region that also experience flooding) became predominantly focused on the current Island residents as individuals, regardless of tribal identity. According to a fact sheet circulated by the office:

> Phase II of the state's NDRC application specifically references the Isle de Jean Charles Band of Biloxi-Chitimacha-Choctaw. This reference was made under the belief that all inhabitants of the Island affiliate with this tribe. There are apparently also members of the United Houma Nation living on the Island, and there may be Island residents who don't

[4]The experiences and aims of the United Houma Nation's Tribal government and citizens are beyond the scope of the research that informs this chapter.

> affiliate with any tribe. As such, specific tribal membership will not be a requirement for inclusion in the resettlement, as the state's objective is the resettlement of all willing members of the Isle de Jean Charles community, irrespective of any familial, cultural or tribal affiliation. (quoted in Crepelle 2019, p. 28)

The Office of Community Development began referring to the Tribal Council as "stakeholders" rather than partners. They imposed a more conventional top-down, rather than tribal-driven or participatory, planning process beginning with a needs assessment predominantly based on land-use data collection on the Island and outreach with individual households who remained on the Island rather than the collective needs of the Tribe as a community (LDOA 2017b). Tribal leaders began to question the state's commitment to their efforts and read their actions as divisively exacerbating tensions on and around the Island—what some Tribal citizens referred to as a process of "divide and conquer." Moreover, instead of relying on the existing efforts of the Tribal Council whose planning and outreach was critical to the state being awarded the NDRC award, they slowly began to roll out new leadership structures to govern the planning process, such as a project steering committee, composed of indigenous and nonindigenous Island residents, representatives from both the Isle de Jean Charles Biloxi-Chitimacha-Choctaw and United Houma Nation, and state employees to give input in the planning process (LDOA 2019a). The steering committee met six times over a 9-month period in 2018 (LDOA 2019a). The meetings then stopped abruptly confusing at least some steering committee members who began to hear about resettlement activities in the press rather than directly from the state planners themselves (Jessee Forthcoming).

State planners and their subcontractors have also rendered the extensive work of the Isle de Jean Charles Biloxi-Chitimacha-Choctaw Tribal Council has devoted to the process invisible in planning materials and public statements. From after the award announcement in 2016 until at least late 2018, the state's resettlement website displayed a project timeline with no reference to any pre-NDRC resettlement work that the Tribe had conducted or their extensive contributions to the successful NDRC application (LDOA 2018a). Timelines posted at public community planning meetings also omitted the tribal-driven history of resettlement planning (Jessee Forthcoming). Additionally, in one article written for an online planning publication, Louisiana's Resilience Policy and Program Administrator makes no mention of the Tribe's name or the indigenous heritage of the Island (Sanders 2018). Instead, he referred to the people of the Island as "pioneers"—an odd phrase to describe Native Americans given the historical role of European and American pioneers in the occupation and dispossession of indigenous peoples' lands. The article, entitled *Don't Label Them Climate Change Refugees, says a Louisiana Planner, They're Pioneers*, importantly rejects the paternalistic and problematic "climate refugee" discourse explicitly. It does so, however, while reproducing an ahistorical reframing of the resettlement that erases the indigenous heritage of the Island and the Tribal leaders who have advanced resettlement as a way of saving the Island's heritage. According to the article, "Many of the Island's residents are leaning in by collaborating with a project team of state officials, planners, engineers and architects to plan the look, feel, function and composition of the new community." However, it was Louisiana state planners who "leaned in" to collaborate with an existing Tribal Council who established and maintained a set of existing partnerships that produced clear plans for a tribal community resettlement—which was funded.

In early 2019, 3 years after tribal leaders began to feel excluded from the state's planning process—and nearly 3 years after Chief Naquin initially suspected the state was trying to transform their tribal resettlement into a subdivision—the State announced an action plan amendment to the US Department of Housing and Urban Development for a "project narrative clarification" and "introduction of new activities," formalizing a shift that began almost immediately following the award announcement. The proposed amendment replaces the section of the funded NDRC application that commits to "supporting and enhancing tribal identity, sovereignty, and dignity" and the tribal community center (LDOA 2019b). Despite their stated goal to "create an opportunity" for former residents to join the resettlement, the state's outreach has been focused predominantly on current Island residents, and only as of early 2019 did they begin to publicize the basic terms under which former Island residents can claim a spot in the new location. According to the proposed amendment, eligibility of former Island residents to claim a plot of land in the resettlement location would be restricted to those who currently own a home and "can demonstrate the financial ability to build a new home" (LDOA 2019b). The state's program amendment would enable anyone "who lived in a Hurricane Isaac federally-declared disaster parish on Aug. 28, 2012" to receive a lot in the new location and would allow the state to publicly auction lots that do not get claimed by former island residents (LDOA 2019a, p. 10). Additionally, there remain questions as to how those who could not afford to relocate on their own will afford the new costs of owning and maintaining a home, insurance, taxes, utilities, etc. given the parameters established by the Office of Community Development. Three years after the award, there is immense uncertainty as to how one of the primary goals of the Tribal leadership over the years of resettlement efforts—to reunite scattered Tribal citizens in order to ensure the Island's traditional cultural survival and growth—will be achieved.

The redefinition of the scope of the resettlement and lack of commitment to the existing tribal vision has estranged Tribal leadership from the planning process and limited the potential resources for the resettlement garnered through their advocacy. The NDRC was one of many mechanisms Tribal leaders hoped would support their long sought-after resettlement goals. In addition to working with the state's Office of Community Development toward the NDRC funding, Tribal leaders and their resettlement team secured planning support from multiple programs including the Citizens Institute for Rural Design, the Environmental Protection Agency's Healthy Places for Healthy People, the Smithsonian's Recovering Voices grant, and the National Academy of Sciences to enhance the planning process, directed toward ensuring a number of plan elements including a health clinic, honoring traditional ecological knowledge, and building their organizational capacity after years of struggling with the strains of displacement. Tribal leaders remain uncertain as to whether or not the outcomes of these efforts will be realized, because the state of Louisiana has retained ownership over the land they purchased with NDRC funds for the resettlement and have not committed to the incorporation of these other Tribal-driven initiatives on that land. Moreover, while the Tribe envisioned owning and governing communal spaces of the resettlement site, the recent proposed amendment states that infrastructure and recreational spaces will be "maintained by a governing nonprofit or unit of local government" (LDOA 2019b, p. 10), exacerbating questions as to how the Isle de Jean Charles Biloxi-Chitimacha-Choctaw Tribe as a distinct community might plan for their future on the lands purchased for their relocation with a grant they contributed extensively to garnering. One tribal member summed up these concerns during an informal interview after hearing in the newspaper about a state planning meeting to be held the next day, despite their serving on the state-created project steering committee:

> "This ain't no Road Home Program. We wanted them to help us, not take over the whole thing. My spirit aches knowing they are going to try to get people to sign up for a spot in their resettlement, while not explaining what they will have to pay since the Tribe is getting pushed out. It ain't free. We don't know what they tell people when they talk on their porches. We don't know if the people hear from the state that they will need to pay taxes or flood insurance on this new house, and it will be more than they pay now on the Island. My spirit aches because the whole tribe and tribal leaders were not invited to their meetings, and the state people do not trust us. We don't know if we should even advise our people. This was for our children. We want what is best for the future of our people as a whole. But now we are stuck between a rock and a hard place of not doing anything… But who's going to be there after 2022 when this grant ends?"

The state's approach to administering NDRC funds has produced immense uncertainty and threatened to alienate and undercut some of the most compelling advocates for the Island people—who also happen to be arguably the most effective advocates for coastal resettlement in Louisiana for nearly two decades. What does that do for future coastal planning? What will other community and tribal leaders learn when they reflect on how this process has unfolded? Will they have to choose between adaptation and identity? Between relocating out of harm's way and maintaining their collective self-determination? Or between safety from coastal flooding and justice? At the time of this writing, Isle de Jean Charles Biloxi-Chitimacha-Choctaw Tribal leaders are concerned that public perception will be they have benefitted from the $48 million investment in their resettlement and therefore do not need continued or future support. In reality, the state's process has been incredibly painful for tribal leaders and has led to new kinds of uncertainty for the Tribe's future.

Scholars and journalists are beginning to raise questions as to the ability of public-philanthropic-sponsored design competitions to grapple with the social and historical complexities amplified by disasters and climate change. While competitions may offer resources for some jurisdictions but do not support others, they run the risk of creating divisions rather than solidarity among communities struggling to adapt (Spanne 2016). I hope that this section has highlighted that the concerns, plans, and visions advanced by tribal leaders or activists embraced by government agencies in design stages of these competitions may not transfer throughout implementation. First, the state's ambiguous approach to community—

articulating both geographic notions inherited from the CPRA plan and risk reduction conventions that highlight physical exposure to hazards as well as tribal notions of community inherited from existing Isle de Jean Charles resettlement efforts—throughout their application and implementation processes along with regional politics have derailed the commitment of state actors to some of the most important social and cultural aspects of the initial application. Additionally, the innovation of resilience design competitions seems primarily focused on design with minimal attention to the innovations needed in governance. Local jurisdictions may draw on conventional approaches to spending federal dollars, constraining the innovation harnessed during the development of applications to those competitions. Design is not a replacement for governance, and the problems raised in this chapter should demand government agencies find new ways to support the initiatives of those already working to support community-oriented and indigenous resilience. I often wonder how things would be different today, with regard to the Isle de Jean Charles resettlement, if state policy-makers and planners had remained committed to the values and ideas embraced in initial application and their initial partnerships when they realized the social complexity of the Island, even if it meant pursuing additional funding and developing additional partnerships to accommodate other tribes and those whose visions of the future departed from the one long-advanced by the Isle de Jean Charles Biloxi-Chitimacha-Choctaw Tribe?

## 6.5 Beyond Beneficiaries: Resourcing Community and Tribal-Driven Resettlements

Given the historical and contemporary tensions described in previous sections of this chapter, policy-makers, planners, and funders should concentrate on adjusting planning conventions, grant implementation processes, federal regulations, and expectations in order to engage in participatory processes that ensure community leaders—or in the case of tribal resettlement, tribal leadership—remain in control of their complex and dynamic adaptation processes. What should the Tribe have been told about their rights to maintain ownership over the resettlement ideas that they had been honing for about 20 years when they shared them with the state, thinking the state was going to support their existing efforts to lead their resettlement? What role should the state have taken to support those tribal leaders already mobilizing to resettle their communities *en groupe?* Once tensions between notions of community and goals for resettlement emerged, what would have been a better process for harmonizing divergent adaptation aims?

One resource for policy-makers, scholars, and practitioners engaged with those already working on resettlement planning processes is *The Peninsula Principles on Climate Displacement Within States*—a policy document produced by Displacement Solutions in 2013 based on consultation with international experts in human rights, international law, refugees, and migration. The principles highlight the agency and dignity of those experiencing displacement and frame the role of state agencies as resource providers and advocates more so than leaders of resettlement processes. According to the principles, national policies should institutionalize the government's role as providers of resources, assistance, and protection against human rights abuses. Mackinnon and Derickson (2013)'s "politics of resourcefulness" is also useful. Their work highlights the need to distribute resources with the goal of addressing inequalities, to institutionalize and commit to democratic and participatory processes and capacity building within communities, to value traditional indigenous knowledges (ecological and other forms of expertise), and to ensure the political and cultural recognition of communities and tribes. As indicated earlier in the chapter, the United Nations Declaration on the Rights of Indigenous Peoples is also a critical resource.

There is also a need for planners and policy-makers to enhance organizing capacity when it is not fully developed. Though many community and tribal leaders, like the Isle de Jean Charles Biloxi-Chitimacha-Choctaw Tribal Council, have long begun planning their resettlements to ensure a future for their people in the wake of climate-related hazards and disaster, they likely also need various kinds of support to fully realize their visions and in dealing with the effects of existing displacement. Government representatives have access to many resources that might be useful to resettling communities including knowledge of agencies, laws, and procedures; people to serve as staff; letters of support; and grant writing skills. State or federal agencies could also provide financial compensation for time spent on resettlement activities, office space for community or tribal use, and guidance for people to gather additional resources themselves. Providing material support to existing community-to-community and tribe-to-tribe knowledge exchanges and to existing relationships of solidarity among community-based organizations would enhance resilience and might also supplement governmental deficiencies (Lazrus et al. 2016; Maldonado et al. 2015). Access to professional networks, like the American Planning Association, the Society for Applied Anthropology, Rising Voices, and the Rural Sociological Society, are also potentially valuable.

Outcomes of any knowledge-generating processes related to resilience or resettlement planning are hot commodities at the moment. There is widespread demand for publications of "best practices" or "lessons learned" of resilience and adaptation processes. Indeed, one of the goals of governmental support for the Isle de Jean Charles Biloxi-Chitimacha-Choctaw resettlement was to develop a conduit for future learning about resettlement processes and sustainable development. The Tribal leadership and their team of partners had embraced that role using their resettlement to create a teaching and learning community for sustainable development. The state of Louisiana has indicated they envision a resettlement process which is a "scalable" and "replicable" model (LDOA 2018b). Louisiana state planners have presented their work on this resettlement at numerous conferences including the American Planning Association, American Association of Geographers, and others around the world without including Tribal leaders and their knowledge of the long road to resettlement. For more just processes of resettlement knowledge sharing, it is critical that local activists and Tribe-based partners have real power and authorship in the activities through which their experiences, struggles, and activism are represented, objectified, commodified, and circulated for multi-societal benefit. These kinds of presentations and any reports produced should be crafted in collaboration and co-authored with community-based organizations and Tribal leaders, and the resources acquired from the circulation of documented outcomes should directly benefit those community-based organizations monetarily and through recognition. This is especially important during Tribal resettlements, where intellectual property issues and representations of identity, ancestry, and social organization are so often conditioned by historical colonial social relations and can be used to threaten cultural survival or political recognition.

## 6.6 Conclusions

In this chapter I have argued that historical atrocities and intergenerational experiences of injustice influence the lived experiences of contemporary adaptation policy and resettlement planning efforts. The refusal of risk reduction professionals to adequately address past injustices and their social and environmental legacies—what I refer to as ahistorical adaptation—can disrupt efforts to build meaningful partnerships between communities or tribes and government agencies. A lack of awareness and explicit acknowledgment of local and national histories of race and racism, the politics of disaster, or struggles for self-determination, for example, might be felt as an invalidation of peoples' experiences, suffering, perseverance, and wisdom. Without meaningfully addressing broader social, cultural, and historical contexts, climate change adaptation policies may serve as just another site for the continued oppression of already marginalization peoples.

In order to avoid ahistorical adaptation, I call on risk reduction professionals, state planners, scholars (myself included), and others engaged in hazard mitigation and environmental adaptation—especially community resettlement processes—to conceptualize and conduct their work through wider timelines and notions of risk and adaptation than they may be accustomed. Many of the state planners I have spoken with during the course of my research who are working on administering National Disaster Resilience Competition funds for the Isle de Jean Charles resettlement acknowledge the histories discussed in this chapter and their legacies within individual conversations. Also, the final LA SAFE regional adaptation strategy, which was released as this chapter was headed to press, briefly acknowledges the legacies of colonialism and racism and how they contribute to current risk (LDOA 2019c, p 96–97). This evolution from earlier drafts of the framework might demonstrate an increasing sensitivity to the historical production of vulnerability. Future research might explore 1) the suitability of the recommended strategies advanced in that document for addressing the scope of risks that Louisiana residents face, 2) the extent to which the strategies lead to the redress of injustices mentioned in the "Race and Ethnicity" blurb, and 3) the appropriate role of oil and gas industries in resilience and adaptation policy in Louisiana. In order to meaningfully adapt to the socio-ecological risks that threaten coastal habitability in Louisiana, state and federal policy-makers, scholars, planners, and others working on environmental adaptation must reckon with the entangled roots of risk: settler colonization, racism, capitalism, and the environmental destruction caused by industrial development. Inequalities and violence related to gender and sexuality should also be considered in this context though it is regrettably missing from the analysis in this chapter.

Policy and planning approaches must provide adequate resources for the implementation of community-based and tribal-driven adaptation initiatives, especially when, as in the case of the Isle de Jean Charles Biloxi-Chitimacha-Choctaw Tribe, those efforts have been underway for nearly two decades. Resilience and adaptation policies and practice should bolster existing environmental and social justice efforts by finding ways to provide financial and other needed resources to those doing this essential work. Tribal and community resettlement processes evoke disparate legal and policy worlds that relate to grant administration, property law, and the rights of indigenous peoples—making public support for just resettlement processes incredibly complex and potentially impossible without significant social, cultural, legal, and bureaucratic transformation. Institutionalizing the kinds of support most needed will likely require changes in federal and state regulations, budgets, and more serious conversations about holding extractive industries accountable. Providing reparations for legacies of forced displacement, slavery, land grabs, and impacts of deforestation and oil and gas development should be investigated as resilience and adaptation strategies. While the focus of risk reduction professionals and policy-makers may remain transfixed on the future—imagining new relationships to place and habitation in flood zones—we must simultaneously reckon with the past. "Reshaping" coastal Louisiana must include reshaping the colonial social relations that have devastated this landscape and many of the people who have lived, and who currently live, here.

## References

Austin, D. E. (2006). Coastal exploitation, land loss, and hurricanes: a recipe for disaster. *American Anthropologist, 108*(4), 671–691. https://doi.org/10.1525/aa.2006.108.4.671.

Barker, J. (2011). *Native acts: Law, recognition, and cultural authenticity*. Durham: Duke University Press.

Barry, J. M. (1997). *Rising tide: The great Mississippi flood of 1927 and how it changed America*. New York: Touchstone.

Batte, J. (2016, May 11). Tensions arise between local Indian tribes over effort to abandon sinking island. *HoumaToday,* Retrieved from https://www.houmatoday.com/news/20160511/tensions-arise-between-local-indian-tribes-over-effort-to-abandon-sinking-island

Bettini, G. (2017). Unsettling futures: Climate change, migration, and the (ob)scene biopolitics of resilience. In A. Baldwin & G. Bettini (Eds.), *Life adrift: Climate change, migration, critique* (pp. 79–96). Lanham: Rowman & Littlefield Publishers.

Bronen, R. (2011). Climate-induced community relocations: Creating an adaptive governance framework based in human rights doctrine. *New York University Review of Law and Social Change, 35*(2), 356–406. Retrieved from https://socialchangenyu.com/.

Bronen, R., Maldonado, J. K., Marino, E., & Hardison, P. (2018). Climate change and displacement: Challenges and needs to address an imminent reality. In M. Cernea & J. K. Maldonado (Eds.), *Challenging the prevailing paradigm of displacement and resettlement: Risks, impoverishment, legacies, solutions* (pp. 252–272). New York: Routledge.

Burkett, M., Verchick, R. R. M., & Flores, D. (2017). Reaching higher ground: Avenues to secure and manage new land for communities displaced by climate change. *Center for Progressive Reform*. Washington, DC: Center for Progressive Reform. Retrieved from http://progressivereform.org/articles/ReachingHigherGround_1703.pdf

Carter, L., Terando, A., Dow, K., Hiers, K., Kunkel, K. E., Lascurain, A., et al. (2018). In D. R. Reidmiller, C. W. Avery, D. R. Easterling, K. E. Kunkel, K. L. M. Lewis, T. K. Maycock, & B. C. Stewart (Eds.), *Impacts, risks, and adaptation in the United States: Fourth national climate assessment* (Vol. II, pp. 743–808). Washington, DC: U.S. Global Change Research Program.

Cernea, M. M. (1997). The risks and reconstruction model for resettling displaced populations. *World Development, 25*(10), 1569–1587. https://doi.org/10.1016/S0305-750X(97)00054-5.

Cernea, M. M. (2000). Risks, safeguards, and reconstruction: A model for population displacement and resettlement. In M. Cernea & C. McDowell (Eds.), *Risks and reconstruction: Experiences of resettlers and refugees* (pp. 11–15). Washington, DC: The World Bank.

Clipp, A., Gentile, B., Green, M., Galinski, A., Harlan, R., Rosen, Z., et al. (2017). *2017 Coastal Master Plan: Appendix B: People and the landscape*. Version Final. (43 pp.). Baton Rouge: Coastal Protection and Restoration Authority. Retrieved from http://coastal.la.gov/wp-content/uploads/2017/04/Appendix-B_People-and-the-Landscape_FINAL.pdf

Coastal Protection and Restoration Authority (CPRA). (2017). *2017 Coastal master plan: Appendix E: Flood risk and resilience program framework*. Version Final. (pp. 1–66). Baton Rouge: Coastal Protection and Restoration Authority. Retrieved from http://coastal.la.gov/wp-content/uploads/2016/04/Appendix-E_FINAL_10.09.2017.pdf

Cohen, F. S. (2005). *Cohen's handbook of federal Indian law*. Newark: LexisNexis.

Collier, S. J., Cox, S., & Grove, K. (2016). Rebuilding by design in post-Sandy NYC. *LIMN, 7*. Retrieved from https://limn.it/rebuilding-by-design-in-post-sandy-new-york/

Colten, C. E. (2017). Environmental management in coastal Louisiana: A historical review. *Journal of Coastal Research, 33*(3), 699–711. https://doi.org/10.2112/JCOASTRES-D-16-00008.1.

Colten, C. E., Simms, J. R., Grismore, A. A., & Hemmerling, S. A. (2018). Social justice and mobility in coastal Louisiana, USA. *Regional Environmental Change, 18*(2), 371–383. https://doi.org/10.1007/s10113-017-1115-7.

Comstock, D., & Fox, R. (1993). Participatory research as critical theory: The North Bonneville, USA, experience. In P. Park, M. Brydon-Miller, B. Hall, & T. Jackson (Eds.), *Voices of change: Participatory research in the United States and Canada* (pp. 103–124). Westport: Bergin & Garvey.

Couvillion, B. R., Beck, H., Schoolmaster, D., & Fischer, M. (2017). *Land area change in coastal Louisiana (1932 to 2016)* (No. 3381). Reston: US Geological Survey. https://doi.org/10.3133/sim3381.

Crepelle, A. (2019). The United States first climate relocation: recognition, relocation, and indigenous rights at the Isle de Jean Charles. *Belmont Law Review, 6*, 1), 1–1),40. Retrieved from http://www.belmont.edu/law/.

Cronin, V., & Guthrie, P. (2011). Community-led resettlement: From a flood-affected slum to a new society in Pune, India. *Environmental Hazards, 10*(3–4), 310–326. https://doi.org/10.1080/17477891.2011.594495.

Crowell, M., Coulton, K., Johnson, C., Westcott, J., Bellomo, D., Edelman, S., et al. (2010). An estimate of the U.S. population living in 100-year coastal flood hazard areas. *Journal of Coastal Research, 26*(2), 201–211. https://doi.org/10.2112/jcoastres-d-09-00076.1.

Cusik, D., & Aton, A. (2017). Puerto Ricans could be newest U.S. climate refugees. *Scientific American E & E News: Climate*. Retrieved from https://www.scientificamerican.com/author/e-e-news/

Dalbom, C., Hemmerling, S. A., & Lewis, J. (2014). *Community resettlement prospects in Southeast Louisiana: A multidisciplinary exploration of legal, cultural, and demographic aspects of moving individuals and communities*. New Orleans: Tulane Institute on Water Resources Law & Policy. Retrieved from http://www.law.tulane.edu/uploadedFiles/Institutes_and_Centers/Water_Resources_Law_and_policy/Content/Community%20Resettlement%20Prospects%20in%20Southeast%20Louisana.pdf

Dawson, A. (2017). *Extreme Cities: The peril and promise of urban life in the age of climate change*. London: Verso.

Den Ouden, A. E., & O'Brien, J. M. (Eds.). (2013). *Recognition, sovereignty struggles, and indigenous rights in the United States: A sourcebook*. Chapel Hill: UNC Press Books.

Dunbar-Ortiz, R. (2014). *An Indigenous Peoples' history of the United States*. Boston: Beacon Press.

Estes, N. (2019). *Our History Is the Future: Standing Rock Versus the Dakota Access Pipeline, and the Long Tradition of Indigenous Resistance*. New York City: Verso Books.

Faas, A. J. (2016). Disaster vulnerability in anthropological perspective. *Annals of Anthropological Practice, 40*(1), 14–27. https://doi.org/10.1111/napa.12084.

Ferguson-Bohnee, P. (2018, May). *Self determination in a sinking basin - hurdles and hindrances to protecting tribal cultural heritage in Southern Louisiana*. Presentation at the State of the Coast Conference. New Orleans, LA.

Ferris, E. (2012). *Protection and planned relocations in the context of climate change*. Geneva: UN High Commission of Refugees, Division of International Protection. Retrieved from http://www.unhcr.org/5024d5169.html

Fleming, B. (2019). Design and the Green New Deal. *Places Journal*, Retrieved from https://placesjournal.org/article/design-and-the-green-new-deal

Fullilove, M. T. (2005). *Root shock: How tearing up city neighborhoods hurts Americans and what we can do about it*. New York: One World Press.

Gohd, C. (2018, January 5). A wave of climate refugees could soon emerge in the US. *Futurism*. Retrieved from https://futurism.com/wave-climate-refugees-soon-emerge-us/

Hardy, R. D., Milligan, R. A., & Heynen, N. (2017). Racial coastal formation: The environmental injustice of colorblind adaptation planning for sea-level rise. *Geoforum, 87*, 62–72. https://doi.org/10.1016/j.geoforum.2017.10.005.

Hardy, D., Lazrus, H., Mendez, M., Orlove, B., Rivera-Collazo, I., Roberts, et al. (2018). *Social vulnerability: Social science perspectives on climate change*, Part 1. Washington, DC: USGCRP Social Science Coordinating Committee. Retrieved from https://www.globalchange.gov/content/social-science-perspectives-climate-change-workshop

Hauer, M. E., Evans, J. M., & Mishra, D. R. (2016). Millions projected to be at risk from sea level rise in the continental United States. *Nature Climate Change, 6*, 691–694. https://doi.org/10.1038/nclimate2961.

Hobor, G., Plyer, A., & Horwitz, B. (2014). *The coastal index*. New Orleans: The Data Center. Retrieved from https://s3.amazonaws.com/gnocdc/reports/TheDataCenter_TheCoastalIndex.pdf.

Jankowski, K. L., Törnqvist, T. E., & Fernandes, A. M. (2017). Vulnerability of Louisiana's coastal wetlands to present-day rates of relative sea-level rise. *Nature Communications, 8*, 14792. https://doi.org/10.1038/ncomms14792.

Jerolleman, A. (2019). *Disaster recovery through the lens of justice*. Springer Nature.

Jessee, N. (2015). Can't Stop the Water Jason Ferris and Rebecca Marshall Ferris, dirs. 40 min. Philadelphia: Cottage Films, 2013. *American Anthropologist, 117*(4), 808–810. https://doi.org/10.1111/aman.12376.

Jessee, N. (2016, April 22). Hope for "just resilience" on Earth Day. *EnviroSociety*. Retrieved from www.envirosociety.org/2016/04/hope-for-just-resilience-on-earth-day

Jessee, N. (Forthcoming). *Reshaping Louisiana's Coastal Frontier* (Unpublished doctoral dissertation). Temple University.

Katz, M. (2003). Staying afloat: How federal recognition as a Native American tribe will save the residents of Isle de Jean Charles, Louisiana. *Loyola Journal of Public Interest Law, 4*(1), 126. Retrieved from http://law.loyno.edu/loyola-journal-public-interest-law.

Keene, E. (2017). Lessons from relocations past: Climate change, tribes, and the need for pragmatism in community relocation planning. *American Indian Law Review, 42*(1), 259. https://www.jstor.org/stable/10.2307/26492279.

Koenig, K. A., & Stein, J. (2013). State recognition of American Indian Tribes: A survey of state-recognized tribes and state recognition processes. In A. E. Den Ouden & J. M. O'Brien (Eds.), *Recognition, sovereignty struggles, and indigenous rights in the United States: A sourcebook* (pp. 115–148). Chapel Hill: UNC Press Books.

Koslov, L. (2016). The case for retreat. *Public Culture, 28*(2), 359–387. https://doi.org/10.1215/08992363-3427487.

Kuh, K. F. (2016). Agnostic adaptation. In R. K. Craig & S. R. Miller (Eds.), *Contemporary issues in climate change law and policy: Essays inspired by the IPCC* (pp. 167–180). Washington, DC: ELI Press.

Louisiana Office of Indian Affairs (LA Indian Affairs). (2018). Federal and State Tribal Contact List – 2018. Baton Rouge: Louisiana Office of Indian Affairs. Retrieved from http://gov.louisiana.gov/assets/Programs/IndianAffairs/LouisianaTribes.pdf

Laska, S., Peterson, K., Alcina, M. E., West, J., Volion, A., Tranchina, B., et al. (2010). *Enhancing Gulf of Mexico coastal communities' resiliency through participatory community engagement.* CHART Publications. Paper 21. New Orleans: UNO-CHART. Retrieved from https://scholarworks.uno.edu/chart_pubs/2

Laska, S., Peterson, K., Rodrigue, C., Cosse', T., Philippe, R., Burchett, O., et al. (2015). "Layering" of natural and human caused disasters in the context of anticipated climate change disasters: The coastal Louisiana experience. In M. Companion (Ed.), *The impact of disasters on livelihoods and cultural survival: Opportunities, losses and mitigation* (pp. 225–238). Boca Raton: CRC Press.

Lasley, C. B. (2012). *Catastrophes and the role of social networks in recovery: A case study of St. Bernard Parish, LA, residents after Hurricane Katrina* (dissertation). New Orleans: University of New Orleans Theses and Dissertations. (1504).

Lazrus, H., Gough, B., & Maldonado, J. (2016). The rising voices: Building bridges between scientific and indigenous communities. *Natural Hazards Observer, 40*(4), 16–20. Retrieved from https://hazards.colorado.edu/natural-hazards-observer/volume-xl-number-8.

Lindemuth, J. (2017). *Re: Legal status of tribal governments in Alaska.* Anchorage: Department of Law, Office of Attorney General. Retrieved from https://s3.amazonaws.com/arc-wordpress-client-uploads/adn/wp-content/uploads/2017/10/21021026/AG-Opinion-re-Tribal-Government-in-Alaska.pdf

Logan, J. R., Zhang, W., & Alba, R. D. (2002). Immigrant enclaves and ethnic communities in New York and Los Angeles. *American Sociological Review, 67*(2), 299–322. https://doi.org/10.2307/3088897.

Louisiana Act 102 House Bill 660 (2018).

Louisiana Budget Project. (2018, September 13). Poverty and income inequality continue to plague Louisiana. *Louisiana Budget Project.* Retrieved from http://www.labudget.org/2018/09/release-povertyand-incomeinequality-continue-to-plague-louisiana/

Louisiana Department of Administration (LDOA). (2015a). *Resettlement as a resilience strategy and the case of Isle de Jean Charles.* Office of Community Development Disaster Recovery Unit (OCD DRU). Baton Rouge: OCD DRU. Retrieved from http://www.doa.la.gov/ocddru/ndrc/idjc_prospectus_final_27oct15_updated_logos.pdf

Louisiana Department of Administration (LDOA). (2015b). *National Disaster Resilience Competition Phase II Application.* Baton Rouge: Louisiana Department of Administration.

Louisiana Department of Administration (LDOA). (2015c). *National Disaster Resilience Competition Phase II Application, Attachment D: Consultation Summary Public Comment Summary & Responses.* Office of Community Development Disaster Recovery Unit (OCD

DRU). Baton Rouge: OCD DRU. Retrieved from https://www.doa.la.gov/OCDDRU/NDRC/AttchD_ConsultationSummary_LA.pdf

Louisiana Department of Administration (LDOA). (2017a). *LA SAFE Louisiana's strategic adaptations for future environments policy framework*. Baton Rouge: Louisiana Department of Administration. Retrieved from https://www.doa.la.gov/ocddru/ndrc/lasafe_Report_Final.pdf

Louisiana Department of Administration (LDOA). (2017b). *The Resettlement of Isle de Jean Charles: Report on data gathering and engagement phase*. Baton Rouge: Louisiana Department of Administration. Retrieved from http://isledejeancharles.la.gov/sites/default/files/public/IDJC-Final-Report-Update.pdf

Louisiana Department of Administration (LDOA). (2018a). *Isle de Jean Charles project timeline*. Baton Rouge: Louisiana Department of Administration. Retrieved from http://isledejeancharles.la.gov

Louisiana Department of Administration (LDOA). (2018b). *A message from Pat Forbes, Executive Director of OCD*. Baton Rouge: Louisiana Department of Administration. Retrieved from http://isledejeancharles.la.gov

Louisiana Department of Administration (LDOA). (2019a). *Resettlement of Isle de Jean Charles-Background and overview*. Baton Rouge: Louisiana Department of Administration. Retrieved from http://isledejeancharles.la.gov/sites/default/files/public/IDJC_Background_Overview_3_20_19

Louisiana Department of Administration (LDOA). (2019b). *Substantial Amendment 5: Introduction of new activities and project narrative clarifications for the utilization of community development block grant funds under the National Disaster Resilience Competition (NDRC) Resettlement of Isle de Jean Charles*. Baton Rouge: Louisiana Department of Administration. Retrieved from https://www.doa.la.gov/OCDDRU/Action%20Plan%20Amendments/NDR/IDJC_Substantial_APA_5_FINAL03272019.pdf

Louisiana Department of Administration (LDOA). (2019c). *Our land and water: A regional approach to adaptation*. Baton Rouge: Louisiana Department of Administration. Retrieved from https://s3.amazonaws.com/lasafe/Final+Adaptation+Strategies/Regional+Adaptation+Strategy.pdf

Louisiana Housing Corporation. (2019). *Louisiana housing needs assessment*. Public Administration Institute E.J. Louisiana State University. Baton Rouge, LA. Retrieved from https://www.lhc.la.gov/hubfs/Document%20Libraries/Housing%20Policy%20and%20Data/RLMA3.pdf

Louisiana Senate Concurrent Resolution 105. (2004). Louisiana State Senate, Baton, Rouge, LA.

MacKinnon, D., & Derickson, K. D. (2013). From resilience to resourcefulness: A critique of resilience policy and activism. *Progress in Human Geography, 37*(2), 253–270. https://doi.org/10.1177/0309132512454775.

Macklin, R. (2009). *Eliminating health disparities: From a grass-roots perspective*. Bureau of Minority Health Access. Retrieved from http://ldh.la.gov/assets/docs/GovCouncil/MinHealth/HealthDisparitiesReport200809.pdf

Maldonado, J. K. (2014). A multiple knowledge approach for adaptation to environmental change: Lessons learned from coastal Louisiana's tribal communities. *Journal of Political Ecology, 21*, 61–82. https://doi.org/10.2458/v21i1.21125.

Maldonado, J. K. (2018). *Seeking justice in an energy sacrifice zone: Standing on vanishing land in coastal Louisiana*. New York: Routledge.

Maldonado, J. K., Shearer, C., Bronen, R., Peterson, K., & Lazrus, H. (2013). The impact of climate change on tribal communities in the US: Displacement, relocation, and human rights. *Climatic Change, 120*(3), 601–614. https://doi.org/10.1007/s10584-013-0746-z.

Maldonado, J. K., Naquin, A. P., Dardar, T., Parfait-Dardar, S., & Bagwell, B. (2015). Above the rising tide: Coastal Louisiana's tribal communities apply local strategies and knowledge to adapt to rapid environmental change. In M. Companion (Ed.), *Disasters' impact on livelihood and cultural survival: Losses, opportunities, and mitigation* (pp. 239–253). Boca Raton: CRC Press.

Manning-Broome, C., Dubinin, J., & Jenkins, P. (2015). *View from the Coast: Local perspectives and policy recommendations on flood-risk reduction in south Louisiana.* [Policy report]. Baton Rouge: Center for Planning Excellence.

Marino, E. (2012). The long history of environmental migration: Assessing vulnerability construction and obstacles to successful relocation in Shishmaref, Alaska. *Global Environmental Change, 22*(2), 374–381. https://doi.org/10.1016/j.gloenvcha.2011.09.016.

Marino, E. (2015). *Fierce climate, sacred ground: An ethnography of climate change in Shishmaref, Alaska.* Fairbanks: University of Alaska Press.

Matz, D., Vogel, E. B., Mattar, S., & Montenegro, H. (2015). Interrupting intergenerational trauma: Children of Holocaust survivors and the Third Reich. *Journal of Phenomenological Psychology, 46*(2), 185–205. https://doi.org/10.1163/15691624-12341295.

McCintire, W. G. (1954). *Prehistoric settlements of coastal Louisiana* (dissertation). Baton Rouge: Louisiana State University Historical Dissertations and Theses. (8099).

Miller, M. (2004). *Forgotten tribes: Unrecognized Indians and the federal acknowledgment process*. Lincoln: University of Nebraska Press.

Nagata, D. K., Kim, J. H. H., & Nguyen, T. U. (2015). Processing cultural trauma: Intergenerational effects of the Japanese American Incarceration. *Journal of Social Issues, 71*(2), 56–370. https://doi.org/10.1111/josi.12115.

Nelson, M., Ehrenfeucht, R., & Laska, S. (2007). Planning, plans, and people: Professional expertise, local knowledge, and governmental action in post-hurricane Katrina New Orleans. *Cityscape, 9*(3), 23–52. https://www.jstor.org/stable/20868630.

O'Keefe, P., Westgate, K., & Wisner, B. (1976). Taking the naturalness out of natural disasters. *Nature, 260*, 566–567. https://doi.org/10.1038/260566a0.

Oliver-Smith, A. (2009). *Development and dispossession: The crisis of forced displacement and resettlement.* Santa Fe: School for Advanced Research Press.

Oliver-Smith, A., & Hoffman, S. (1999). *The angry earth: Disaster in anthropological perspective*. New York: Routledge.

Perry, R. W., & Lindell, M. K. (1997). Principles for managing community relocation as a hazard mitigation measure. *Journal of Contingencies and Crisis Management, 5*(1), 49–59. https://doi.org/10.1111/1468-5973.00036.

Peterson, K., & Maldonado, J. K. (2016). When adaptation is not enough: Between the "now and then" of community-led resettlement. In S. A. Crate & M. Nuttall (Eds.), *Anthropology and climate change: From action to transformation* (2nd ed.). New York: Routledge.

Randolph, N. (2018). License to extract: How Louisiana's Master Plan for a Sustainable Coast is sinking it. *Lateral- Journal of the Cultural Studies Association, 7*(2). https://doi.org/10.25158/L7.2.8.

Rigaud, K. K., Jones, B., Bergmann, J., Clement, V., Ober, K., Schewe, J., et al. (2018). *Groundswell: Preparing for internal climate migration*. Washington, DC: World Bank.

Rivard, C. (2015). Archival recognition: The Pointe-au-Chien's and Isle de Jean Charles Band of the Biloxi-Chitimacha Confederation of Muskogees' fight for federal recognition. *Settler Colonial Studies, 5*(2), 117–127. https://doi.org/10.1080/2201473X.2014.957257.

Sanders, M. (2018, August 23). Don't label them climate change refugees, says a Louisiana planner, they're pioneers. *Common Edge*. Retrieved from http://commonedge.org/dont-labelthem-climate-change-refugees-says-a-louisiana-planner-theyre-pioneers/

Sassaman, K. E., & Anderson, D. (2004). Late Holocene Period, 3750 to 650 B.C. In R. D. Fogelson (Ed.), *The handbook of North American Indians. Vol. 14, Southeast* (pp. 101–114). Washington, DC: Smithsonian Institution Press.

Simms, J. R. (2016). "Why would I live anyplace else?": Resilience, sense of place, and possibilities of migration in coastal Louisiana. *Journal of Coastal Research, 33*(2), 408–420. https://doi.org/10.2112/jcoastres-d-15-00193.1.

Simon, D. (2008, 22 September). Tribal chief on Isle de Jean Charles says it's time to leave. *Nola.com*. Retrieved from https://www.nola.com/news/2008/09/tribal_chief_on_isle_de_jean_c.html

Sneath, S. (2018, July 26). Louisiana tribes say federal recognition will help to face threat of climate change. *The Times-Picayune*. Retrieved from https://www.nola.com/expo/news/err2018/07/449c2f22d39490/louisiana-tribes-say-federalr.html

Solet, K. (2006). *Thirty years of change: How subdivisions on stilts have altered a Southeast Louisiana Parish's coast, landscape and people* (dissertation). New Orleans: University of New Orleans Theses and Dissertations. (1029).

Spanne, A. (2016, March 23). The lucky ones: Native American tribe receives $48m to flee climate change. *The Guardian*. Retrieved from https://www.theguardian.com/environment/2016/mar/23/native-american-tribes-first-nations-climate-change-environment-indican-removal-act

Starna, W. A. (1992). "Public ethnohistory" and Native-American communities: History or administrative genocide? *Radical History Review, 1992*(53), 126–139. https://doi.org/10.1215/01636545-1992-53-126.

Swyngedouw, E. (2011). Depoliticized environments: the end of nature, climate change and the post-political condition. *Royal Institute of Philosophy Supplements, 69*, 253–274. https://doi.org/10.1017/S1358246111000300.

The Peninsula Principles on Climate Displacement within States. *Displacement solutions*. (2013). Geneva. Retrieved from http://displacementsolutions.org/wp-content/uploads/final-peninsulaPrinciples final.pdf

Tierney, K. (2014). *The social roots of risk: producing disasters, promoting resilience*. Stanford: Stanford Business Books, an imprint of Stanford University Press.

Titz, A., Cannon, T., & Krüger, F. (2018). Uncovering "community": Challenging an elusive concept in development and disaster related work. *Societies, 8*, 71. https://doi.org/10.3390/soc8030071.

Tuck, E., & Yang, K. W. (2012). Decolonization is not a metaphor. *Decolonization: Indigeneity, Education & Society, 1*(1), 1–40. Retrieved from https://www.materialculture.nl/sites/default/files/201902/Decolonization_Is_Not_a_Metaphor.pdf.

Turner, R. E., & McClenachan, G. (2018). Reversing wetland death from 35,000 cuts: Opportunities to restore Louisiana's dredged canals. *PLoS One, 13*(12). https://doi.org/10.1371/journal.pone.0207717.

U.S. Army Corps of Engineers (USACE). (2013). *Revised programmatic environmental impact statement for the Morganza to the Gulf of Mexico project*. Retrieved from http://www.mvn.usace.army.mil/Portals/56/docs/PD/Projects/MTG/FinalRevisedProgrammaticISMtoG.pdf

U.S. Department of Housing and Urban Development (HUD). (2015). *National Disaster Resilience Competition Phase II Fact Sheet*. Washington, DC: U.S. Department of Housing and Urban Development. Retrieved from https://www.hud.gov/sites/documents/NDRCFACTSHEETFINAL.PDF

U.S. Department of Labor, Bureau of Labor Statistics. (2018). Louisiana economy at A glance. Washington, DC: U.S. Department of Labor. Retrieved from https://www.bls.gov/eag/eag.la.htm.

U.S. Geological Survey, Earth Resources Observation and Science Center. (2017). Isle de Jean.

UN General Assembly (2007). United Nations Declaration on the Rights of Indigenous Peoples. *Resolution adopted by the General Assembly on 2 October 2007*. A/RES/61/295. Retrieved from http://www.refworld.org/docid/471355a82.html

Walls, M. L., & Whitbeck, L. B. (2012). The intergenerational effects of relocation policies on indigenous families. *Journal of Family Issues, 33*(9), 1272–1293. https://doi.org/10.1177/0192513X12447178.

Weber, L., & Peak, L. (2012). *Displaced life in the Katrina diaspora*. Austin: University of Texas Press.

Wendland, T. (2018, January 5). Lack of funds keeps Louisiana from buying out coastal residents. *NPR*. Retrieved from https://www.npr.org/2018/01/05/575876626/lack-of-fundskeeps-louisiana-from-buying-out-coastal-residents

Whyte, K. P. (2016). Is it colonial déjà vu?: Indigenous peoples and climate injustice. In J. Adamson & M. Davis (Eds.), *Humanities for the environment: Integrating knowledge, forging new constellations of practice* (pp. 88–105). New York: Routledge.

Wildcat, D. R. (2009). *Red alert!: Saving the planet with Indigenous knowledge*. Golden: Fulcrum.

Williams, B. (2002). The concept of community. *Reviews in Anthropology, 31*(4), 339–350. https://doi.org/10.1080/00988150214749.

Wilmsen, B., & Webber, M. (2015). What can we learn from the practice of development-forced displacement and resettlement for organized resettlements in response to climate change? *Geoforum, 58*, 76–85. https://doi.org/10.1016/j.geoforum.2014.10.016.

Wilson, S. G., & Fischetti, T. R. (2010). *Coastline population trends in the United States: 1960 to 2008, U.S.* U.S. Census Bureau Publication No. P25–1139. Washington, DC: U.S. Census Bureau. Retrieved from https://www.census.gov/prod/2010pubs/p25-1139.pdf

Wisner, B., Blaikie, P., Cannon, T., & Davis, I. (2004). *At risk: Natural hazards, people's vulnerability, and disasters* (2nd ed.). London: Routledge.

Wolfe, P. (2006). Settler colonialism and the elimination of the native. *Journal of Genocide Research, 8*(4), 387–409. https://doi.org/10.1080/14623520601056240.

# Chapter 7
# Sojourners in a New Land: Hope and Adaptive Traditions

**Kristina J. Peterson**

> ***Somos el barco***, *somos el mar,*
> *Yo navego en ti, tu navegas en mi*
> *We are the boat, we are the sea, I sail in you, you sail in me*
> *The boat we are sailing in was built by many hands*
> *And the sea we are sailing on, it touches every land- chorus*
> *So with our hopes we set the sails and face the winds once more*
> *And*
> *with our hearts we chart the waters never sailed before - chorus*
> *Lorre Wyatt (1983)*

## 7.1 Introduction

The Louisiana State flag has a mother Pelican caring for and protecting the next generation. Like the pelican on the flag, the task in Louisiana is to provide care for and protection of generations yet to come. The task facing us now, however, is rejuvenating rural communities that are suffering from inadequate physical and social structures. This rejuvenation will benefit and heal the current communities as they prepare to become areas of hospitality for those who are and will become sojourners from Louisiana coastal and riverine areas. At the core of hospitality of receiving the stranger are the acts of welcoming, supplying food, storytelling, and relationship and community building (Buber 1970). Culture, which includes "art and literature, lifestyles, ways of living together, value systems, traditions, and beliefs," is an overlooked element in rebuilding cities and rural areas, ravaged by disasters, war, and other forms of distress, according to a joint World Bank and UNESCO report (UNESCO 2018). This chapter explores the role of culture in the resettlement inland of coastal Louisianans.

---

K. J. Peterson (✉)
Lowlander Center, Gray, LA, USA
e-mail: Kristina@lowlandercenter.org



S. Laska (ed.), *Louisiana's Response to Extreme Weather*,
Extreme Weather and Society, https://doi.org/10.1007/978-3-030-27205-0_7

Humans are now in a new and critically different relationship with the world (Morton 2013). Called the "new anthropocene era," it is a relationship where human activity impacts the world at a planetary level seen most clearly in both slow and sudden weather events contributing to drastic shifts in population (Hawken 2017). In this new era, a radical extension of hospitality will be a necessity (Newman 2007). This new and radical hospitality will involve hosting and showing appreciation for all life, not just humans (Ogletree 2003; Morton 2017).

The 2016 floods confirmed that most of Louisiana is at risk to riverine, extreme rain events, and coastal flooding. As the population of the state works to adapt to the changing geography of and increased risks of such extreme weather, there will be shifts in living patterns and reworked understandings of dwelling place or home place (Brown 2011). The relocation of Louisiana coastal residents is already taking place (Dalbom et al. 2014). The research that has been done helps to understand internal post-disaster displacement but has focused mostly on urban areas. Little attention has been given to internally displaced rural communities and to the rural receiving communities (Hauer 2017).

Historically, individuals, households, neighborhoods, and communities do not "just pick up and move." Relocation is usually spurred by some critical need or set of circumstances that ends in the difficult choice of leaving a home place to become landless and placeless (Fullilove 2004). To experience the extremes of diaspora, or sojourning, it is most likely that all other resources or options for staying have been exhausted before leaving. The out-migration is likely to happen in stages as serial events prompt such personally critical decisions at each juncture. The lack of justice or the transgressions against justice are often connected with the impetus for going in many migration situations. Leaving becomes a hope that the situation might change and provide a just future in the new locale. It is imperative that the sojourner not be considered an object of charity, but a full and just participant in the new community. Notably for people in a deeply religious state (Cooperman 2015), the Bible addresses very similar circumstances. The principle from Judaic law that the sojourner must not be deprived of justice and be paid fair wages is apropos for Louisiana (Smith-Christopher 2002). Thus, for the receiving community to be ready to extend hospitality, it must first be healed or made whole by having obtained justice and economic security itself (Rom-Shiloni 2013; Escobar 2008). This chapter explores the possibilities for "just" sojourning and radical hospitality. Addressing immediate needs while understanding long-term implications of current choices and decisions involves creative living, hope, and a massive scale of collaboration and cooperation, brought together under a comprehensive state master plan of rejuvenation. A visionary plan will call upon the work of the various communities in the state to create real, appropriately aspirational rejuvenating designs. After all, we, through our history, were all once sojourners (Brasseaux and Davis 2017).

The first part of this chapter will highlight the challenges, or so-called wicked problems (Rittel and Webber 1973), facing people who live in receiving regions. Communities receiving the newcomers face not only the problems of extending services to the new arrivals but doing so in the context of chronic and layered problems that already make their home places precarious, which the arrival of more people can exacerbate. The challenge of the inward migration can also present opportunities for the in-migrants bringing skills and abilities to these new locations (Woods 2018). Planners will be challenged by the magnitude of potentially a million people displaced from the coast over the next 50 or so years (Hauer 2017). It is predicted that a majority of coastal Louisiana residents will migrate within the bounds of the state (Hauer 2017). Yet, the carrying capacity of some of the inland rural regions of the state is already stretched to collapse, racial tensions still exist, and fear of diminished quality of life likely make hospitality to receive more people difficult (Brown 2011; Derrida 2000; Kearney and Semonovitch 2011; Ogletree 2003).

The second part of the chapter suggests that both faith and culture can be utilized as assets for a comprehensive master plan that stresses innovation, collaboration, and inclusion of all regions of the state. Rejuvenation of neglected rural areas can take place through judicious work of boundary organizations[1] and cross-boundary networks utilizing social and cultural assets (Hufford 1994). The rejuvenation of cultural arts and social assets, food, faith, music, stories, and family networks can retain and share historical knowledge to those who are leaving flooded areas, the sojourners, and those who are receiving the sojourners. Collaboration and relationship building can contribute to problem-solving and can be addressed through cultural networks (UNESCO 2018). The benefits of utilizing cultural assets within the restoration and rejuvenation of a people and place during and following social and physical disruptive events are increasingly being used by governments and institutions (CIRD 2018).

This discussion is focused on a rural to rural relationship of "historied" communities and "communities-of-place" who understand themselves as part of their surroundings and as responsible for the shared commons of their home, a sense of shared community that Tonnies famously called *gemeinschaft* (Tonnies 2001). Rural communities, in many cases, are self-organizing or managed and are built around shared lifeways and an ethical code of caring for and sharing with each other and their shared dwelling place (Low and Lawrence-Zuniga 2003). By exploring the ways faith, tradition, food, music, and storytelling intersect within rural areas of the state that have different religious and historical traditions, models of hospitality can be discovered and constructed. Through cross-boundary work and trust networks, the task can be the building of new community through co-learning (Maida 2007).

## 7.2 "Wicked Problems" and Racial Divides

The term "wicked problem" was coined by Horst Rittel to describe the challenges and complexities of problematization, the search for root causes and the necessary planning required for social issues that are out of the ordinary (Freire 1996; Fullilove 2004). Two defining characteristics of wicked problems that seem appropriate to

[1]A boundary organization is one that helps collaboration, information, and facilitation between diverse entities such as research disciplines, public policy, and community participants.

Louisiana's current situation are the endless varieties of actions and approaches that could be taken, and the reality that any actions taken are likely to have irreversible consequences. If these actions generate more, new problems instead of addressing the existing ones, "wicked problems" will be exacerbated rather than mitigated. Because every action has an impact, best knowledge and principles have to be applied with each intervention, preferably in anticipation of the change (Rittel and Webber 1973).

The process of locating areas or regions in the state conducive for population growth is complex. Rural folks, for the most part, want to be with rural folks with similar shared social values.

> When my people moved from Plaquemines to Amite all they wanted was a place to call home and a welcoming church and not to be afraid.[2]

The racial divides that have kept neighbors from knowing neighbors is still present in the state. Pushback from communities to resist "the other" happens all too frequently. In conversations with people in several of these locations, there is a sense of fear that permeates from stereotypes seen on the news and in entertainment. The *Southern Poverty Law Center* tracks changes in hate group behavior and has reported a rise in hate-related activity in Louisiana toward minority groups (SPLC 2018).[3] Racial tensions can increase with additional people and diminished resources. Without a counter-narrative to highlight hospitality and creative interventions, there will be resistance to the in-migration. Following the 2016 flooding in New Roads, Louisiana, state recovery officials identified a predominately Black community that had flooded severely. They worked with the residents who wanted to move together to a less flood-prone area in a different section of town. The initiative experienced resistance from white residents who opposed their moving near them (Jones 2018). More recently, the community of Schriever was critical of the state's planning of the HUD-NDRC-IdJC Tribal[4] grant, not wanting an open door policy of who could live in the Schriever community (DeSantis 2019). Another example occurring in the same area is the construction of slab-on-grade subdivisions in the Gray, Louisiana, area for residents moving from closer to the coast, with opponents claiming that the development is turning the quiet rural community into suburban "sprawl." Their argument is that the development is contributing to changing the density and identity of the receiving community and increasing the original residents' flooding.

A male resident whose family dates back to the original Acadian settlers stated:

> My house and yard now flood with all the houses they stuck to the ground; they even took down the trees the kids played in. Those trees helped stop the water. It ain't right to do that to us. We can't even sit on the front porch…[5]

---

[2] Personal communication with Rev. Tyronne Edwards, March 28, 2019. Deep appreciation is extended to Rev. Edwards and the other friends who offered stories, advice, editing, and critique to hone this chapter.

[3] https://www.splcenter.org/hate-map

[4] Housing and Urban Development—National Disaster Resilience Competition Award for Isle de Jean Charles Tribe.

[5] Personal communication with a Bayou Blue resident, April 22, 2018.

Whether it is racial or economic fears or flooding from increased development, the outcomes are similar: Those who are at risk and need to relocate have fewer options to make informed choices of safe and healthy living places for their children because of the real and feared risks of the original residents.

The concerns expressed by rural communities about "newcomers" contain multiple issues. The landscape, the sounds, the way of life have forever changed and restructured the social networks of the historied families and the way their land was used for essential lifeways (Low and Lawrence-Zuniga 2003). One oft-voiced concern is the lack of protection of farms and forestlands that put rural lifeways, and all whom they serve with food, at risk. Common concerns heard from the Bayou Blue area are:

1. More flooding with the increased development (as quoted above)
2. More noise, less silence, quiet, and peace
3. More issues of road safety, not able to either walk or ride bikes
4. "Can't let the chickens run free" as farm life diminishes
5. Higher property taxes
6. Fear of crime and drugs as more people move into the area[6]

## 7.3 Development Impacts on Rural Lifeway

People moving in incremental steps trying to stay close to their former residence and extended family are attracted to the new development on former rural tracts of land. The demand for new housing is placing a demand on all available properties. Speculation of land by investors and realtors of larger parcels of rural/farmland property is occasionally causing friction between neighbors and in the families of commonly held land. Family members wanting to retain the land within their families can be in conflict with other family members who would like to sell for financial reasons. Because of increased land values, family members who would like to retain the land for family often do not have the financial means to "buy out" their extended family members. Conflict arises when the property sold becomes a convenience store, gas station, or development, contributing to a diminished rural lifestyle of the older homestead family next door.[7]

Rural areas in Louisiana are being gentrified by new housing developments that have changed the landscape and challenged the lifeways of rural families. "No place to park when gathering for crawfish boils at Mama's home" is not just an inconvenience. It is a significant change of family dynamics and community celebrations. Feeling left out of the conversation in preserving or negotiating their way of life has created harsh feelings not only toward the Parishes and the State but

[6] Self Development of People, Presbyterian Church (PCUSA), Bayou Blue correspondence/meeting, April 23, 2017.

[7] Self Development of People, Prebyterian Church (PCUSA), Bayou Blue correspondence/meeting, April 23, 2017.

also toward the newcomer. For example, when a local farmer went to the Parish Council meeting to raise concerns of the amount of building and the cutting down of cypress trees causing additional flooding, he was told that the developer knew what he was doing and it was okay. There is now chronic flooding in the location of concern.[8]

Some families who are being hemmed in or overwhelmed by the new developments and wanting to maintain their rural lifeways have encouraged their children and grandchildren to move to rural areas further inland and obtain land before it becomes too costly or is bought for new development. This process of moving from one rural spot to the next is not new. Over generations as the coast has slowly migrated inland, there has been a reconfiguration of fishing villages and small enclaves that meander up the bayou onto the next small village or town (Hemmerling 2017). Like many rural lifeway areas, in order to stay "rural," a bit of land is needed to grow food for an extended family and subdivide or share space for children as they grow up and live on the family enclave. In coastal Louisiana a tradition dating back to French colonial times, long narrow pieces of land measured by "arpents," were given to families so they could access the river or bayou as well as the woods and agricultural land.. In order to find space large enough to continue to engage in the giving of resources, food, and mutual aid, families now have to leapfrog over the towns and head further north and inland. As people move north, there are new sets of problems, many unexpected.

Louisiana families from coastal areas who headed inland following 16 years of disaster events from the time of Katrina believed they were making sound choices to avoid flooding. The people already living in the inland areas and the newcomers were not told that flood insurance was needed; thus, they assumed the area was safe. They had land and resources to share when their extended family needed a place to stay, grow crops, and raise animals. Some who purchased "safe" land found themselves flooded by the 2016 events. Believing they were moving to protect family and their future assets, they again were severely impacted by an "unexpected flooding event." Now separated from their original home base and again having their subsistence way of life taken by flooding, they are finding it more difficult to recover. The move inland brings with it other possible risks new residents will experience. One sojourner now returns to the bayou to sustain his inland family by helping others with shrimp, crab, and oyster harvesting. Others who have moved inland frequently return to maintain family connections, to participate in the life of the community, and to share food resources, such as taking sweet potatoes and kale to the bayou and taking shrimp back to the inland families.[9]

Many inland rural regions, as potential receiving regions, are coping with limited resources and bearing the burden of socially and economically constructed vulnerabilities with little attention to them, or public awareness from the more populated areas of the state. Information regarding the challenges to various parts of

[8] First Peoples' Conservation Council meeting notes March 162,018.

[9] Community conversations with Placquimines Parish residents December 2016.

the state are not often part of public discourse and thus the needs are overlooked or not understood. There has been however, over the past several years, a growing public awareness as the online *Lens* and the *New Orleans Advocate* newspaper have highlighted failing infrastructures in the central and northern rural areas of the state such as the investigative reports on the condition of water systems (Ballard 2017; Wright 2019). The non-profit *Together Louisiana*[10] has helped share and advocate as a voice with rural areas (Jones 2018).

The failing inland infrastructures have not been mitigated nor the risks of flooding conveyed adequately. The current flooding patterns caused by climate-change induced extreme weather may not follow historical patterns of events thus the local knowledge to predict and respond may be dimish. A Tallulah resident shared that her community was well aware of the infrastructure issues but lacked a voice or the resources to address them, so they "make do together" with what is on hand.[11] Awareness of and attention to the risk conditions that exist in many rural areas, both on the coast and inland, have to be justly addressed. The carrying capacity of many parishes is at its limit, contending with bad water and overall failed health standards and flooding in their own right. These "layered" social, environmental, historical traumas and physical infrastructure deficiencies can be understood as a "wicked problem" (Laska et al. 2015). Environmental and social justice principles need to be applied in addressing the substandard conditions.

The displacement and transition of population, business, industry, and social functions from the coast is sometimes a hope to get to higher, drier, and safer places. Lasley has explored the organic self-transition of people from Saint Bernard to St. Tammany following Katrina utilizing social networks (Lasley 2012). As some parts of the community resettle to a particular locale, others in the social network including restaurants and businesses are drawn to the same area to continue old neighborhood patterns. As these new self-directed enclaves develop, such as in St. Tammany, they can either be stigmatized as outsiders or be socially merged into cultural fabric of the new location.

When one part of the population moves "up the bayou," it leaves a hole in the fabric of the community it left behind. Fewer people mean fewer relationship resources and declining population numbers threaten the doors of schools, churches, community buildings, and retail facilities (Lowlander Center 2017). Small businesses are stretched to doing "self-subsidizing" of their own businesses in order to serve their extended network of customers, family, and friends.[12] Often a local void in services, goods, and community connections happens when a rural business closes placing extra hardships on the population remaining.

---

[10] *Together Louisiana*, www.togetherla.org

[11] Personal correspondence with Tallulah resident and church leader, March 2019.

[12] Group discussion with Appalachian College Students, bayou residents, and business owners, March 14, 2019.

## 7.4 Carrying Capacity of Coastal and Inland Communities

In many rural areas of the state, be they inland or coastal, communities are experiencing sporadic or continual disruptions to the systems that impact their well-being.[13] Coastal communities find themselves at the end of the road or bayou for services as the coast erodes their familial home and life-place. Inland communities are experiencing dysfunctional water systems and lack of social services with limited capacity to address the diminished services impacting their population (Ballard 2017; Wright 2019). For both the inland and coastal communities there are similar issues with dissimilar settings, inland and coastal, communities are reaching or are past their carrying capacity, and thus are approaching their tipping points (The Sustainable Scale Project 2003). The carrying capacity refers to the ability to maintain social, environmental, and physical infrastructure for a population.

Strangely, because of the magnitude of problems facing Louisiana, the state is at the forefront of problematizing the many risks and vulnerability issues, and thus has a critical leadership role to play in being visionary. The choice is dramatic and not overstated—the people, institutions, and communities of Louisiana can provide critical aspirational and visionary leadership by example or become a whole sacrificial state even beyond the coastal sacrificial zone (Freire 1996; Bowen 2009; Maldonado 2018).

- Resources for holistic problem-solving and planning in rural areas are not as prevalent as they are for small towns and urban regions, thus increasing rural areas' vulnerability (Cutter et al. 2016). As illustrated in Chap. 11 of this book by Alessandra Jerolleman, resources are not plentiful and capacity is lacking to apply and execute grant programs in rural areas. The lack of capacity is sometimes as basic as having no or limited access to Internet. Specialized rural organizations have taken the lead to assure rural areas are not left behind but their efforts are not always supported with equal political clout or voice to secure adequate proactive funding. Philanthropy is reluctant to fund rural programs especially if they do not see a strong outcome to their investments. One small family foundation shared that there is a stereotypical image of rural folks that tends to keep funders and resources away (Swierzewski 2007). Often these overlooked places will need long-term resources that support capacity building, staff support, and operational program funding. Lack of initial capacity or size should not be a factor in the allocation of funds.

With the state's loss of population in 2018 (Hauer 2017) and the anticipated future loss of population due to climate and economic patterns, needed revenue sources will be scarce. Rural and small communities already suffering from the downturn in the oil fields and agriculture industry will have to make cuts in services affecting the well-being of their immediate jurisdictions and the surrounding rural enclaves. The loss of revenue accompanying the loss of population will be intensified

[13] The interactions between human choice, physical infrastructure, and environmental stressors are key determinants for tipping points leading to systems collapse.

with the next big hazard event, stretching recovery dollars and burdening the state further. Communities of color will be the most impacted in rural areas where political agency is ignored (Doherty 2004).

Additionally, changes in environmental conditions will impact all varieties of wildlife and their movements and habitats. Human and non-human habitat adaption to weather and climate stressors will become increasingly challenging to traditional agricultural practices and ecological preservation. The "community" that is in transition includes all the wildlife and plants. Planning for transition is for the entire population, human and non-human. Hunters and fishers, a large economic recreational industry, are paying close attention to these changes and are often advocates in preservation practices. Decision-making processes become networks or a web of life that is complex (Morton 2017). In short, how the lifeworld and natural resources are hosted and carefully managed in the "retreat" from the coast will have a direct impact on all species native to the entire area.

### *7.4.1 Health as Indicator*

The *Robert Wood Johnson Foundation* health report gives Louisiana a low score on overall wellness and well-being. The map (Fig. 7.1) highlights the severity of health conditions in Louisiana (University of Wisconsin Population Health Institute 2018). The conditions for determining the health score include individual choices in health behavior such as exercise, tobacco, diet, drug use, and social and economic factors such as access to limited quality care, education, employment, social support, community safety, and income and the physical environment attributes that include air and water quality, housing, and transit (University of Wisconsin Population Health Institute 2018). Inadequate housing, weatherization, and inability to maintain and repair housing contribute to vulnerability (Tierney 2014). Mitigation for deteriorating housing stock is challenged by lack of rural funding (Horney et al. 2017). Physical health indicators reveal that the state ranks toward the bottom of all states in many functions critical for a vital population (Horney et al. 2017).

Indeed, Concordia Parish, north of Baton Rouge on the border of Mississippi, has been singled out as having the worst health conditions in the country (Belluz 2016). The Parish is lacking in many of the essentials including access to clean water, sanitation, affordable healthy foods, and safe housing (Doherty 2004). Other inland communities south along an 85-mile stretch on the Mississippi River between Baton Rouge and New Orleans, often referred to as "Cancer Alley" (Bullard 1994; Blackwell et al. 2017), are surrounded by 25 percent of the nation's petrochemical production (Jackson and Chapple 2018). The map (Fig. 7.1) shows toxic release and superfund sites that have a bearing on health factors of the surrounding communities (Bullard 1994). Communities of color and those of lesser means are the ones who suffer most with the contamination in many places causing abnormally high cancer rates, asthma, miscarriages, and other health-related issues (Bullard 1994). Jonathan

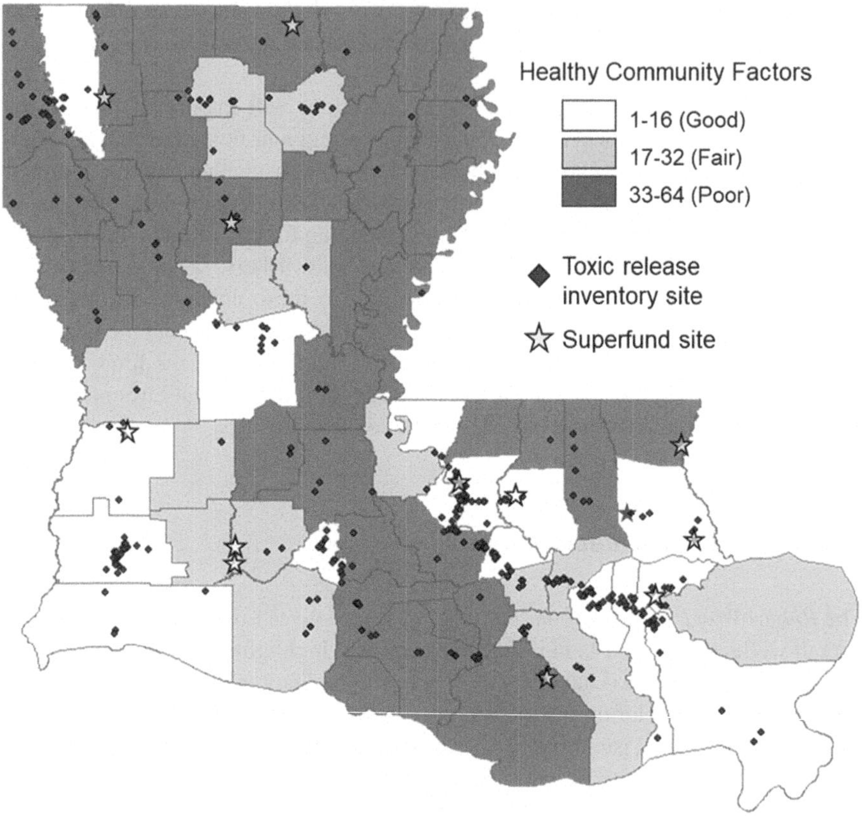


**Fig. 7.1** Map showing health factors rankings by parish and locations of toxic release inventory and superfund sites in Louisiana. The ranks are based on four types of measures: health behaviors, clinical care, and social and economic and physical environment factors. (Data retrieved from *County Health Rankings & Roadmaps* and the Environmental Protection Agency (EPA))

West, a UNO student, described being weary from the grief of having seen cancer in every one of his friends' families in Norco, LA.[14]

## *7.4.2 Water Quality*

The conditions that make for either a vulnerable or a sustainable population include the capacity of different communities to address the long-term and chronic everyday disasters they face (Bolin and Bolton 1986). Families need safe water to drink, cook, and bathe with. Purchasing water is not an option for people living in areas with high unemployment and underemployment. People want to be generous but it

[14] Personal communications at University of New Orleans, May 18, 2006.

is difficult with limited resources. Ms. Ancar, a resident of Grand Bayou, Louisiana, often said to visitors "Come join me for some coffee and if I don't have coffee we will have water" (personal communication April 9, 2003). For inland communities desiring to give hospitality, some residents have no water to share.

The quality of water is threatening the health of the most vulnerable, elderly and children, sick. The physical infrastructures vital for water delivery in the central part of the state are close to collapse (Ballard 2017). In 2013, the American Society of Civil Engineers gave Louisiana an infrastructure grade of "C," after finding that 62 percent of the state's roads were of mediocre to poor condition, 3815 bridges were structurally deficient or functionally obsolete, 33 dams were considered "high hazard," and drinking water and wastewater systems were in need of $10.9 billion in improvements during the next 20 years. Many of these drinking water and wastewater systems are located in rural areas (Schliefstein 2013). The combined conditions of inadequate storm water drainage, potable water, and sewage, accompanying nuisance flooding, will soon make coastal communities uninhabitable (Ruppert 2019).

> "I turn on the tap and the water is sludge," a Tallulah resident shared with the author. " I had to move my mama since that is all she had to drink."[15]

Safe clean water is clearly essential to basic health standards (Steingraber 1998), yet water required for daily use may be one of the state's biggest challenges (Adeniyi et al. 2016). The now-closed England Air Force Base in Alexandria, Louisiana, is an EPA superfund site that has the highest detected deadly PFAS[16] levels in groundwater of any PFAS contamination site in the nation. Water from the site has 10,970,000 parts per trillion (ppt), l million times more than the risk level determined by Harvard public health and that the *Agency for Toxic Substance Disease Registry* (ATSDR) recommends as safe (Reed 2018). The *Environmental Working Group* (EWG) report on drinking water by zip codes showed that potable water in all of the Parishes in Louisiana contained either carcinogenic chemicals or metals. Despite the EWG report of carcinogenic and other contaminants in potable water, the water districts are still in compliance with legally mandated federal standards (EWG 2015). The map (Fig. 7.2) delineates cities and development, 2018 coastal predicted flooding moderate risk scenario with no action, rural farmland under cultivation, open territories either forested, or held by government entities and in gray, the water infrastructures in immediate need of repair.

> My baby said his eyes were burning when he brushed his teeth. I told him to turn off the water. Your eyes shouldn't burn from the tap water when brushing your teeth!.[17]

This and the previous quote come from communities in distant parts of the state from locations that will have to relocate or receive families. Water quality in the state for many people is a present concern for their health and an economic hardship in obtaining alternative safe water for drinking, food preparation, and personal hygiene.

---

[15] Personal communication with Tallulah resident, March 16, 2019.

[16] Perfluoroalkyl substances www.health.re.gov/water/about/pfas

[17] Personal communication with Chief Shirell Dardar at Knights of Columbus Hall, March 14, 2019.

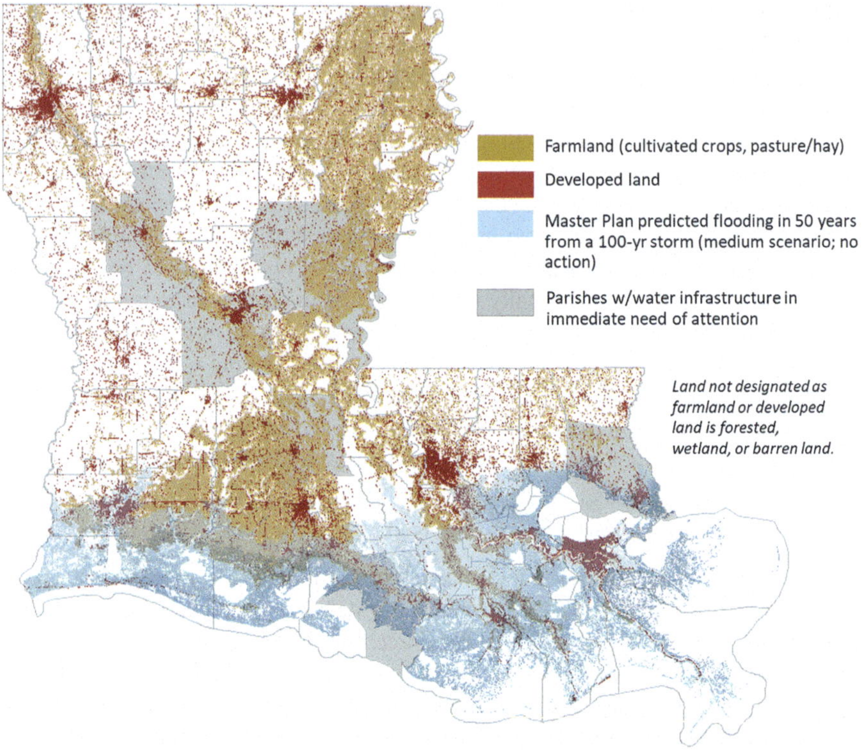


**Fig. 7.2** Map showing where select land cover categories in Louisiana overlap with dysfunctional water infrastructure and which areas would be at risk of flooding in the near future. (Data retrieved from *National Land Cover Database (NLCD) 2011 edition; Rural Water Infrastructure Committee (2018), & Coastal Protection & Restoration Authority CIMS application*)

### *7.4.3 Resilience and Tipping Points*

> How do you know when you reached a tipping point, only when you passed it (Ann Claire Phillips, Rear Admiral, USN (Ret.) October 2017).

Resilience is the ability of a system to absorb shocks and adapt. In ecological terms, when the stressors have reached the tipping point, the system enters into a new regime (Walker and Salt 2012). Some rural areas within the state have surpassed their ability to absorb additional shocks. It is similar for rural communities that have endured chronic and sudden social impacts where a tipping point happens and the rural community either dissolves or has to reinvent itself through creative and radical actions (Nored 1999; Akuno et al. 2017). Examples of such radical holistic remakes come from Kali Akuno's work in Jackson, MS, and the work of Rev. Nored in a rural enclave outside of Birmingham, AL. In order for high-risk areas of the state to avoid becoming a "sacrificial zone," radical hospitable plans of rejuvenation must become priority activities (Maldonado 2018).

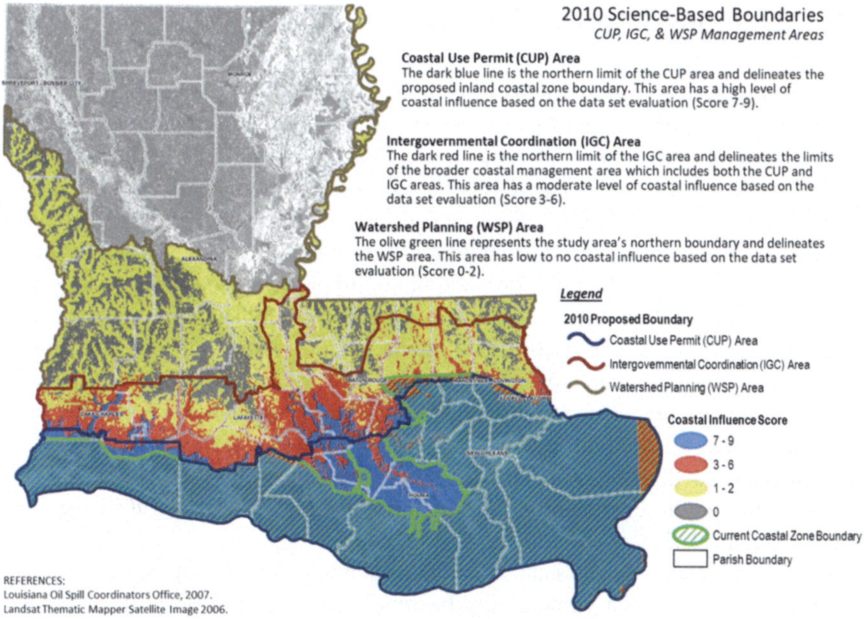


**Fig. 7.3** Map showing updated definitions of Louisiana's coastal zone. (Image source: *Louisiana Department of Natural Resources, Office of Coastal Management*)

Moving from rural place to rural place is difficult and made more complicated by the unknowns of risks and climate. This map (Fig. 7.3) shows the delineation of water management areas. Each area will have different regulations in accordance with federal agencies and with them the availability of agency resources. The rate of acceleration, which the coast is moving inland, has been increasing faster than predicted and will probably continue to do so in the near future (Goodell 2017). As seen in the maps portrayed in Figs. 7.1 and 7.2, most of Louisiana is rural. The serenity and beauty of the rural areas have provided forest products, potatoes, peanuts, cotton, poultry and poultry products, soybeans, strawberries, and sugarcane. It would seem that finding high ground and open area would be easy enough when considering locations for sojourning populations. But it is not so simple to find a good safe place, as seen in the first two maps that indicate land loss, health concerns, and crumbling infrastructures (Curth 2018). The first two maps show existing stressors to health and safety, while the third map shows the new coastal management changes that will require different permitting within the zones. With the inclusion of more inland areas within the coastal permitting area, it too will add time and burden for the extra expense of permitting. Given the gravity of challenges to all regions of the state, it begs the question of how and where to start.

## 7.5 Desperate Times Call for Creative Measures

The development of comprehensive inclusive statewide frameworks and working models is required to address the problems. Contemporary planner and political philosopher Iris Young (1990) argues that the paradigm of "development justice" is through people's capacities for free and autonomous decision-making in inclusive democratic engagement and processes (Harvey 1996). Ideally, the inclusion of the entire community with all of its social differences together must have a voice in the problematization and planning process even if the community members/families do not move together as a community. The community, especially those with local and traditional knowledge, can give voice and representation to the non-human inhabitants that will also be migrating. The trajectory Young is suggesting is in alignment with the work of planners over the past 100 years such as Benjamin Marsh, Arturo Escobar, and Dennis Goulet who have all posited that planning has to be for the benefit of the people and the health of the whole community, human and non-human (Marsh 1909; Escobar 2008; Goulet 1995).

To democratize the process and the work of planning, the ones most impacted by the decision-making, the vulnerability-bearers must be included (Krajeski 2018). By involving those who are most likely to bear the consequences of extreme events, the visioning of the future can be inclusive and the quality of the planning will benefit from a variety of knowledge, backgrounds, and experiences. The inclusion of vulnerability-bearers in problematization and problem-solving often leads to creative alternatives and healthier outcomes as well as a broader base of support for the project. The "down the road" game plan for possible relocation teeters between the imagined scenarios that will be possible and the triggered event. Each formal or informal planning process, given the limits of people's time and the resources of a community, must solve the most critical issues impacting people as defined by the community of vulnerability (West et al. 2008).

Wisner et al. (2004) posit that the disaster following on a hazard event is directly related to the amount of socially constructed vulnerability, the capacity of the jurisdiction, and the extent of mitigation ("adaptation" in the climate change vocabulary). The disaster that follows a hazard event can be minimal or extensive depending on these three variables, all of which can be addressed through human interventions that include enhancing the capacity of a community and eliminating historical and socially constructed vulnerabilities. These types of interventions for a healthy community are not enough unless they infuse adaptation into the physical and social fabric of the communities (Highfield et al. 2014). Without the interventions, both the sojourner and the receiving community will have increased risk and diminished social health conditions.

## 7.6 Counter-Narrative Frameworks and Problem-Solving Approaches

This section of the chapter will suggest counter-narrative frameworks from historical models that achieved innovation in spite of grave challenges (Scott 1990). It will also explore current problem-solving approaches that place citizen engagement at the core of a local and larger state model. By utilizing core strengths and assets of each region of the state and by collaborative and cross-boundary networks, the work can start building relationships with the already existing networks of family, faith, food, and music groups within the state (Malsale et al. 2018). Starting with what Friere calls problematization—finding root causes and doing critical thinking of problems—collaboration by kindred groups and communities-of-interest can build alternative futures through co-creation of knowledges (Freire 1996; Foucault 2009). Problematization helps understand the underside of wicked problems, as they are symptoms of other problems (Rittel and Webber 1973).

Addressing population relocation through shared cultural experiences opens conversations in a "safe place" that builds relationships (Woods 2018). In this visioning process, as conversations about transitions of populations are happening within the faith community, similar discussions are taking place in government (top down) and non-profit agencies (grasstops), but not necessarily taking place between and with community members at the grassroots. There will be many sojourners within the state, and there will be many who will receive the stranger. It should be a slow, long, and intentional process through which trust and relationships will be forged. And modeling of successful efforts is shared among communities. By building relationships of trust first, with and between the populations that are most impacted, difficult issues and topics can be broached with openness for problem-solving (Park et al. 1993). Giving attention and support first to the vulnerability-bearers is necessary for building capacity and for the community's engagement with those who can provide support, for example, faith communities, universities, agencies, or non-profits (Krajeski 2018). Vulnerability-bearers will be both in the moving and in the receiving populations.

Building on community connections and strengths alone will not be sufficient to rectify the complex issues that have caused and maintained the wicked problems outlined in the beginning of the chapter. A robust master plan for the state is desperately needed that takes on the qualities of a New Marshall Plan, understood like the WWII Marshall Plan, as an effort to rebuild infrastructure lost, remove contamination, and rekindle a safe economy and agriculture. Louisiana's current plight needs nothing less than such a robust plan. A New Marshall Plan can embrace asset-based planning with imagination to harness every possible resource available within the receiving regions from the cultural strengths and inclusive of the faith, NGO, and business and governmental communities. It could counter the lack of carrying capacity of the state with a new ethic of a *caring capacity*. The key is complex integrated thinking and resourcing (Thomas 2014). The plan would be a large-scale model for the state, much like what Bob Becker, CEO of New Orleans

City Park, used for post-Katrina recovery and redevelopment of one of the largest parks in the country: have a vision, get it into workable components that augment each other, and attract resource providers to complete those outlined components as through interest in supporting the plan develops and grows (Pope 2013). The vision itself draws resources; it must be innovative with a strong commitment to a caring capacity, one that rejuvenates and restores community with community.

And, because Louisiana is "ahead of the curve" in needing such a large-scale plan, that the state is an innovator on this New Plan, it will attract donors.

If they are done well, innovations in society have the capacity to gather a following because they are visionary, they can give hope, they can be just, and they offer a way for people to be involved (Rogers 2003). The innovations also often have the quality of large thinking even though their implementation begins small. The innovation as experienced in intentional planned communities offered needed hope and justice in difficult times of segregation, housing needs, economic exploitation, and challenged agricultural land use (Howard 1965). Models are shared from the faith community in the following section. They were and are models that defied the impossible. They can give hope and understanding for the tremendously large task ahead for Louisiana.

### *7.6.1 Faith-Inspired Models for Change*

People in Louisiana continue to have strong faith connections, tied with Alabama, with the highest percentage of people in the USA who are linked to faith organizations (Cooperman 2015). The map on religious preference from the 2010 US Religious Census data shows a clear distinction of faith preferences in Louisiana (Fig. 7.4). Roman Catholic and Southern Baptist are the two dominant faith groups, followed by the United Methodist Church (ASARB 2012).

A sharp geographic dividing line between Catholic and Baptists is based on historical migration patterns. Historical colonial dominance of the coastal area by Spanish missionaries and Catholic French and Catholic Canadian[18] settlers still influences the southern coastal region. The northern region was influenced by Protestant itinerate preachers and missionaries and settlers from the central and east coast once Louisiana became a state and has been designated by many as part of the Bible Belt (Tweedlie 1978; Zelinsky 1972). As populations shift, so will the religious mix in the central and northern parts of the state. Catholic and protestant congregations will need to find ways to carry their traditions as they find places to establish congregations, schools, and other faith-based facilities in new areas. They will be sojourners in a new land and will look for hospitality in the place they reside and their faith practiced. Both the sojourner and those extending hospitality will

[18] The predominance of French rule and the Catholic Church manifest themselves into the current-day Napoleonic Law still practiced throughout the state and the county units of government, called *parishes*.

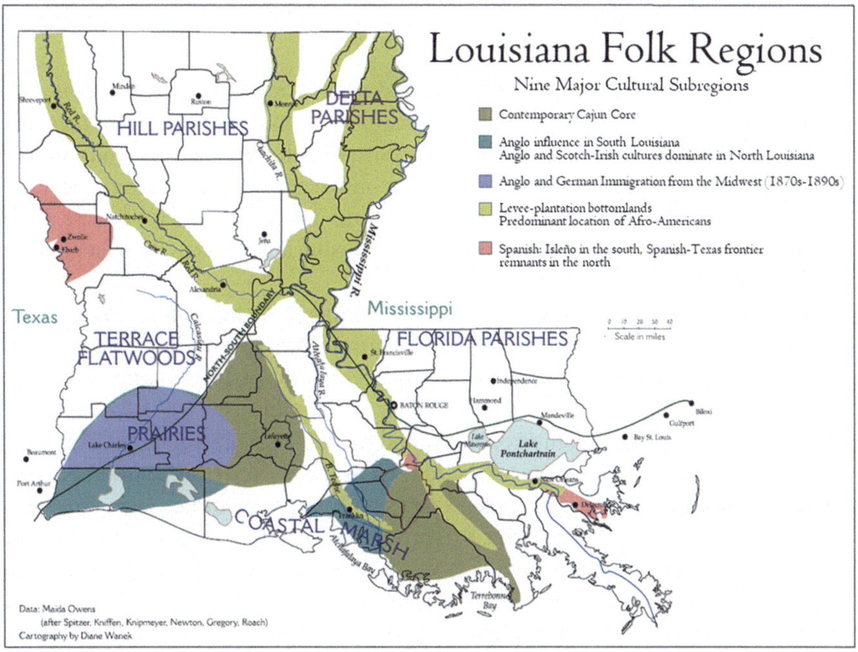


**Fig. 7.4** Map showing Louisiana's main folk regions and nine cultural subregions. (Image source: www.louisianafolklife.org)

find a rich tapestry of history intertwined with traditions of music, stories, and outreach ministries of service that can benefit the new communities.

Protocols of faith practice are different between Protestants and Catholics including the judicatory structures and modes of worship such as liturgy and the adherence to Holy Days. Denominational leadership will be essential in helping to create a welcoming situation for all faiths. Ecumenical organizations like the *Louisiana Interchurch Conference* and *Church Women United* can be instrumental in finding common ground through shared faith values and faith stories.

Over the years, faith-based initiatives and faith-based work have birthed innovation during times of distress and turbulence offering hope and alternatives. These initiatives include leaders such as Rev. Abraham Lincoln Davis, founder of *Southern Christian Leadership Conference.*[19] Others also helped usher in new ways of being within community. Not waiting for government to take action, the faith community has often taken the lead to advocate and/or provide education, health, housing, and environmental witness and social justice in general.[20]

---

[19] https://kinginstitute.stanford.edu/encyclopedia/southern-christian-leadership-conference-sclc, Rev. Davis from Bayou Goula was the first Vice President of SCLC.

[20] National Council of Churches: www.nationalcouncilofchurches.us

The following are some illustrations of faith-based initiatives that can be helpful to the work on leaving and receiving communities.

- *Koinonia Farms* in Plains, Georgia, started as a racially integrated community land trust in 1942. It was the utopian dream of Clarence Jordan, a farmer and Baptist preacher, devised to promote sustainable agriculture and animal husbandry. It helped poor rural residents move away from cotton, the crop equated with slavery, and have a healthy life. The ecologically diverse farm experimented in various forms of affordable housing and water system techniques (Jordan 1972).
- Through Millard Fuller's experience at the cooperative, he founded *Habitat for Humanity*. The experiment outlasted the cross burning and threats on others connected to a radical idea of social and economic justice. Jimmy Carter, a neighbor, was inspired by the achievements of *Koinonia*.
- While Jordan was addressing economic and racial justice at Koinonia, Dan West was developing what has come to be known as the *Heifer Project* International. It started with a simple idea to send cows to Europe, restoring farms that had been destroyed during World War II. Through partnerships with US government agencies, a steady supply of animals of all varieties was provided to European farmers. The first born of every animal was gifted to another farmer, leading to a slow but continual recovery (Kettering 2003). Following Katrina, the Heifer Project helped restore the fishing fleet at Grand Bayou, an indigenous community in Lower Plaquemines Parish, and extend crop resources to Afro-American farmers in the north central part of Louisiana.
- Likewise, *Serrv*, another innovation of the era, still flourishes today. It addressed the postwar economy of local families in Western Europe to help displaced refugees trade their handcrafts for income. Today, fair trade has become a norm. *Serrv* is a founding member of the World Fair Trade Organization and the Fair Trade Federation and employs 8000 artisans and farmers in 24 countries. Both ideas were visionary for the time (Serrv n.d.)
- Grand Bayou Native Village located in Plaquemines Parish initiated a Participatory Action Research project in 2002 following Hurricanes Lili and Isadore through the support of *Church World Service* and *PCUSA*. After establishing their vision of "saving their culture and land," the project partnered with many experts from various disciplines and backgrounds, being awarded a National Science Foundation Grant to do the cross-boundary work necessary to rebuild their village and establish and enhance their political and cultural voice. It was heralded by the then disaster program director of NSF, a cutting edge success (Peterson 2011).

Each of the above programs were not only innovative but they also responded to an immediate human crisis with effective problem-solving. Each program also created a network of collaborators that helped support the vision that is necessary to have long-term sustainability. Following in the tradition of dialogue, the *Louisiana Interchurch Conference* (LIC) is in an excellent position to help bridge geographic and religious boundaries. Its board of directors spans the state, representing leadership of both Catholic and Protestant traditions. LIC can and has been a boundary and facilitating organization addressing issues pertinent to most Louisiana residents, such as housing, farms, children's welfare, disasters, and the environment. Many of the faith leaders involved with LIC are involved with ecumenical and interfaith projects that contribute to the well-being of the state and become a link between stakeholders to address issues. For example, Rob Gorman from Terrebonne Parish, previous chair of the LIC Environmental Taskforce, helped in the creation of the Coastal Coalition and has been a coastal leader for many years.

A recent sign of hope of a faith-based initiative in Louisiana is the work of *Mary Queen of Viet Nam Church*. Following Hurricane Katrina, *Mary Queen of Vietnam Catholic Church* (MQVN) under the leadership of Fathers Vien Nguyen and Luke Dung[21] became the hub for recovery of the New Orleans East area. The efforts of the congregation resulted in the development of a master plan for the immediate community that includes community gardens, economic endeavors, elder housing, a Community Development Corporation (CDC), and health services.[22] Youth and elderly were a critical part of the recovery and of the continuing work of New Orleans East.

## Faith Influenced Land Trusts to Address Social and Environmental Justice

Inspired by Dorothy Day and the *Catholic Worker Movement* (Coles 1987) that promoted equality and justice, Chuck Matthei envisioned farm settlements and land trusts as an innovative tool that can be used to address distressed rural communities (Matthei 1992). Community and conservation land trusts can be used in creative ways to heal environmentally distressed sites and act as protection against gentrification for traditional and historied communities. While community land trusts are primarily focused on human community, the conservation land trust focuses on the non-human environmental setting. Both types of trusts are necessary, one to protect land from gentrification and the other to provide necessary buffer and healing for green space. In Louisiana, one possibility is to have conservation trusts that buffer areas that have been heavily contaminated over the years from oil, gas, and farming practices. The conservation trusts can help in the rejuvenation from the impacts caused by the by-products of those practices. Conservation trusts can help restore healthy water systems, air, and habitat for outdoor activity that is loved throughout the state (Davis 2010).

Community land trusts provide protection from gentrification for communities that are rejuvenating themselves.[23] Creative uses of community land trusts have been utilized in communities of color to preserve and protect community, cultural and heritage, that sometimes gets disrupted by various development strategies such as economic renewal projects. The Albany, Georgia, land trust is an excellent

---

[21] https://www.npr.org/2011/08/29/140036221/turning-toward-faith-during-hurricane-aftermath

[22] Mary Queen of Viet Nam Community Development Corporation, mqvncdc.org

[23] See, for example, New Orleans' Central City Community Land Trust that serves that exact purpose: https://www.ccclt.org/how-we-work/

example of an established community that has weathered time and endured racial injustice (Davis 2010). Louisiana could create a trust to take on an innovation of "climate land trust," land held for restoration and renewal until it is needed by a transitional community or transitional families for resettlement. The innovation of such a land trust can lend itself to the regeneration of damaged land and water helping to address the wicked problems mentioned earlier in this chapter.

## 7.7 Folk Traditions as Rallying Points

Louisiana's population reflects the state's historically diverse economy. Being on the Mississippi River and having a major port on the Gulf, the access to the state resulted in various diversities. The state's connection with sugar in the Caribbean and bananas and coffee from Central America gave variety to music and foods. The horrific years of the slave market centered in New Orleans brought knowledge and rich traditions from Africa. The multiple derivatives of places of French-speaking people brought traditions and foods from around the world. Figure 7.5 shows the cultural distribution of people who carried their traditions that helped create the diversity of festivals, food, faith, and music.

The map illustrates the predominance of Euro-American and Afro-American population in the central and northern regions of the state while the southern region has a greater ethnic diversity. Notice that the historied Cajun and German enclaves are not well represented in the northern regions of the state (Brasseaux et al. 2005). There are many cultural groups that are not indicated on this map such as the significant Vietnamese population in New Orleans and southeast delta region. The map does not reflect the 79 self-identified indigenous tribes within the state. It is not an exhaustive representation but gives an indication of the larger ethnic populations.

### 7.7.1 *Power of Cultural Festivals*

The state of Louisiana is one of the most popular tourist destinations in the country. Contrary to the low ratings on social health factors, the state excels in offering food, hospitality, happiness, music, and faith (Newport 2017; Cooperman 2015; Morton and Lagrave 2018). The people and communities of the state offer a celebratory culture that is sought by visitors and relished by its residents. With over 520 festivals listed within the state in 2018, there were themes that helped build knowledge while highlighting the strengths of the state's cultural diversity.

Louisiana Office of Tourism's "Feed Your Soul"[24] advertising campaign highlights the joys of multiculturalism, food, festivals, and music. The listings of monthly activities throughout the state are evidence of the wealth of cultural talent

[24] www.Louisianatourism.com

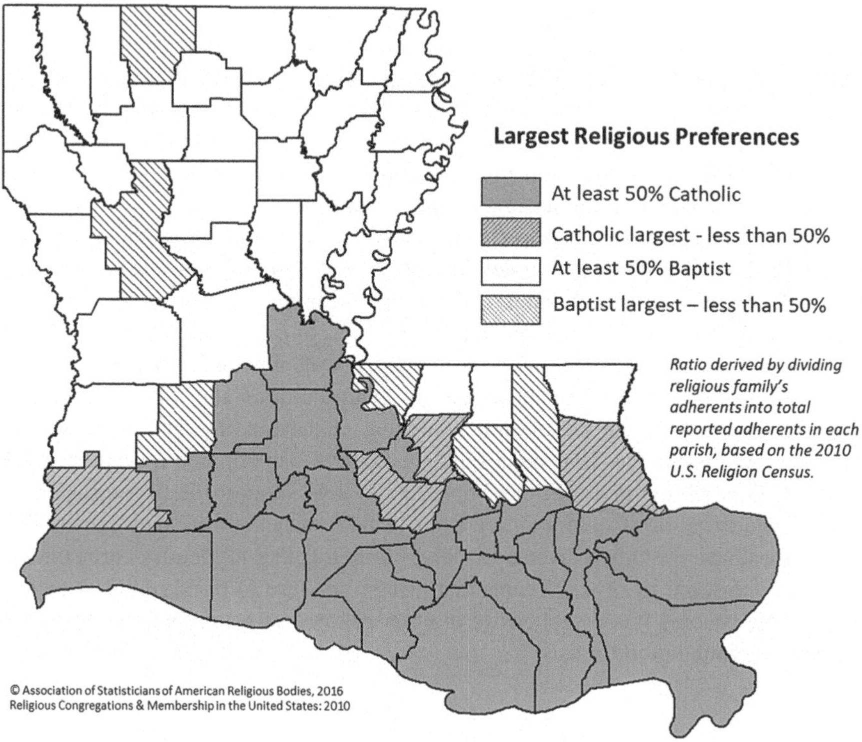


**Fig. 7.5** Map showing the prevalence of Catholicism and Baptism by parish in Louisiana. (Adapted from *2016 U.S. Religion Census: Religious Conregations & Membership Study (Association of Religion Data Archives 2016)*)

and ingenuity garnered by local and regional groups. Utilizing the natural convergence of food, music, festivals, and faith at work in the state organically builds relationships and trust.

The festivals are testimony to the capacity, imagination, and hard work of the communities. The organizational coordination and resources that are expended to host the festivals often take a year of work. Krewes for Mardi Gras coordinate and raise their own funds, the wooden tinder bonfire structures along the Mississippi River in Lutcher for Christmas Eve beacons to St. Nick take months to build and are works of art.[25] Even a local congregation's fundraiser for a cancer patient can take immense work and coordination. Self-organization is a skill and asset seen in the festivals and community celebrations. The altruism that Ostrom (Ostrom and Walker 2003) speaks of within communities is witnessed in these festival events. Grassroots mobilization and altruism can be quick and swift when there is a disaster as seen in groups like the Cajun Navy. Grassroots ability and capacity are present in the state, often in informal networks.

[25] Christmas Eve Bonfires: www.festivalofthebonfires.org

Although there is capacity to mobilize within the state, there are other interpersonal dynamics that will restrict successful relocation such as racial challenges, regional biases, and prejudice. Recently in April 2019, three Afro-American Churches in St. Landry Parish northwest of Baton Rouge were burned as hate crimes. These issues cannot be ignored, including when relocation is being considered. Addressing them must be part of the needed reconciliation and healing as changing inland communities reconfigure themselves.

Prejudice is real and experienced in the daily lives of people throughout the state but can be confronted and addressed through shared cultural connections. Through the work of youth and local congregations, for example, proms in northern parish high schools have moved beyond outdated codes of segregation as recently as 10 years ago and are now integrated.[26] Another example is found in Arizona. *Tucson Meet Yourself* was created to showcase the many cultures and faiths residing in Tucson. The event began for cultural awareness, understanding, and dialogue that can overcome prejudice. Louisiana Folklore Society has initiated a project called Bayou Culture: Moving Traditions Forward[27] to help support the bayou region to discuss what traditions communities want to take with them as they move. The discussions have brought a diverse working group together to identify core cultural values people want to save.[28] Events that bring a mixture of backgrounds together help people become less apprehensive of other cultures, customs, and practices and lay a strong foundation for trust.[29]

## 7.8 Cross-Boundary Work: Trust and Relationship Building

Elinor Ostrom, Nobel Laureate, examined the concept of trust and conducted extensive research on the topic to determine if people were apt to be trusting, especially in a relationship of reciprocity and mutual caring. Her work was explored across geographic and cultural boundaries and found that the more understanding people had of each other, the greater the levels of trust. The research showed that generosity and good will were extended even at a minimal level of interpersonal knowledge (Ostrom and Walker 2003). Her interest in mutual support was fundamental to her understanding of the "commons" and how we can restore a society and economy to be equitable and just. This understanding of "commons" is seen in the mutual aid and stability of the population within the rural areas of the state and in its communities' self-care, generosity, and hospitality (Peterson & Krajeski 2011). Many folks in the bayou region refer to their community as a "giving" lifeway based on a web of life that is both sacred and joyous and full of

---

[26] Jonathan Coats, May 1, 2008, Outdated separation ends with integrated high school prom. www.NOLA.com

[27] Louisiana Folklore Society, www.louisianafolklore.org

[28] www.Tucsonmeetyourself.org

[29] Louisiana Folklore Society, www.louisianafolklore.org

awe. It is situated or place-based knowledge, as in "the shrimp are giving" and "we are giving to family" (Regis 2019). The experienced knowledge of the rural communities supports the concepts of Peter Park, and the development of relational knowledge to build networks and collaboration for problem-solving, as well as Ostrom's concepts of altruism (Park et al. 1993).

Projects that emphasize the co-creating of place, knowledge, and a healthy future can and will build trusting relationships between the displaced and the receiving community (Hillier and Rookby 2005). As trust is gained and projects honed, resources to bring them to fruition can be secured through the efforts of a collaborative team. Especially for small rural communities, teaming small communities together gives more clout or voice in the securing of resources. It serves to maximize human resources as well as people are often overtaxed and wearing many hats.

In Louisiana in the early twentieth century, an innovation brought together best practices held by government agencies at the time to create the Terrebonne Project, a planned community in Terrebonne Parish (Conn 2009). It was one of the hundred communities created all over the country, including Alaska, planned by the *Federal Resettlement Administration*. It addressed the massive displacement of tenant farmers and small business owners as a result of the Great Depression, the great flood of 1927, and the Dust Bowl. Using ideals set by Sir Ebenezer Howard's "garden city" movement, Rexford Tugwell used all available resources in President Franklin Delano Roosevelt's Cabinet, from agricultural, health, arts and folk life, housing, and industry to create resettlements for displaced families (Lash 1988). Using best knowledge of sustainability from around the world, new communities were birthed and families were given the means of sustainability with a small farmstead and involvement in a larger community enterprise.

The formal Resettlement Administration was short lived due to the need of resources for the impending war. However, in spite of the lack of continued funding and support, many of the communities survived by the participants' initiative to make the planned community work (Holley 1940). The popularity of healthy planned greenbelt communities of the New Deal continued to grow, as seen in the work of James Rouse in places like Columbia; Greenbelt, Maryland; and Green Hills, Ohio (Chappelle et al. 1986).

Cross-boundary organizing can help facilitate linking programs like Louisiana Watershed Initiative[30] and EPA's Sustainable Healthy Communities Initiative[31] to support projects that reduce natural hazards while emphasizing water quality benefits. Quality of life can be lifted up as a starting point for rejuvenation of a community. EPA's program encourages cross-boundary work including the faith and cultural community as active planning participants in community projects.

One at risk community suggested that a watershed festival hosted by communities distressed by water issues could coincide with the watershed planning that the state has initiated. The festival could link communities along a watershed for

[30] Louisiana Watershed Initiative, www.watershed.la.gov

[31] https://www.epa.gov/aboutepa/about-sustainable-and-healthy-communities-research-program

storytelling, food, and faith traditions while celebrating the common source of water. Telling the water's story together is a way to start the relationship building between different people and the trust needed to work on other issues that emerge from the stories. Examples of local community problem-solving through the convergence of people's experienced lifeways, stories, and problematization are abundant (Peterson 2011). Networks organically built around extended family and their practices of food, faith, celebrations, and music provide the building of the connectional web to establish relationships (Park et al. 1993).

Collaboration and asset building work is neither a top-down nor a bottom-up approach to planning and usually multidimensional in scope. The in-between entities sometimes called boundary organizations can act as facilitators, interpreters, and resource advocates. They can help facilitate the linkages for collaborative work and can serve as connectors of resources between and within grassroots, grasstops, and formal organizations, be they academic, philanthropic, or government (Taylor et al. 2012; NASEM 2018). When inter–/intra-organizational connections and relationship are developed and collaboration is established with a non-hierarchal structure, each organization increases in strength and effectiveness. All participating entities can claim the co-ownership of the work and the co-learning process (Laska and Peterson 2011).

A New Plan for Louisiana is to get creative with resources, to think with ingenuity while addressing the issues at hand. Using a process similar to what Bob Becker used for the New Orleans City Park recovery post Katrina could result in creative acquisition of resources for projects and to have projects prepositioned for grants when they are available (Borrup 2006). Addressing a particular issue such as clean water links people of place with each other as an affinity group, regardless if some are natives to the area and others are recent arrivals. The affinity group's interest in a particular topic that is critical to the group's well-being can generate the interest for learning and action. The EPA Healthy Community Program and the Citizen Institute on Rural Design (CIRD)[32] are national entities that can help in resourcing a community's vision in addressing an issue. Each helps the community build their capacity for well-being. Using a similar strategy as Becker's City Park planning and recovery model would help link these efforts with and to an overall State plan to prioritize and monitor all the components for evaluation and achievement.

Developing "new communities" or rejuvenating fading communities can offer positive outcomes for people still in place and those who want to remain rural in their living orientation, connection to the land. Relying on old ideologies and practices has placed the state on the bottom rung for education, health care, potable water quality, as well as other physical, economic, and social indicators (US News and World Report 2019). It is critical now with the coastal and inland threats to have a vision and to be aspirational; it is possible to have *caring capacity* through *climate land trusts* and the full utilization of our cultural assets. Cross-boundary networking

[32] CIRD: www.arts.gov/partnership/citizens-institute-rural-design

can be enhanced and increased through relationships of established cultural networks to achieve both of these goals.

## 7.9 Conclusions

The approaches shared in this chapter, utilizing the faith and cultural richness of the state, can enhance and celebrate the depth of shared and appreciated values of music, food, family, and faith. The deep-rootedness of various cultural communities has contributed to the food, music, and festival traditions that make Louisiana attractive to tourism. Thus, the biblical notion of sojourner and the reciprocal one of giving of hospitality are keys to how the migration can be framed to achieve a positive acceptance by the receiving communities.

Geographical population shifts can be an opportunity for the state to address dire social, health, and human dimension issues while in the mire of coastal land loss and forced displacement. If all issues are not addressed together, the consequences will likely repeat former patterns of built vulnerability and diminished human capacity (Taylor-Ide and Taylor 2002).

The coordination, knowledge, expertise, and skill sets for creating and hosting a festival of any type "takes a village." The skills used are the same as those used in community planning and simply in community functioning. They can be harnessed as valuable resources in problem-solving. For an inclusive discourse in holistic planning, the participants need to go beyond non-profits, agencies, academics, and business communities to include the arts and faith communities and essential at the core—the people themselves, especially the youth (Taylor et al. 2012; Esteva and Prakash 2014). As witnessed in the work of the Isle de Jean Charles Biloxi-Chitimacha-Choctaw Tribe resettlement planning process, children and youth were eager to participate in what will be their future (Hart 1997).

Strong cultural and faith traditions are assets often overlooked for problem-solving larger issues such as water, contamination, and relocations (Aldrich and Meyer 2015). Asset-based planning uses community-driven practices, racial equity, transparency, and participation. It also applies accountability in governance of the work process (West et al. 2008). Through formal and informal networks of family and festivities, these affinity groups can help create relationships/friendships and bonds while bringing joyful celebration into problem-solving (Kretzmann and McKnight 1993).

A *New Plan for Louisiana* must be inclusive and multidimensional, using best principles and wisdom from every knowledge discipline and practice. It must have goals that bring the state's health and well-being up to an exemplar of good living for its people, a *caring capacity.* And concurrently the plan will need to consider land usage and its benefit for restoring the health of the environment through *climate land trusts*, while giving a future location for future displaced populations in resettlement.

## References

Adeniyi, O., Hernandez, A., LeBlanc, M., King, J., & Janes, M. (2016). Quantitation of pesticide residue in water and food in Louisiana, USA. *Journal of Water Resource Protection., 8*, 1145–1157. https://doi.org/10.4236/jwarp.2016.812089.

Akuno, K., Hall, S., King, B., & Nangwaya, A. (Eds.). (2017). *Rising, Jackson: The struggle for economic democracy and black self-determination in Jackson, Mississippi*. Montreal, Canada: Daraja Press.

Aldrich, D. P., & Meyer, M. A. (2015). Social capital and community resilience. *American Behavioral Scientist, 59*(2), 254–269. https://doi.org/10.1177/0002764214550299.

Association of Statisticians of American Religious Bodies (ASARB). (2012). Religious diversity in the United States, 2010. [map] Retrieved from http://www.usreligioncensus.org/images/004.jpg

Ballard, M. (2017, April 8). Across Louisiana, crumbling infrastructure threatens small town water supplies. The Advocate. Retrieved on February 6, 2019 from https://www.theadvocate.com/acadiana/news/article_8b57ab2c-19fc-11e7-b596-3b474ce3c816.html

Belluz, J. (2016, December 20). We visited one of America's sickest counties. We're afraid it's about to get worse. *VOX*. Retrieved from https://www.vox.com/2016/12/7/13673802/louisiana-healthcare-concordia-sickest-counties-in-america

Blackwell, V., Drash, W., & Lett, C. (2017 October 20). Toxic tensions in the heart of "Cancer Alley." *CNN*. Retrieved from https://www.cnn.com/2017/10/20/health/louisiana-toxic-town/index.html

Bolin, R., & Bolton, P. (1986). *Race, religion, and ethnicity in disaster recovery. Program on environment and behavioral science*. Boulder, CO: University of Colorado.

Borrup, T. (2006). *The creative community builder's handbook: How to transform communities using local assets art, and culture*. St. Paul, MN: Wilder Publishing Center.

Bowen, W. R. (2009). *Engineering ethics: Outline of an aspirational approach*. London: Springer.

Brasseaux, C., & Davis, D. (2017). *Ain't there no more: Louisiana's disappearing coastal plain*. Jackson, MS: University Press of Mississippi.

Brasseaux, C. A., Brasseaux, R. A., & Bienvenu, M. (2005). *Stir the pot: The history of Cajun cuisine*. New York, NY: Hippocrene Books.

Brown, L. R. (2011). *World on the edge: How to prevent environmental and economic collapse*. New York, NY: Earth Policy Institute. W.W Norton & Company, Inc.

Buber, M. (1970). *I and thou*. New York, NY: Charles Scribner's Sons.

Bullard, R. D. (1994). *Unequal protection: Environmental justice & communities of color*. San Francisco, CA: Sierra Club Books.

Chappelle, S. E., Baker, J. H., Esslingen, D. R., Ridgway, W. H., Schultz, C. B., & Stiverson, G. A. (1986). *Maryland: A history of its people*. Baltimore, MD: The John Hopkins University Press.

Citizen's Institute on Rural Design (CIRD). (2018). "About CIRD". Retrieved from https://www.rural-design.org/about

Coles, R. (1987). *Dorothy day: A radical devotion*. Reading, MA: Addison-Wesley.

Conn, S. (2009). Origins: Guess what? The new deal worked! Retrieved from The Ohio State University and Miami University website: http://origins.osu.edu/history-news/guess-what-new-deal-worked

Cooperman, A. (2015). U.S. public becoming less religious. Pew Research Center. Retrieved from https://www.pewforum.org/2015/11/03/u-s-public-becoming-less-religious/

Curth, K. (2018, November 12). LA's aging infrastructure: Nearly 2,000 structurally deficient bridges, no money for repairs. *Fox8Live*. Retrieved from http://www.fox8live.com/2018/11/13/las-aging-infrastructure-nearly-structurally-deficient-bridges-no-money-repairs/

Cutter, S. L., Ash, K. D., & Emrich, C. T. (2016). Urban-rural differences in disaster resilience. *Annals of the American Association of Geographers, 106*(6), 1236–1252. https://doi.org/10.1080/24694452.2016.1194740.

Dalbom, C., Hemmerling, S., & Lewis, J. (2014). *Community resettlement prospects in Southeast Louisiana*. New Orleans, LA: Tulane University Law School, Tulane Institute on Water Resources and Policy.

Davis, J. E. (Ed.). (2010). *The community land trust reader*. Cambridge, MA: Lincoln Institute of Land Policy.

Derrida, J. (2000). *Of hospitality*. Stanford, CA: Stanford University Press.

DeSantis, J. (2019, February 26). Isle de Jean Charles relocation project moves forward; local tribe pushes back. The Times. Retrieved from https://www.houmatimes.com/news/isle-de-jean-charles-relocation-project-moves-forward-local-tribe/article_1869106e-3a4e-11e9-bcf3-f35bfcbcc033.html

Doherty, G. W. (2004). Crisis in rural America: Critical incidents, trauma and disasters. *Traumatology, 10*(3), 145–164. https://doi.org/10.1177/153476560401000302.

Edwards, T. (2019, March 28). Personal communication.

Environmental Working Group (EWG). (2015). Tap Water Database. Retrieved from https://www.ewg.org/tapwater/

Escobar, A. (2008). *Territories of difference: Places, movements, life, redes*. Durham, NC: Duke University Press.

Esteva, G., & Prakash, M. (2014). *Grassroots post-modernism: Remaking the soil of cultures*. London: Zed Books.

Foucault, M. (2009). *History of madness*. New York, NY: Routledge.

Freire, P. (1996). *Pedagogy of the oppressed*. New York, NY: Continuum.

Fullilove, M. T. (2004). *Root shock: How tearing up city neighborhoods hurts America, and what we can do about it*. New York, NY: One World Press.

Goodell, J. (2017). *The water will come: Rising seas, sinking cities*. New York, NY: Little, Brown and Company.

Goulet, D. (1995). *Development ethics: A guide to theory and practice*. London: Zed Books.

Hart, R. A. (1997). *Children's participation in planning*. New York, NY: Taylor and Francis.

Harvey, D. (1996). *Justice, nature and the geography of difference*. Oxford: Blackwell.

Hauer, M. E. (2017). Migration induced by sea-level rise could reshape the US population landscape. *Nature Climate Change, 7*, 321–325. https://doi.org/10.1038/nclimate3271.

Hawken, P. (2017). *Drawdown: The most comprehensive plan ever proposed to reverse global warming*. New York, NY: Penguin Books.

Hemmerling, S. A. (2017). *A Louisiana coastal atlas: Resources, economies and demographics (the natural world of the Gulf south)*. Baton Rouge, LA: Louisiana State University Press.

Highfield, W. E., Peacock, W. G., & Van Zandt, S. (2014). Mitigation planning: Why hazard exposure, structural vulnerability, and social vulnerability matter. *Journal of Planning Education and Research, 34*(3), 287–300. https://doi.org/10.1177/0739456X14531828.

Hillier, J., & Rookby, E. (2005). *Habitus: A sense of place* (2nd ed.). Aldershot, England: Ashgate.

Holley, D. (1940). *Uncle Sam's farmers: The New Deal communities in the lower Mississippi Valley*. Chicago, IL: University of Illinois Press.

Horney, J., Nguyen, M., Salvesen, D., Dwyer, C., Cooper, J., & Berke, P. (2017). Assessing the quality of rural hazard mitigation plans in the southeastern United States. *Journal of Planning Education and Research, 37*(1), 56–65. https://doi.org/10.1177/0739456X16628605.

Howard, E. (1965). *Garden cities of to-morrow*. Cambridge, MA: The MIT Press.

Hufford, M. (Ed.). (1994). *Conserving culture: A new discourse on heritage*. Champaign, IL: University of Illinois Press.

Jackson, J. B. C., & Chapple, S. (2018). *Breakpoint: Reckoning with America's environmental crises*. New Haven, CT: Yale University Press.

Jones, T. L. (2018, August 5). Black community in flood-prone New Roads area feels pushback form white neighborhood over relocation site. *The Advocate*. Retrieved from https://www.theadvocate.com/baton_rouge/news/article_ee6b7b88-95db-11e8-a52e-cb57c9ecb7d0.html

Jordan, C. (1972). In D. Lee (Ed.), *The substance of faith and other Cotton Patch Sermons*. New York, NY: Association Press.

Kearney, R., & Semonovitch, K. (2011). *Phenomenologies of the stranger: Between hostility and hospitality*. New York, NY: Fordham University Press.

Kettering, D. D. (2003, June 30). *Dan West Papers*. Series 18. Brethren Historical Library and Archives: Goshen College.

Krajeski, R. (2018, July). *Framing research: Concepts and jargon in disaster research*. Panel presented at Natural Hazards Center workshop: Boulder, CO.

Kretzmann, J. P., & McKnight, J. L. (1993). *Building communities from the inside out: A path toward fighting and mobilizing a community's assets*. Chicago, IL: ACTA Publications.

Lash, J. P. (1988). *Dealers and dreamers: A new look at the New Deal*. New York, NY: Doubleday.

Laska, S., & Peterson, K. (2011). The convergence of catastrophes and social change: The role of participatory action research in support of the new engaged citizen. *Journal of Applied Social Sciences., 5*(1, Spring), 24–36. https://doi.org/10.1177/193672441100500102.

Laska, S., Peterson, K., Rodrigue, C., Cosse, T., Philippe, R., Burchett, O., & Krajeski, R. (2015). Layering of natural and human-caused disasters in the context of sea level rise: Coastal Louisiana communities at the edge. In *Disaster's impact on livelihood and cultural survival: Losses, opportunities, and mitigation*. Boca Raton, FL: CRC Press.

Lasley, C. B. (2012). *Catastrophe and the role of social networks in recovery: A case study of St. Bernard Parish, LA residents after Hurricane Katrina*. Unpublished University of New Orleans. Dissertation.

Low, S., & Lawrence-Zuniga, D. (Eds.). (2003). *The anthropology of space and place: Locating culture*. Oxford: Blackwell Publishing.

Lowlander Center. (2017). *Building the resilience of small coastal businesses*. New Orleans, LA: Entergy.

Maida, C. (Ed.). (2007). *Sustainability and communities of place*. Brooklyn, NY: Berghahn Books.

Maldonado, J. (2018). *Seeking justice in an energy sacrifice zone: Standing on vanishing land in coastal Louisiana*. New York, NY: Routledge Press.

Malsale, P., Sanau, N., Tofaeono, T., Kavisi, Z., Willy, A., Mitiepo, R., Lui, S., Chambers, L., & Plotz, R. (2018, December). Protocols and partnerships for engaging Pacific Island communities in the collection and use of traditional climate knowledge: Traditional weather and climate knowledge remains important for environmental decision-making in the Pacific. *Bulletin of the American Meteorological Society., 99*(12), 2471–2489. https://doi.org/10.1175/BAMS-D-17-0163.1.

Marsh, B. C. (1909). *An introduction to city planning: Democracy's challenge to the American city*. Whitefish, MT: Kessinger Publishing, LLC.

Matthei, C. (1992). *U.S. land reform movements. The community land trust reader*. Cambridge, MS: Lincoln Institute of Land Policy.

Morton, T. (2013). *Hyperobjects: Philosophy and ecology after the end of the world. Posthumanities (series 27)*. Minneapolis, MN: University of Minnesota Press.

Morton, T. (2017). *Humankind: Solidarity with nonhuman people*. London: Verso.

Morton, C. & Lagrave, K. (2018). The best cities in the U.S.: 2018 Readers' Choice Awards. *Conde Nast Traveler.* Retrieved from https://www.cntraveler.com/gallery/best-cities-us

National Academy of Sciences, Engineering, & Medicine (NASEM). (2018). *Understanding the long-term evolution of the coupled natural-human coastal system: The future of the U.S. Gulf Coast a consensus study report*. Washington DC: The National Academies Press. https://doi.org/10.17226/25108.

Newman, E. (2007). *Untamed hospitality: Welcoming God and other strangers*. Grand Rapids, MI: Brazos Press.

Newport, F. (2017). Mississippi retains standing as most religious state. Gallup: Social & Policy Issues. Retrieved from https://news.gallup.com/poll/203747/mississippi-retains-standing-religious-state.aspx

Nored, R. E. (1999). *Reweaving the fabric: How congregations and communities can come together to build their neighborhoods*. Montgomery, AL: Black Belt Press.

Ogletree, T. W. (2003). *Hospitality to a stranger: Dimensions of moral understanding*. Louisville, KY: Westminster John Knox Press.

Ostrom, E., & Walker, J. (Eds.). (2003). *Trust and reciprocity: Interdisciplinary lessons from experimental research*. New York, NY: Russell Sage Foundation.

Park, P., Brydon-Miller, M., Hall, B., & Jackson, T. (Eds.). (1993). *Voices of change: Participatory research in the United States and Canada*. Westport, CT: ABC-CLIO.

Peterson, K. (2011). *Transforming researchers and practitioners: The unanticipated consequences (significance) of Participatory Action Research (PAR)*. University of New Orleans Theses and Dissertations. 129. New Orleans, LA.

Peterson, K., & Krajeski, R. (2011). Grounded in faith, inspired to action: Bayou women own their recovery. In D. Emmanuel & E. Enarson (Eds.), *Women of Katrina: The gendered dimensions of disaster recovery*. Nashville, TN: Vanderbilt Press.

Pope, J. (2013, April 10). Bob Becker, City Park's CEO, named a distinguished UNO alumnus. *NOLA: The Times Picayune*. Retrieved from https://www.nola.com/education/2013/04/bob_becker_city_parks_ceo_name.html

Reed, G. (2018). A toxic threat: Government must act now on PFAS contamination at military bases [Fact Sheet]. Center for Science and Democracy at the Union of Concerned Scientists. Retrieved from https://www.ucsusa.org/sites/default/files/attach/2018/09/a-toxic-threat-pfs-military-fact-sheet-ucs-2018.pdf

Regis, H. A. (2019). Re-thinking subsistence in turbulent times: New contexts, configurations, and intersections with social and environmental justice: Festivals to subsistence and back again. Panel presented at the 79th annual meeting of the Society for Applied Anthropology Conference: Portland, OR.

Rittel, H. W. J., & Webber, M. M. (1973). Dilemmas in a general theory of planning. *Policy Sciences, 4*, 155–169. https://doi.org/10.1007/BF01405730.

Rogers, E. M. (2003). *Diffusion of innovations*. New York, NY: Free Press.

Rom-Shiloni, D. (2013). *Exclusive inclusivity: Identity conflicts between the exiles and the people who remained (6th–5th Centuries BCE)*. The Library of Hebrew Bible/Old Testament Studies Vol. 543. Bloomsbury, NY.

Ruppert, T. (2019, May 7) *Keeping history above water: Politics and policy*. St Augustine, FL.

Schliefstein, M. (2013, March 19). Louisiana's infrastructure ranks C- for poor roads, dams, drinking water systems. *NOLA Times Picayune*. Retrieved from https://www.nola.com/politics/2019/03/broad-street-overpass-bike-lanes-will-feature-new-protections-cyclists-still-wary.html

Scott, J. C. (1990). *Domination and the arts of resistance*. New Haven, CT: Yale University Press.

Serrv. (n.d.). Our Story. Retrieved from https://www.serrv.org/category/about-us

Smith-Christopher, D. (2002). *A biblical theology of exile*. Minneapolis, MN: Fortress Press.

Southern Poverty Law Center (SPLC): Active Hate Groups in the United States 2018 [map]. Retrieved from https://www.splcenter.org/hate-map

Steingraber, S. (1998). *Living downstream: A scientist's personal investigation of cancer and the environment*. New York, NY: Vintage Books.

Swierzewski, R. (2007). *Rural philanthropy: Building dialogue from within*. Washington, DC: National Committee for Responsive Philanthropy.

Taylor, D. C., Taylor, C. E., & Taylor, J. O. (2012). *Empowerment of an unstable planet: From seeds of human energy to a scale of global change*. New York, NY: Oxford University Press.

Taylor-Ide, D., & Taylor, C. E. (2002). *Just and lasting change: When communities own their future*. Baltimore, MD: John Hopkins University Press.

The Sustainable Scale Project. (2003). "Carrying Capacity". Retrieved from http://www.sustainablescale.org/ConceptualFramework/UnderstandingScale/MeasuringScale/CarryingCapacity.aspx

Thomas, E. (2014). *Planning and building livable, safe & sustainable communities: The patchwork quilt approach*. New Orleans, LA: National Hazards Mitigation Association.

Tierney, K. (2014). *The social roots of risk: Producing disasters, promoting resilience*. Stanford, CA: Stanford University Press.

Tonnies, F. (2001). *Community and civil society*. Cambridge: Cambridge University Press.

Tweedlie, S. W. (1978). Viewing the Bible Belt. *Journal of Popular Culture, 11*, 865–876. https://doi.org/10.1111/j.0022-3840.1978.1104_865.x.

UNESCO. (2018). *Reshaping Cultural Policies 2018: 2005 Convention Global Report*. Paris, France. Retrieved from http://uis.unesco.org/sites/default/files/documents/reshaping-cultural-policies-2018-en.pdf

University of Wisconsin Population Health Institute: 2018 County Health Rankings. Retrieved from http://www.countyhealthrankings.org/rankings/data/LA

US News & World Report. (2019). Best states 2019: Ranking performance throughout all 50 states. Retrieved from https://media.beam.usnews.com/69/4f/9ec3a3e94c4080b146d64d27288a/190508-best-states-overall-rankings-2019.pdf

Walker, B., & Salt, D. (2012). *Resilience practice: Building capacity to absorb disturbance and maintain function*. Washington, DC: Island Press.

West, J., Peterson, K., Alcina, M., & Laska, S. (2008). Principles of participation and issues of entry for Participatory Action Research (PAR) in coastal community resiliency enhancement collaboration. *Journal for Community Engaged Research and Learning Partnerships., 1*(1). http://www.manifestationjournal.org/.

Wisner, B., Blaikie, P. M., Cannon, T., & Davis, I. (2004). *At risk: Natural hazards, people's vulnerability and disasters*. London: Routledge.

Woods, M. (2018, November). Precarious rural cosmopolitanism: Negotiating globalization, migration and diversity in Irish small towns. *Journal of Rural Studies, 64*, 164–176. https://doi.org/10.1016/j.jrurstud.2018.03.014.

Wright, T. (2019, January 24) Tough choices ahead for Louisiana's struggling water systems. *The Lens*. Retrieved from https://thelensnola.org/2019/01/24/tough-choices-ahead-for-louisianas-struggling-water-systems/

Wyatt, L. (1983). *We are the boat*. [Recorded by Folk Legacy Records, Inc.] Roots and Branches. BMI. Greenfield, MA.

Young, I. M. (1990). *Justice and the politics of difference*. Princeton, NJ: Princeton University Press.

Zelinsky, W. (1972). *Cultural geography of the United States. (Foundation of Cultural Geography Series)*. Upper Saddle River, NJ: Prentice Hall Publishing.

# Part IV
# Types/Locations of Communities and Their Responses to Extreme Weather: Urban

# Chapter 8
# Post-disaster Development Dilemmas: Advancing Landscapes of Social Justice in a Neoliberal Post-disaster Landscape

**Anna Livia Brand and Vern Baxter**

## 8.1 Introduction

Disasters like Hurricane Katrina generate deep debate about human vulnerability and the viability of built environments. Fundamental questions raised in the aftermath of disaster include *how could they live there, why would they live there?* It is difficult for those facing the impacts of disaster to field unwelcome questions about connections to home and place. We all have connections to the places we live that go beyond rational measurement and plumb the depths of emotion. Well-understood psychological reasons draw people home to an imaginary state that existed before lifeworlds were rendered unrecognizable by disaster (Burley 2010). But there is no going back after Katrina. The storm and ensuing levee failures left an enduring mark on the city. It displaced over a million people, devastated the Gulf Coast, and claimed the lives of over 1000 Louisiana residents (Plyer 2016).

While there is no going back, the road forward is unclear. Post-disaster planning is often framed as a moment from which a new future landscape might be imagined. The wake of disaster invites imaginings of more socially just settlement patterns and inspires radical alternatives to what was there before. Myriad individuals and groups invested time and energy shaping a positive recovery imaginary in New Orleans, from the scale of neighborhood meetings patched together in churches and schools, homes, and front porches to the scale of city-wide meetings that spanned nearly 10 years after Katrina.

A. L. Brand (✉)
Department of Landscape Architecture & Environmental Planning,
University of California Berkeley, Berkeley, CA, USA
e-mail: annalivia@berkeley.edu

V. Baxter
Department of Sociology, University of New Orleans, New Orleans, LA, USA
e-mail: vbaxter@uno.edu



S. Laska (ed.), *Louisiana's Response to Extreme Weather*,
Extreme Weather and Society, https://doi.org/10.1007/978-3-030-27205-0_8

For many, this future landscape was imagined as resilient, green and sustainable, socially just and equal. Yet policy makers and urban designers who raised questions about how, why, and where resettlement *should* take place in the wake of Katrina often excluded the deeply unequal and settled patterns of a racial state, geographies developed, and resisted over different racial formations and political economies of urban development across the city's history. Further, they often failed to imagine how processes like advanced capitalism and racism work geographically in the twenty-first century or might work in a disaster context. Therefore, outside of specific and even narrow instances, the larger post-disaster story fails to meet these hopes, despite moments where more racially and socially just solutions prevail.

Indeed, many questions remain about what the future might bring and what shape vulnerability might take. Hurricane Katrina and more recent environmental disasters (e.g., Hurricanes Harvey, Irma, and Maria) transpired under the hegemonic reign of a neoliberal political economy. Neoliberal doctrine above all advances individual freedoms and social well-being within an institutional framework defined by strong private property rights and free markets, where the state generally assists and facilitates business and private development goals (Harvey 2005, p. 2). A tilt toward private sector solutions to planning and urban development problems pre-dates Katrina and aptly characterizes recovery and rebuilding efforts over the past dozen years. Contemporary human, social, racial, and environmental justice claims are made on the state within a dominant neoliberal paradigm, whose colorblindness (Omi and Winant 2014) makes it difficult to advance the type of equity work needed to address the deeply racialized disparities that characterized the city before the storm. Perhaps more concerning, the social justice limitations of neoliberalism make it imperative that we not only articulate its operation as a foundation for disaster response and future planning but we also unsettle its often unquestioned position as the dominant economic paradigm of our generation.

Deep racial and social inequality and concurrent patterns of geographic expression are not new to New Orleans, but the moment of Katrina and ensuing years of redevelopment reflect compounding development dilemmas that make it hard to advance social justice claims. The larger political economy within which a disaster unfolds helps shape possible and probable responses. Disasters also unfold on racial landscapes that unequally affect people of color. Finally, a deeper environmental development history contextualizes both disaster and rebuilding. Separately, these development dilemmas are at odds with goals of a more socially, racially, economically, and environmentally just landscape. Together, they solidify a rebuilding based on elitism and reproduction of privilege carried over from long historical trajectories of racism, political economy, and a skewed environmental landscape on which claims for social justice are lodged.

This chapter explores these interrelated and compounding development dilemmas and argues that these dilemmas, both on their own and together, impede the advance of social justice as a normative claim on urban planning and the state. The three development logics that dominate post-Katrina New Orleans—neoliberalism, racialization, and land/environmental development—set the stage for how planners imagine and rationalize the future trajectory of the city. Although post-Katrina development was also shaped by resistance to these frameworks, we explore how social justice claims were fundamentally limited by these dominant logics.

In the following sections, we explore how particular strategies of a neoliberal political economy and neoliberal ideology shaped planning and recovery in New Orleans after Katrina. Next, we argue that racialized landscapes and the long history of race as development logic have also shaped planning and reenvisioning the city's future. We then turn to a discussion of how the longer history of unequal environmental vulnerability presents specific post-disaster development dilemmas. We then discuss how these development dilemmas compound one another to shape normative ideas of state action and equity in a post-disaster moment and argue that to envision a more just landscape requires further articulation of how the three dilemmas work in concert to reproduce inequality.

## 8.2 Neoliberal Disaster Landscapes

More attention is warranted to the ways neoliberal ideologies prevailed in the urban planning process before and after disaster and, therefore, to one of the main forums through which recovery ideas, representations, and development agendas are shaped. Raising questions about how neoliberal ideologies work in recovery planning requires that we think through ideological shifts at work in late capitalism and how these shifts emerge in the disciplines and logics of urban development and planning. Understanding the embeddedness of neoliberal ideologies like personal responsibility, entrepreneurialism, and primacy of private over public solutions to collective problems, as plausible responses to disaster and post-disaster urban planning, requires that we resituate the moment of Katrina as it unfolded from a much longer trajectory of unequal development.

The neoliberal moment in post-disaster New Orleans was foreshadowed by an earlier tilt toward private sector solutions to planning and urban development problems that created a context for disaster capitalism and rebuilding after the storm (Klein 2007). Manufacturing decline and the oil crisis of the late 1980s brought a deep recession, high unemployment, and incarceration rates of mostly black residents that reached the highest in the world, while a narrative of individual failure accompanied an attack on public housing as the site of social problems and urban decline. The oil bust, coupled with the larger retrenchment of the social welfare state (Peck and Tickell 2002), further solidified an urban geography of white population loss, poorly performing and racially segregated public schools, and segregated and deteriorating housing stock (Lewis 2003). While certainly these shifts were resisted at every turn (Woods 2017), racial geographies of the city calcified prior to Katrina through a complex renaissance of downtown development and abandonment (Campanella 2006) that had gentrified much of the historic higher ground. By the 1990s, advance of the city's tourist economy was marked by a hotel boom downtown and increased gentrification on high ground (Campanella 2006).

The city's long stigmatized and ignored public housing developments invited market-based alternatives and public-private partnerships, both before and after the storm, to replace publicly funded housing. Much like other cities, the federal Department of Housing and Urban Development (HUD) failed to adequately maintain public housing projects in New Orleans as part of a decade-old strategy to engender crisis and justify demolition of public housing. Since the 1980s, the federal government supported solutions to low-income housing problems with a series of programs (e.g., Business Improvement Districts, Community Development Block Grants, New Market Tax Credits, Low Income Housing Tax Credits (LIHTC)) committed to private initiatives, while moral entrepreneurs trumpeted individual solutions to problems of the poor (Peck 2006; Wacquant 2010). An ethic of self-sufficiency and entrepreneurialism accompanied creation, in 1974, of the Section 8 housing voucher program and the use of LIHTC and Hope VI programs to support demolition of public housing and subsidies for low-income homebuilders instead of low-income housing residents (Hackworth 2007, pp. 47–8). In New Orleans, the St. Thomas housing project met the wrecking ball in 2002 to clear land in the Lower Garden District for gentrification (Arena 2012). Tax Incremental Financing (TIF) helped fund the new mixed-income River Gardens Apartments and Walmart store at the former St. Thomas site as a blueprint for the demolition of public housing after the storm (See Fig. 8.1).

Like other U.S. cities, a retrenchment of social welfare provisions and publicly funded health care in New Orleans before Katrina accompanied growth of a disadvantaged population reliant on non-living wage jobs in an increasingly tourist-dependent city (Gotham 2007). Additionally, increased privatization of public facilities and institutional infrastructure preceded Katrina, as did other forms of devaluation and destruction (Brenner and Theodore 2002). The transfer to the state's general fund for 10 years before the storm of federal money (under Disproportionate Share program) appropriated to maintain and upgrade Charity Hospital is an example of degrading public facilities in preparation for privatization and the roll out of gentrification (Ott 2012). Plans to abandon Charity Hospital and develop a biomedical district in the Mid-City neighborhood adjacent to downtown were proposed prior to Katrina. Like public housing, maintenance on the hospital was deferred, which helped create a crisis in public sector health care. Minimal flooding during Katrina enabled Louisiana Governor Kathleen Blanco to close the hospital and appoint Louisiana State University to oversee it. Blanco's successor, Governor Bobby Jindal, hired private contractors in 2013 to manage the rest of the state's public hospitals, while the federal government agreed to spend $625 million for a new Veteran's Administration Hospital to anchor gentrification and poverty removal across the Mid-City neighborhood (see Fig. 8.1).

Privatization also impacted the City's public-school system prior to Katrina. Just before the storm, poor overall performance on standardized tests by students in the New Orleans public school system prompted a shift in control of the system from the local School Board to a state contractor, the private accounting firm of Alvarez

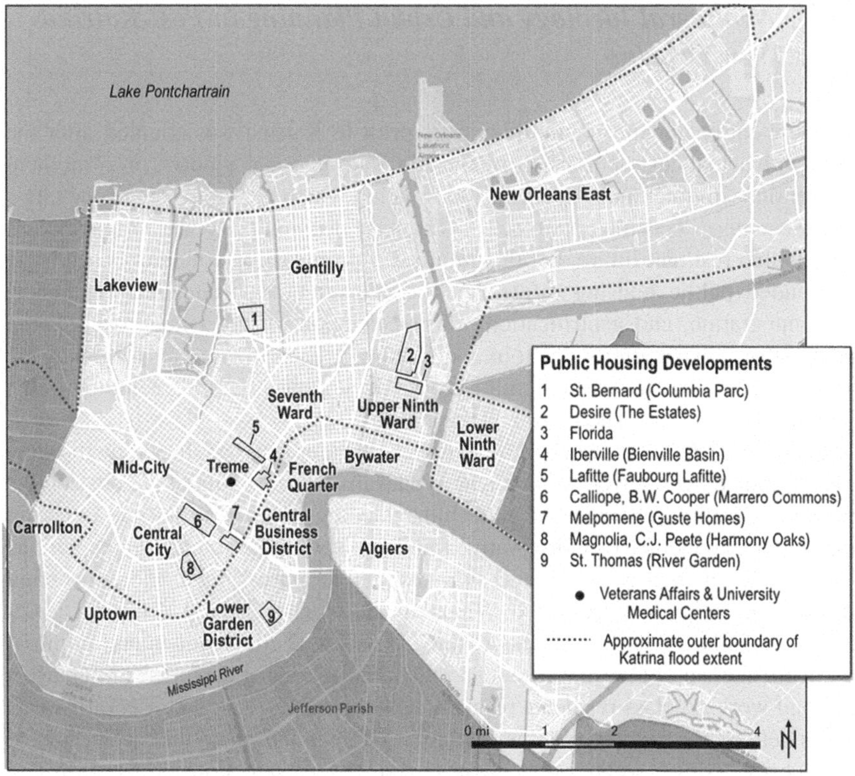


**Fig. 8.1** Map of Orleans Parish, LA, showing neighborhoods and the locations of major public housing developments as they were before Hurricane Katrina. (Data retrieved from City Planning Commission of New Orleans and Google Maps. Base map courtesy of Esri, DeLorme, HERE, MapmyIndia, © OpenStreetMap contributors, & the GIS community)

and Marcel. Days after the storm shuttered the city's schools, 7500 public school teachers were laid off, their unions broken, and their schools reopened as charter schools or as private schools funded by state vouchers (Brown 2015; Akers 2012). More flexible employment conditions now prevail and tenure protection was removed for teachers in a decidedly neoliberal transformation of New Orleans public education.

The advance of private profit and individual solutions to collective problems in the years prior to Katrina effectively calcified deep social, economic, and environmental inequalities and normalized what would be a flood of neoliberal policy measures advanced in the wake of Katrina. While racial and environmental equity and sustainability were fought for locally, particularly by African-American residents in the years preceding Katrina, "the disaster before the disaster" laid the groundwork for what would unfold later.

### 8.2.1 *Neoliberal Ideology and Urban Planning in Post-Katrina New Orleans*

The slow entrenchment of multiple crises prior to Katrina was coupled after the storm with actual destruction. The moment of Katrina was a window, or a "moment of creation" (Brenner and Theodore 2002) within which reconfiguring of the city's land-use patterns became possible. This formerly declining city was reimagined as a city with a resilient future, but this future was promised by a market-based logic that guided public housing reform, public education, environmental resilience, racial integration, and gentrification. The localization of larger neoliberal spatial and privatization paradigms that took root in New Orleans in the years following the storm reshaped the city through multiple sectors and mechanisms, including urban planning (Brenner and Theodore 2002; Johnson 2011).

As Peck (2006) notes, much of the post-Katrina moment was shaped by the slow pace of immediate recovery efforts and by narratives formed about the city and its residents. Despite adherence to color-blind ideology, post-Katrina narratives, in both the media and across conservative think tanks, were dominated by depictions of the black, urban poor as "savages who resorted to murder, rape, and destruction during a time of crisis" (Peck 2006; Whitehall and Johnson 2011, p. 60). This depiction extended larger narratives about New Orleans as a crime-ridden and immoral city in decline, but it also played out an individualized racist depiction of the urban poor and working-class residents of many neighborhoods. These conservative narratives defined the city's problems (and its residents) and, from this perception of delinquency, crime, welfare-dependency, and government corruption, proposed policy orientations for rebuilding the city that centered the private market, limited federal initiative, and demanded individual accountability in exchange for government support.

The advance of a "common sense" neoliberalism at the local level informed redevelopment policies at all scales and influenced which neighborhoods were deserving of recovery efforts and dollars (Johnson 2011). In the months following the storm, then New Orleans Mayor Ray Nagin initiated a planning process when he created the Bring New Orleans Back Commission (BNOB 2006). The BNOB urban planning committee promoted a limited return of residents to the city and proposed a reduced geographic footprint at the same time that they recommended deployment of resources to areas and residents least affected by flooding rather than to areas and people hardest hit by the storm (Brand 2007; BNOB 2006). Although the Mayor backed away from his own committee's recommendations, the lingering focus on neighborhood scale redevelopment occupied a central place in subsequent planning efforts, as did unresolved concern about who could return home after the storm. Of course, a lack of citizen participation in this planning effort and its limited analytical rigor limited the plan's capacity to address the uneven racial geography of the city, not to mention the city's environmental vulnerability. However, subtle support for removal of unwanted residents (areas proposed for return to green space were racially mixed or predominantly black) was rationalized as a more suitable and sustainable development plan. The BNOB logic of prioritizing rebuilding based on depth of flooding and likelihood of residents to return reveals how the neoliberal discourse of individual market power can extend to redevelopment plans that prioritize the upper middle class and their right to the city over that of poor residents of color (see also Chap. 9 of this volume).

Indeed, residents of many neighborhoods were told that they had to prove their viability to rebuild and were forced to compete for public resources such as public schools and street repairs. Each of the city's subsequent major planning efforts promoted individualization of neighborhood recovery and reinforced market-driven redevelopment agendas. Much of the post-disaster planning and development process proceeded with a competitive approach that pitted neighborhood against neighborhood, combined with targeted recovery zones meant to incentivize private investment. Community-level planning was one way that residents engaged this dilemma and resisted limitations of a neoliberal recovery paradigm (Brand 2007). However, while neighborhood-level input and participation were critical to the city's master planning and zoning efforts, what remains missing from city-led planning is a deeper consideration of the distributional consequences of recovery spending in each neighborhood and how residents' ideas about equity reflected their various privileges (Brand 2015; Reardon et al. 2009).

### *8.2.2 Post-Katrina Development Outcomes*

Policies and ideas associated with neoliberal urban reform and a new form of uneven development define redevelopment strategies and landscapes after Katrina. A further focus on entrepreneurism in the post-disaster city reinforced individual responsibility for recovery while also supporting a free-market redevelopment landscape. In addition to a full-scale restructuring of public education and public/affordable housing, transformations of the city also include massive land clearance, the redevelopment of a Bio-Medical district in Mid-City, and a reworked racial and economic landscape that privileges gentrification, expansion of racial privilege, and new forms of exclusion and abandonment.

Thousands of public housing residents were displaced without access to their belongings when slightly flooded and historically undermaintained public housing projects were closed after Katrina. The revalorization of public housing sites in post-Katrina New Orleans relied not only on discursive strategies that devalued and stigmatized these properties but also on neoliberal urban development solutions designed to privatize public property (Weber 2002). The federal department of Housing and Urban Development (HUD) and local Housing Authority of New Orleans (HANO) announced plans in June, 2006, to demolish the "Big Four" housing developments (C.J. Peete, Lafitte, St. Bernard, B.W. Cooper—see Fig. 8.1), home to 4500 people before the storm (Tulane School of Architecture 2007). Local residents and preservation activists struggled mightily to convince authorities to repair and reopen public housing, but in a dramatic 7–0 vote on December 1, 2007, the New Orleans City Council endorsed demolition.

Public housing projects demolished after the storm were replaced by an expansion of Section 8 low-income housing vouchers and mixed-income developments that featured a combination of market rate, subsidized, and a reduced number of public housing units. A federal Gulf Opportunity Zone (GoZone) designation enabled allocation of New Market Tax Credits (NMTC) and other public funds to private firms like Capital One, Enhanced Community Development, Whitney New Markets Fund, Crescent Growth Capital, and other "tax credit equity investors" that could partner with private developers like HRI, Columbia Residential (St. Bernard Development), Providence Community Housing, L + M Development Partners (Lafitte Development), and others to build mixed-income developments where most units command market rents (Novogradac 2015). Before the storm, the St. Bernard Public Housing Development contained 1464 public units; its replacement, Columbia Parc, has 221. C.J. Peete public housing contained 1403 public units; its replacement, Harmony Oaks, has 198. Guste public housing contained 993 public units; its replacement has 577 (Webster 2016a). The over 3300 fewer families now living in public housing were low-income, mostly black residents displaced by neoliberal post-Katrina housing policies to make way for gentrification of large swaths of Central and Mid-City, Treme, and the Lower Garden District (see Fig. 8.1).

The same publicly subsidized private companies that built mixed-income housing on sites made vacant after demolition of public housing often built off-site private housing near the vacated sites (e.g., Enterprise Community Partners and L + M Development built near former Lafitte public housing development (see Fig. 8.1)). These companies also did commercial developments nearby (e.g., L + M developed Whole Foods in Mid-City) to claim increased rents available from poverty removal and gentrification. A prominent example of publicly subsidized private commercial and community service property development is Crescent Growth Capital, which lined up NMTC financing for New Orleans Culinary Hospitality, Ursuline Academy, NOMMA School, NOCCA, Myrtle Banks, Carver Theater, Joy Theater, St. Margaret's Daughters Home, Belleville Assisted Living, New Orleans Healing Center, Transportation Consultants at Port of New Orleans, and others—all part of the neoliberal strategy of using the tax code rather than government grants as a market-driven approach to incentivize private sector investment in low-income communities (Crescent Growth Capital 2018).

The Section 8 voucher program was expanded after Katrina to replace public housing and meet housing needs of other low-income residents. In 2005, Housing Authority of New Orleans (HANO) served 14,129 families—8981 received vouchers and 5148 lived in public housing. In 2016, HANO served 19,175 families and only 1820, or 9%, lived in public housing (Webster 2016a). Higher rents and a shortage of rental housing meant more residents needed housing assistance after the storm. The Section 8 waitlist currently stands at 20,000 and is closed to additional residents in need of housing assistance (HANO 2019). Available federal assistance under the Section 8 program pays landlords favorable rents and places additional cost burdens on poor residents. Section 8 housing costs are higher than public housing costs because residents must pay deposits, utilities, and up to 30% of their income in rent (Webster 2016a). Most Section 8 vouchers pay for housing in low-income black neighborhoods, many in the large residential area called New Orleans East, and most are located away from higher-ground neighborhoods along the river. One report concluded that 82% of landlords, most in more affluent neighborhoods, refused to accept vouchers or placed unreasonable demands on tenants, which suggests that the Section 8 program reinforced racial residential segregation after the storm (Webster 2016a).

While both the closure of public housing and precariousness of public education and public transit make it harder for working-class or poor families to thrive in the city, these moves are rationalized by past failures of the state and the continuing criminalization of the poor. As expected, neoliberal urbanism is reflected in a new urban form that, as discussed above, includes large-scale redevelopment, new urban housing markets, and an intensification of gentrification (Barrios 2010; Brenner and Theodore 2002). New forms of gentrification and uneven development intersect with the appropriation of black sociospatial culture and environmental sustainability. It is important to point out here that the frontiers of development in post-Katrina New Orleans have been unevenly distributed across neighborhoods. Affluent, white, but heavily flooded Lakeview received instant endorsement for rebuilding after the storm while post-Katrina environmentalism prompted near abandonment of black neighborhoods in New Orleans East and the Lower Ninth Ward. Gentrification and attraction of new white residents to "authentic" black areas like Treme were accompanied by near abandonment by publicly organized rebuilding efforts of other predominantly black, less gentrifiable neighborhoods (Brand and Bailey 2017). Cultural gentrification and tourism-based promotion of black culture in the historic Treme neighborhood may bring benefits to individual residents, but it simultaneously displaces producers of local culture (Kelley 2003). We move next to locate the neoliberal transformation of post-Katrina New Orleans within the re-racialization of the city's inhabitants and landscapes.

## 8.3 Racial Landscapes, Racial Processes, and Racialization

It is unquestionable that pre-Katrina New Orleans was not a socially just or equal landscape. This does *not* imply that residents in the city's most neglected neighborhoods did not have strong connections to place or that their sociocultural and spatial practices did not signify alternative interpretations and imaginations about their built and social landscapes (Brand 2018). It *does* mean that residents long suffered from failing public schools, poor access to living wage employment opportunities, and urban development histories that left many poorer and predominantly black urban areas particularly vulnerable to flooding, environmental/developmental racism, and disinvestment. Although the emergent city can be understood in a neoliberal framework, the cementing of market logic and individualized solutions to collective problems must also be framed both as an extension of deeper historical geographies and as (re)racialization of the city's landscapes. The landscape within which Katrina unfolded was unequal by nearly every metric that urban planners use to measure a place and its people,[1] presenting a post-disaster development dilemma that, while not new or even unique in the larger American context, reflects the depth and expansiveness of racial/spatial formation. The sheer historical depth of racial geographic expression in the United States, now exacerbated by disaster and a political economy that limits claims based on racism, presents specific development dilemmas for those wishing to attend to racial, spatial, and socioeconomic justice.

In this section, we rely on a conceptualization of racial processes as inherently spatial (McKittrick 2006). The spatial formation of race can be tracked and understood just as racial processes can be tracked and understood through their ideological and practical shifts over time (Omi and Winant 2014). An inquiry into the racial landscapes of New Orleans requires resituating our analytical/temporal framework to grasp the rise of neoliberal space making and the "as is" essentialism and reduction of black lives that informed post-Katrina narratives (McKittrick and Woods 2007, p. 6). The time and space of race at the moment of Katrina are therefore complexly arranged geospecific contexts represented not by their settled contours but by their shifting enactments of ongoing racial processes. The New Orleans racial landscape, like that of most Southern American cities, is rooted in the formation of the city and its longer history of colonization, slavery, and post-emancipation segregation. Patterns set forth by both slavery and freedom are set deeply into neighborhoods like Central City, the Lower Ninth Ward, Treme, and New Orleans East, where the city's majority black population lives.

### *8.3.1 An Historically Racialized Landscape*

The landscape of New Orleans is thoroughly racialized. White settlement of the historic high ground between the Mississippi River and Lake Pontchartrain required removal of indigenous Native Americans and forced migration of enslaved Africans into the city. Enslaved Africans were bound both to the land and to their masters in a primitive state of incarceration. This was a racial dictatorship protected by religious ideology that prevailed for more than 250 years (Omi and Winant 1994, p. 65). The toil of enslaved bodies transformed the landscape and embodied wealth that engendered profit, status, and credit for the enslaver. Antebellum historians largely

[1]Part of the dilemma for urban planners are the metrics used to measure place and people. As McKittrick and Woods (2007) argue, while identifying and mapping racial inequalities is critical to the task of naming racism, this effort, when not accompanied by efforts to see and understand how black geographies are made through struggle and resistance (6), reifies and naturalizes racialized inequality and uneven racial geographies. While a fuller discussion of this dilemma is out of the scope of this chapter, it should be noted that many of the measurements served up as evidence of the city's failures were used to further neoconservative and neoliberal redevelopment agendas (see Peck 2006, etc.) in the wake of Katrina.

recognize the combination of enslaved labor and collateral as fundamental to development not only of the landscape of New Orleans but of global industrial capitalism and the industrial revolution (Baptist 2014; Du Bois 1962 [1935]; Johnson 2013).

Enslaved Africans in antebellum New Orleans were concentrated uptown along the Mississippi River near white plantation owners, while free people of color mostly lived downriver and "back of town" in Treme and the Seventh Ward (Morse 2008; see Fig. 8.1). Renegade enslaved Africans, Native Americans, and other outcasts occupied the back swamps of the city. After reconstruction, scientific theories of racial inequality legitimated race-based segregation and most black people were segregated into squalid dwellings in low-value, flood-prone swampland like Gert Town and Broadmoor (see Fig. 8.1).

The invention of the wood screw pump in 1917 permitted drainage of the back swamps of Mid-City and the Lakefront area which prompted white flight to these early suburbs and reinforced the location of whites in better drained parts of the city (Colten 2005, p. 77; Spain 1979). While shaped by the rationality of progressive era approaches to public works, and public health concerns of planners, development of the city's drainage and sewer system during the Jim Crow era of racism meant that newly expanded urban "territories" were racially and economically segregated as they emerged from the swamps (Colten 2002, p. 239, 2005; Lewis 2003). Black areas were cordoned off by commercial and industrial corridors (e.g., Dryades Market) laid out in the first Comprehensive Zoning Ordinance of 1929 and by restrictive covenants that blocked blacks from white neighborhoods (Hastings 2004, p. 300, 326; Morse 2008). Chronic housing shortages plagued the black community after World War II when passage of the Federal Housing Act of 1949 extended the federally funded public housing program to construct segregated housing in cities across the country (Von Hoffman 2000). The location of federally funded housing developments built for blacks only in New Orleans in the 1940s and 1950s further concentrated the black population in the Second (B.W. Cooper/Calliope), Seventh (St. Bernard), Ninth (Desire), and Eleventh (Magnolia) Wards, creating a black residential belt off the river, much of which was formerly back swamp (Spain 1979).

Civil rights struggles and the emergence of a more democratic racial ideology failed to challenge the racialization of space in New Orleans. The official collapse of the separate but equal doctrine that justified Jim Crow segregation until the 1954 Brown v. Board of Education decision of the US Supreme Court coincided with construction of separate but equal suburban housing in 1955 at Pontchartrain Park—walled off from surrounding white suburbs by canals and a railroad line, one road in and one road out (Baxter and Casati 2019). Segregation was further reinforced by construction of the Desire housing project in 1956 in the Upper Ninth Ward on low ground next to the Industrial Canal. The Ninth Ward served generally as a landscape for blacks displaced from their homes by construction of Armstrong Park, Union Passenger Terminal, City Hall, the Superdome, and Interstate 10 (Spain 1979; Crutcher Jr 2010). Initially supported by construction of the I-10 expressway along Claiborne Avenue (through the city's historically black Treme and seventh Wards), the construction of lower-income subdivisions and later white flight from formerly integrated neighborhoods during the oil crash and recession of the late 1980s made

New Orleans East yet another black area located on "reclaimed" swamp land.

As previously noted, the geography within which Katrina unfolded was critically and indelibly marked by generations of urban development paradigms that were themselves inherently raced. The deeper history was of course always marked by struggle and the resistant spatial ontologies of communities of color, as Woods (2017) argues in his historical analysis of the pre-Katrina city. Yet, while a nascent neoliberal bent would harden through post-Katrina development, the colorblindness of neoliberalism would recalibrate the city's racial geographies through new modes of abandonment, asset-stripping, and gentrification (Woods 2017).

### *8.3.2 Newly Racialized Landscapes*

Post-Katrina racialization takes place through multiple sectors: including tourism, the demolition of public housing, public education reform, gentrification, public transportation investments, and tourist-centered redevelopment. It also included hyperdevelopment and abandonment. This framework accounts for abandonment of the Lower Ninth Ward and much of New Orleans East, as well as for unprecedented levels of development and gentrification pressure in Treme, Mid-City, and high-ground neighborhoods along the river such as Bywater (Brand and Bailey 2017). Population shifts after Katrina are one indication of changing racial demographics. Between 2000 and 2016, the city lost over 95,000 black residents (U.S. Census 2000; 2010; U.S. American Community Survey 2016). Geographically, this loss is represented in black population decline in areas along the river and adjacent to downtown, particularly in high-ground neighborhoods like Bywater, the Marigny, Central Business District (CBD), and Lower Garden District (see Fig. 8.1).

Pre-Katrina land tenure and valuation trends recalibrated in many ways after Katrina. Several majority black housing projects were constructed on lower ground and anchored low-income black neighborhoods with low rents and low house values. The historical context of racialized neighborhoods is that blacks are less able than whites to obtain home mortgages either because they are excluded from housing markets or because they have, on average, much less money and less stable jobs than whites. That means black residents are more likely than white residents to rent as opposed to purchase a home. This left many black residents without a real voice in a recovery process after Katrina that was heavily tilted toward homeowners (Reed 2006; Morse 2008). Residents of poorer black neighborhoods (typically on lower ground) with more renters and lower house values were also considered less likely to return and rebuild after the storm than residents of wealthier white neighborhoods like Lakeview.

Land formerly occupied by public housing sites (St. Bernard, Iberville, Lafitte, St. Thomas), and by majority black residents, was coveted by developers and potential white residents because of its elevation and proximity to employment downtown. Indeed, the advance of mixed-income and market-rate approaches to redeveloping sites on historic and largely unflooded high ground significantly reduced black population. The storm allowed closing of this racially based rent gap and displacement of low-income black renters so gentrification could proceed, leaving "symbolic" numbers of black residents within the subsidized portion of new federally backed housing built in the wake of the storm.

Beyond renters, black homeowners were also short-changed as the Road Home program funded by HUD and the State of Louisiana to help homeowners not covered by the National Flood Insurance Program initially allocated grants based on the value of the house pre-Katrina rather than the cost of repairs to the house after the storm. That meant houses in Lakeview neighborhoods occupied by whites received much larger average rebuilding grants than houses in Gentilly and the Lower Ninth Ward occupied by blacks. This inequity was remedied by a later court case, but most black homeowners were never made whole given that the court decision was not retroactive for property owners who had already made claims.

Asset-stripping describes well the allocation of disaster funding after the storm (Woods 2017). The statute that authorized Community Development Block Grant (CDBG) funding for New Orleans originally required that 70 percent of all money be awarded to support persons of low to moderate income. The State of Louisiana obtained waivers from the federal government after Katrina and ended up granting only 34 percent of CDBG money to low- and moderate-income residents. Homeowners received 79 percent of the grant money while the 34 percent of residents identified as renters received only 20 percent of CDBG grant money (Morse 2008). CDBG funding was also used for projects such as the Lafitte Greenway and Crescent Park, supporting public space expansion in areas experiencing rapid land revaluation and green gentrification (Anguelovski, Connolly and Brand 2018). This exemplifies the compounding dilemma of market-based economic logic rooted in deeply racialized systems of spatial inequality.

Snapshots of reinvigorated racial inequality include the continued lack of a grocery store in the Lower Ninth Ward and lack of sustained redevelopment in New Orleans East. Both areas continue to exhibit the slowest rate of return after Katrina. The slow rates of return reveal the intertwined workings of racialized capitalism and development given the direct prohibitions on residents' return (particularly in the Lower Ninth Ward) but also the ways that racial geographies unfold along with new contexts of geographical abandonment, continued inequalities in access to resources, and market-based development ideologies that typically avoid lower-income communities of color.

Racial processes are also evident in the return of middle-class and white elites to the city through production of luxury housing stock on high ground near the river and in the revaluation of property in the CBD and Mid-City neighborhoods. It takes place through the reconfiguration and aspatial reorganization of public education delivered by charter schools, the focus on tourist-oriented transit development (Lowe et al. 2018), and further calcification of the city's tourist economy (Gotham 2007). It also takes place through the abandonment of a dedicated public hospital for the city's poor and low-income residents as disaster recovery money was spent in the biomedical district to facilitate economic redevelopment in Mid-City.

A final point on racial landscapes that is underexamined in urban planning, including post-disaster planning, is the continued emphasis on economic opportunity, which obfuscates how white supremacism and privilege operate spatially. Many planning scholars advance platforms for racial equity and highlight claims on the state and on our collective urban imaginary for more racially just landscapes, and these views go above and beyond economic opportunity (Bates et al. 2018). Our point is that more work is needed in urban planning on how racial geographic structuring and white supremacism work in and through our discipline.

### *8.3.3 Racialized Space as Development Dilemma*

The endurance of the city's racialized geography poses specific development dilemmas often overlooked and ill-conceived in the realm of post-disaster urban planning. The seeming intractability of the city's unequal racial geographies was used as fuel to advance neoconservative spatial and fiscal ideologies in the months and years following Katrina (Peck 2006). Despite the hegemony of racial neoliberalism as the city's redevelopment paradigm, urban planning and numerous participatory processes in the years following Katrina promised possible avenues for eliciting more just visions for the city's future. Yet, when racialized space is so deep, it cannot be simply solved by existing urban planning paradigms (Omi and Winant 2014; Peck and Tickell 2002; Brand 2015).

This latter point is critical because urban planning, despite its normative commitments to social justice, lacks a clear framework for equity in the late twentieth and early twenty-first centuries (Zapata and Bates 2015). The types of policy making and redistribution of power, wealth, and spatial arrangements required to ameliorate a history of inequality experienced by the city's black population raise critical claims for the state and urban planning policy makers in the neoliberal era (Reardon et al. 2009). There are, to be sure, planners whose work highlights these claims and who work to ameliorate intersecting conditions of racialized/spatialized inequality. Even if urban planning re-embraced policies oriented toward equity and redistribution, as many planners and policy makers did in the 1970s and 1980s (Krumholz 2003), market-based hegemony makes these types of claims less tenable. In this sense, we must understand New Orleans's racial geographies as both evidence of the ongoing work of a racial state and racial formation and also as geographies resynthesized under the guise of neoliberal space-making (Omi and Winant 2014).

Racial geographies are reworked in the neoliberal era, but they are not new. The not-newness of racial geographies underlines the hegemony of a racial state whose development and planning has always been enacted through space. Although many scholars point to this "fatal coupling" (Gilmore 2002) of race and space, the central development dilemma posed by racial geographies is not only that they are not new but that they are normalized in moments such as Katrina and alternatives proposed by black residents have been largely muted under neoliberal colorblindness (Reardon et al. 2009.) Yet, the effects of political economies and racial landscapes cannot be extracted from the development modalities that produced the pre-Katrina moment and the deep environmental injustices that existed prior to the storm. The environmental and development landscape underlies how racial justice was spatialized and it is to this development dilemma that we now turn.

## 8.4 Unjust and Unequal Environmental Development

New Orleans is an "impossible but inevitable city" (Lewis 2003, p. 19) born of the logics of economic expansion and the management of nature, both of which coincide with a racialized development history. The city is located on an unstable deltaic plain that features the dangerous convergence of sea-level rise and shrinkage of the land (subsidence). This convergence is made more dangerous by reduced rates of sediment supply to rebuild land due to the construction of an extensive levee and canal system to protect the city from flooding (Blum and Roberts 2012). Put another way, construction of levees and drainage canals increases subsidence and flood risk and therefore contains an environmental contradiction because they dry out land and reduce the ability of the Mississippi River to deposit sediment required to make new land (Baxter 2014). Consequently, further drowning of land in the deltaic plain of New Orleans is inevitable, which means every citizen of New Orleans is at considerable risk of future flooding (Blum and Roberts 2012).

Not every citizen of New Orleans confronts an *equal* risk of exposure to future flooding because of the city's history of unequal and uneven development. Structural patterns of inequality and racism created and sustained unequal power relations and increased disaster vulnerability for poor black citizens shunted into more flood-prone parts of the city; this is a condition that has not abated post-Katrina (Susman et al. 1983; Freudenberg et al. 2008).

### *8.4.1 A History of Environmental Vulnerability in New Orleans*

New Orleans is a flood-prone city where, after the demise of slavery, blacks, indigenous peoples, criminals, and outcasts largely populated marginal lands with inferior drainage near industrial areas and in back swamps located beyond the thin teapot or, "sliver by the river," that lies above sea level (Lewis 2003; see Fig. 8.1). It is well documented how the city expanded toward Lake Pontchartrain when invention of the wood screw pump enabled drainage of former back swamps (Lewis 2003; Colten 2005; Campanella 2008). Subdivisions sprang up across the center and back of town, some of which carried racial covenants to keep out blacks and other so-called undesirables. The city became reliant on pumping floodwaters out of these flood-prone areas whenever there was a hard rain. However, pumping stations and drainage canals were not enough to protect the city from storm surge associated with a major hurricane. That required federal intervention.

Congress authorized federal involvement in levee protection for New Orleans in 1955 when it ordered the Army Corps of Engineers to develop plans for a Lake Pontchartrain and Vicinity Hurricane Protection Project. Substantial funding for the project became available in 1965 when Congress allocated $56 million for hurricane protection in New Orleans after Hurricane Betsy flooded 6600 homes, many in eastern New Orleans (Seed and Bea 2006, p. 4–22; Shallat 2006). Political wrangling continued for decades between real estate investors, environmental activists, and other community groups, over the extent and type of levee protection to construct. A "high-level" barrier plan that required raising, strengthening, and extending existing levees was finally adopted in 1985 and produced the hurricane protection system that failed catastrophically when Hurricane Katrina struck the city (Seed and Bea 2006).

Construction of a hurricane protection system for New Orleans suffered with the roll out of the neoliberal policy regime in the 1980s. Sporadic funding and cost-cutting private contractors delayed completion of the levee system as corners were cut on site preparation and building materials, especially on levees built along major outfall canals that drain water out of the city and along the Industrial Canal (Seed and Bea 2006). The mark of neoliberalism was also evident in rollbacks of environmental regulations after the election in 1980 of President Ronald Reagan consolidated government power in favor of financial and land-use deregulation to stimulate real estate investment as a cornerstone of national economic policy (Hackworth 2007). Federal agencies like the Federal Emergency Management Administration (FEMA) and the National Flood Insurance Program provided insurance that softened oversight of local land-use policies and increased federal responsibility for flood risk, which empowered real estate developers and resulted in a 50 percent increase over 30 years in floodplain construction across the country (Burby 2001). New Orleans was no exception as construction of subdivisions flourished on ill-prepared swamp land in New Orleans East and on the West Bank of the Mississippi River (Baxter 2014).

Katrina flooded eighty percent of New Orleans and the damage raged across social class and racial lines, though low-income black areas were most likely to be damaged by floodwaters (Logan 2006) and black residents were the most impacted. Post-World War II migration of whites to Lakeview, Gentilly, and parts of New Orleans East placed them in slab houses built at grade below sea level, so failure of the Army Corps to construct adequate levees, and inadequate land-use regulations, made these affluent areas vulnerable to extreme flooding as well. While many affluent whites suffered flood damage from Katrina, the white population had had a significant out-migration from New Orleans East before Katrina, thus reducing their vulnerability. White residents also dominated the high ground along the river and the river bend in Algiers, the French Quarter, Bywater, and Marigny neighborhoods that largely escaped the flood as well as the artificial high ground created at the Lake's edge by using lake dredge (see Fig. 8.1). If we recall that many black neighborhoods are located in a belt located off the river (Central City, Mid-City, Gentilly,

Upper Ninth Ward) that severely flooded, racial inequities in exposure to flooding are clear (Logan 2006).

### *8.4.2 Race and Environmental Justice in Post-Katrina New Orleans*

Concerns about environmental justice in post-Katrina New Orleans are multi-scaled and multifaceted. The development dilemma posed by the city's daunting environmental future is twofold. First, the environmental development history of the city created sociospatial vulnerability to large-scale events such as Hurricane Katrina *and* to everyday flooding events like the rain-event flood in August, 2017, which dumped 8–10 inches of rain on the city within 4 hours. These concerns are not equally dispersed across the city nor are residents equally situated to respond to environmental crises, making non-white and lower-income residents more vulnerable than others. Second, rational responses to recovery that attempted solutions abstracted from larger sociopolitical and racialized contexts largely failed to attend deeper racial and political economies. Not only were residents geographically unequally vulnerable prior to the storm; those vulnerabilities were exacerbated by solutions posed for rebuilding different areas of the city after the storm.

Environmental vulnerability is coupled closely with racialized environmental injustice in New Orleans. Toxic sites such as landfills and industrial plants are more likely sited in the city's black geographies. The Ninth Ward is bisected by the Industrial Canal, constructed in the 1920s to improve shipping along the Gulf Coast. On one side of the Industrial Canal lies the Agriculture Street landfill, a toxic site that festered for decades under the HUD funded Press Park housing development, constructed for low-income blacks in the late 1960s (Webster 2016b). The local school, Moton Elementary, also sat above the landfill, and the former Desire public housing development was not far away. Before the storm, low-income black homeowners on the east side of the Canal in the Lower Ninth Ward were protected by an inferior sheet pile levee built only 3 feet above grade on a small berm of land that was overwhelmed by Katrina's tidal surge (Seed and Bea 2006).

### *8.4.3 Post-disaster Redevelopment and Environmental Justice*

Katrina's levee breaks not only devastated tens of thousands of properties; they exposed the city's deep and multi-scaled environmental dilemmas. Despite calls by activists across the city, the massive work required to rebuild levees after Katrina and to analyze the city's environmental vulnerability remains incomplete. The city is reliant on a drainage system with outdated pumping technologies and an unreliable pumping system. It also relies on a levee system and coastal restoration programs whose locus of control lies outside the scale of the city, which reveals the multifaceted scales of hurricane protection and seemingly intractable problems associated with land loss and climate change.

A proposal advanced by planners to move entire communities, such as the Lower Ninth Ward, to higher ground through land swap programs designed to keep residents together failed to gain real traction after the storm. What has gained traction are efforts to utilize urban planning to catalyze stormwater management alternatives across the city. This is mainly contained in the city's master plan, comprehensive zoning, and water planning efforts. The Master Plan, completed in 2010, and the Comprehensive Zoning Ordinance, completed in 2015, handle stormwater management by private development incentives. The Water Plan, which the city largely adopted as its resiliency plan, is primarily focused on subsidence and "living with water" via new stormwater management systems, such as bioswales and blueways. As Fisch (2014) shows, the water plan has largely failed to address underlying racial equity concerns and, outside of raising property values in a city already plagued by affordable housing issues, has failed to consider the alignment of stormwater management issues associated with racialized landscapes and a right to place. For instance, one of the proposals in the Water Plan would replace a large swath of the Lowe Ninth Ward with a blue way in an area where some residents have already rebuilt their homes. The failure here is not necessarily in thinking that big changes are necessary for more environmentally just landscapes, but in not recognizing the ways that these landscapes intersect with racial processes and other forms of dispossession that Lower Nine residents have faced.

The city's environmental injustice geography has deepened along racial fault lines since Katrina. For instance, in the wake of the storm, the city pushed to open a new landfill next to the Vietnamese community of Village de l'est in New Orleans East (See Fig. 8.1). In addition, the Army Corps of Engineers renewed planning efforts to redesign the Industrial Canal lock, despite continued opposition from Lower Ninth Ward residents who, citing the repercussions of disrupting toxic canal mud and then dumping newly cut fill into Bayou Bienvenue, continue to fight the lock redesign after decades of doing so.

Environmental injustice concerns connected to the historical development of the city, subsidence, and increased vulnerability to climate change and land loss present an imposing set of physical and environmental dependencies that are difficult to address in a political climate where climate change denial prevails at high levels of government. Centrality of an environmentally just landscape to our thinking about social justice agendas and how to actualize them is also hindered by ongoing racial projects at work in redevelopment that fail to account for the nimble ways racialization is deployed and redeployed spatially. Much like Colten's (2002) analysis of the mutual evolution of racial and geographic landscapes, and Hardy et al.'s (2017) understanding of racial landscapes and vulnerabilities as always in formation, we think of rational approaches to post-Katrina resettlement (e.g., BNOB Commission, Urban Land Institute, Rockefeller-backed resiliency plans) as bound up in contemporary "color blind" racial processes and structures. These rational approaches to green redevelopment largely fail to deal with the political economy of disasters and the deeper racial terrain within which new modes of "environmentally just" landscapes are promoted.

Rational policy proposals to focus development on higher ground that is less prone to flooding and extreme environmental events do not actualize racially equal landscapes unless all metrics of access to these high-ground neighborhoods are worked out. It is evident that increased green gentrification along high-ground areas, including rising land values and an increase in green amenities, has reproduced a geography where white and affluent residents are better protected and black residents are more vulnerable (Anguelovski et al. 2018; Brand and Bailey 2017). Without mechanisms that center racial equity and access to affordable housing and community development support, market-led approaches to high-ground resettlement can lead to less affordable housing and increased displacement of economically vulnerable residents (Brand and Bailey 2017).

## 8.5 Discussion: Cumulative and Compounding Dilemmas

It is impossible to extract environmental vulnerability from the larger social systems within which it takes place. As Laska and Morrow (2006, p. 16) argue, social vulnerability "comes not so much from exposure as from the manner in which human-environment systems are coupled." More recent work by Hardy et al. (2017, p. 63) argues further that "racial formations are also environmental" and that the "socio-ecological relations of racialized space" are produced over time through racial projects. What we see in the post-disaster city are a new series of interdependent elements of a twenty-first century racial project built on a veneer of rationality and color-blind decision-making.

Each of the development dilemmas discussed above enables a general perpetuation and intensification of the types of social and environmental injustices prevalent in New Orleans prior to Katrina. The onset of a racial state committed to the reproduction of white privilege after the demise of slavery concocted various strategies to isolate black and other minority residents on lower, less well-drained land in New Orleans. Efforts to remedy racial residential segregation in the late twentieth century largely failed as market-oriented and color-blind spatial logics ostensibly replaced race with wealth and income as criteria to allocate home mortgages and rebuilding funds. Promotion of color-blind, market-oriented governmental interventions, and a slow withdrawal of the state from social services, helps some residents while abandoning others. At the end of the day, life is harder for economically and socially disadvantaged residents that lived in New Orleans prior to the storm and whose efforts to return were often blocked.

Neoliberalism can be examined as racial structuring that reinvigorated and transformed racialized landscapes that existed prior to the storm, but not in ways that attenuate racial injustice. Put simply, poor African-American residents that made up the majority of the city's population before Katrina are faced with daunting obstacles, not only in terms of day-to-day living in an unequal city but also in terms of the complex dimensions of resettlement and claims making on the state that could center the reality of racialized politics and contexts of life before and after Katrina. Color-blind racism and individualized solutions to collective problems post-Katrina are further complicated by tensions between the history of environmental development, environmental justice, and future development plans.

We find Colten's (2002) framing of "mutually evolving patterns" useful for thinking about how these dilemmas compound one another. Neoliberalism marked recovery and rebuilding of the city, yet this color-blind approach is itself blind to underlying racial structures. Historically, we see this in Colten's analysis of progressive era approaches to constructing the City's drainage systems where, despite racially progressive development goals to extend drainage to all residents, land reclamation intersected with and created new, racially segregated geographies toward the lake. The rational approach, at the time, extended the geographies of racial residential segregation and uneven development. This pattern repeats post-Katrina, but in particular ways that reproduce a geography of racialized environmental vulnerability in an era of increasing climate change vulnerability. Here, we suggest that whiteness and white supremacist systems are reworked geographically in the post-Katrina city along environmental, developmental, and racial fault lines.

Claims for racial justice and equity made by residents in the Lower Ninth Ward and New Orleans East are not easily settled by rational approaches to planning or by decisions about where development should or should not happen. Lower-lying areas such as the East and the Lower Ninth Ward are more vulnerable to flooding, but so too are the Lakeview and Gentilly areas of the city. The point is that the longer history of who settled where reveals more complicated histories that require attention if we are to imagine and promote more socially, racially, economically, and environmentally just landscapes.

## 8.6 Conclusion

This chapter rests within a volume which argues that Louisiana is ahead of the curve regarding what it can teach us about post-disaster redevelopment. New Orleans is 14 years past a devastating human-made catastrophe that brought existential threats to the city. The incredibly hardworking and engaged efforts of residents and supporters, aided by billions in disaster aid and insurance money, rebuilt the city, albeit in ways that largely reinforced preexisting inequities and development dilemmas.

Perhaps, if New Orleans is ahead of the curve, it is in the demonstrative confrontation with and illumination of the types of development dilemmas that we outlined above. Scholarship on neoliberalism, racial geographies, and environmental justice abounds, but at the nexus of the three lie specific entanglements that make dreams of a socially and racially just and resilient city incredibly difficult to realize—if attention to racialized geographic formations and reparations for these formations are not centrally attended to in our future imagining and day-to-day decisions. If New Orleans is ahead of the curve, it is ahead in its illumination of the depth of these dilemmas and the ease with which racial inequality is reproduced through disciplines committed to equity.

Despite ominous warnings, we do not wish to squash dreams of a more resilient, green, sustainable, socially just and equal landscape, nor are we convinced of the impossibility of these claims and visions. Lessons learned from rebuilding New Orleans after Katrina contain many seeds for hope. However, what we do suggest is that the paradigms within which we operate matter in terms of shaping disaster response. Further, what post-Katrina development shows is that the moment of Katrina provided an opening through which a highly racialized neoliberalism powerfully framed what happened to the city and what should happen next. The velocity and multidimensionality of this redevelopment trajectory itself poses dilemmas for enacting more racially just landscapes unless these paradigms are called to task.

## References

Akers, J. M. (2012). Separate and unequal: The consumption of public education in post-Katrina New Orleans. *International Journal of Urban and Regional Research, 36*(1), 29–48. https://doi.org/10.1111/j.1468-2427.2011.01062.x.

Anguelovski, I., Connolly, J., & Brand, A. L. (2018). From landscapes of utopia to the margins of the green urban life: For whom is the new green city? *City, 22*(3). https://doi.org/10.1080/13604813.2018.1473126.

Arena, J. (2012). *Driven from New Orleans: How nonprofits betray public housing and promote privatization*. Minneapolis, MN: University of Minnesota Press.

Baptist, E. E. (2014). *The half has never been told: Slavery and the making of American capitalism*. New York, NY: Basic.

Barrios, R. E. (2010). You found us doing this, this is our way: Criminalizing second lines, Super Sunday, and habitus in post-Katrina New Orleans. *Identities: Global Studies in Culture & Power, 17*(6), 586–610.

Bates, L. K., Towne, S. A., Jordan, C. P., & Lelliott, K. L. (2018). Introduction: What shakes loose when we imaging otherwise. *Planning Theory & Practice, 19*(2), 254–256. https://doi.org/10.1080/14649357.2018.1456816.

Baxter, V. (2014). Rent, real estate, and flood mitigation in New Orleans East. *Antipode, 46*(4), 1014–1031. https://doi.org/10.1111/anti.12080.

Baxter, V., & Casati, M. (2019). Building black suburbs in New Orleans. In T. Adams & M. Sakakeeny (Eds.), *Remaking New Orleans: Beyond Exceptionalism & Authenticity* (pp. 203–233). Durham, NC: Duke University Press.

Blum, M. O., & Roberts, H. (2012). The Mississippi Delta region: Past, present, and future. *Annual Review of Earth & Planetary Sciences, 40*, 655–683. https://doi.org/10.1146/annurev-earth-042711-105248.

Brand, A. L. (2007). Rebuilding the right to return: Toward a framework of social and spatial justice in New Orleans. *Critical Planning, 14*, 71–89.

Brand, A. L. (2015). The politics of defining and building equity in the 21st century. *Journal of Planning Education & Research, 35*(3), 249–264. https://doi.org/10.1177/0739456X15585001.

Brand, A. L. (2018). The duality of space: The built world of Du Bois' double-consciousness. *Environment and Planning D: Society and Space, 36*(1), 3–22. https://doi.org/10.1177/0263775817738782.

Brand, A. L., & Bailey, H. (2017, April). *High ground, low ground: The new topography of racial development in the 21st century.* Presentation at the annual meeting of Association of American Geographers Conference, Boston, MA.

Brenner, N., & Theodore, N. (2002). Cities and the geographies of actually existing neoliberalism. *Antipode, 34*(3), 349–379. https://doi.org/10.1111/1467-8330.00246.

Bring New Orleans Back Commission (BNOB). (2006). *Action Plan for New Orleans: The New American City.* BNOB Urban Planning Committee. New Orleans, LA: BNOB Commission.

Brown, E. (2015, September 3). Katrina swept away New Orleans' school system, ushering in new era. *Washington Post.* Retrieved from https://www.washingtonpost.com/news/education/wp/2015/09/03/katrina-swept-away-neworleans-school-system-ushering-in-new era/?noredirect=on&utm_term=.6c3e8beb9857

Burby, R. J. (2001). Flood insurance and floodplain management: The U.S. experience. *Journal of Environmental Hazards, 3*(3), 111–128. https://doi.org/10.1016/S1464-2867(02)00003-7.

Burley, D. (2010). *Losing ground: Identity and land loss in coastal Louisiana.* Oxford, MS: University Press of Mississippi.

Campanella, R. (2006). *Geographies of New Orleans: Urban fabrics before the storm.* Lafayette, LA: University of Louisiana Lafayette.

Campanella, R. (2008). *Bienville's dilemma: A historical geography of New Orleans.* Lafayette, LA: University of Louisiana Lafayette.

Colten, C. E. (2002). Basin Street blues: Drainage and environmental equity in New Orleans, 1890–1930. *Journal of Historical Geography, 28*(2), 237–257. https://doi.org/10.1006/jhge.2001.0400.

Colten, C. E. (2005). *An unnatural metropolis.* Baton Rouge, LA: Louisiana State University Press.

Crescent Growth Capital Staff. (2018). "Past Projects." Retrieved from www.crescentgrowthcapital.com

Crutcher, M. E., Jr. (2010). *Tremé: Race and place in a New Orleans neighborhood* (Vol. 5). Athens, GA: University of Georgia Press.

Du Bois, W. E. B. (1962). *Black reconstruction in America: Toward a history of the part which black folk played in the attempt to reconstruct democracy in America, 1860–1880.* New York, NY: Russell and Russell.

Fisch, J. (2014). *Green infrastructure and the sustainability concept: A case study of the Greater New Orleans Urban Water Plan* (Master's thesis). University of New Orleans Thesis and Dissertations. 1869.

Freudenberg, W. R., Gramling, R., Laska, S., & Erikson, K. T. (2008). Organizing hazards, engineering disasters? Improving the recognition of political-economic factors in the creation of disasters. *Social Forces, 87*(2), 1015–1038. https://www.jstor.org/stable/20430901.

Gilmore, R. W. (2002). Fatal couplings of power and difference: Notes on racism and geography. *The Professional Geographer, 54*(1), 15–24. https://doi.org/10.1111/0033-0124.00310.

Gotham, K. F. (2007). *Authentic New Orleans: Tourism, culture, and race in the Big Easy.* New York, NY: New York University Press.

Hackworth, J. (2007). *The neoliberal city.* Ithaca, NY: Cornell University Press.

HANO. (2019). https://www.hano.org/housing/hcvp.aspx. Accessed August 2019.

Hardy, R. D., Milligan, R. A., & Heynen, N. (2017). Racial coastal formation: The environmental injustice of colorblind adaptation planning for sea-level rise. *Geoforum, 87*, 62–72. https://doi.org/10.1016/j.geoforum.2017.10.005.

Harvey, D. (2005). *A brief history of neoliberalism.* New York, NY: Oxford University Press.

Hastings, D. (2004). *Early neighborhood development in New Orleans: Neither new south nor old* (Unpublished Dissertation). College of Urban & Public Affairs, University of New Orleans. New Orleans, LA.

Johnson, C. (Ed.). (2011). *The neoliberal deluge: Hurricane Katrina, late capitalism, and the remaking of New Orleans.* Minneapolis, MN: University of Minnesota Press.

Johnson, W. (2013). *River of dark dreams.* Cambridge, MA: Harvard University Press.

Kelley, R. D. G. (2003). Disappearing acts: Capturing Harlem in transition. In *Introduction in Alice Attie: Harlem on the verge.* New York, NY: Quantuck Lane Press.

Klein, N. (2007). *The shock doctrine: The rise of disaster capitalism.* New York, NY: Metropolitan Books.

Krumholz, N. (2003). Equitable approaches to local economic development. In S. S. Fainstein & S. Campbell (Eds.), *Readings in planning theory* (pp. 224–236). Malden, MA: Blackwell.

Laska, S., & Morrow, B. H. (2006). Social vulnerabilities and Hurricane Katrina: An unnatural disaster in New Orleans. *Marine Technology Science Journal, 40*(4), 15–26. https://doi.org/10.4031/002533206787353123.

Lewis, P. F. (2003). *New Orleans: The making of an urban landscape* (p. 42). Santa Fe, NM: Center for American Places.

Logan, J. R. (2006). *Impact of Katrina: Race and class in storm damaged neighborhoods*. Providence, RI: Brown University Publication. Retrieved from https://s4.ad.brown.edu.

Lowe, K., Brand, A. L., & Hall, E. (2018, October). *Off track? The racial dimensions of the modern streetcar. Presentation at the annual meeting of Association of Collegiate Schools of Planning Conference*, Buffalo, NY.

McKittrick, K. (2006). *Demonic Grounds: Black women and the cartographies of struggle*. Minneapolis, MN: University of Minnesota Press.

McKittrick, K., & Woods, C. (2007). *Black geographies and the politics of place*. Boston, MA: South End Press.

Morse, R. (2008). *Environmental justice through the eye of Hurricane Katrina*. Washington, DC: Joint Center for Political and Economic Studies, Health Policy Institute.

Novogradac, M. (2015). Lessons from Katrina include value of tax credit programs. *Novogradac*. Retrieved from https://www.novoco.com/notes-from-novogradac/lessons-katrina-include-value-tax-credit-programs

Omi, M., & Winant, H. (1994). *Racial formation in the United States*. New York, NY: Routledge.

Omi, M., & Winant, H. (2014, October 8). How colorblindness co-evolved with free market thinking. *The Public Eye*. Retrieved from https://www.politicalresearch.org/2014/10/08/how-colorblindness-co-evolved-with-free-market-thinking/

Ott, K. B. (2012). *The closure of New Orleans' Charity Hospital after Hurricane Katrina: A case of disaster capitalism* (Unpublished master's thesis). University of New Orleans. New Orleans, LA.

Peck, J. (2006). Liberating the city: Between New York and New Orleans. *Urban Geography, 27*(8), 681–713. https://doi.org/10.2747/0272-3638.27.8.681.

Peck, J., & Tickell, A. (2002). Neoliberalizing space. *Antipode, 34*(3), 380–404. https://doi.org/10.1111/1467-8330.00247.

Plyer, A. (2016, August 26). Facts for features: Katrina Impact. Retrieved from https://www.datacenterresearch.org/data-resources/katrina/facts-for-impact/

Reardon, K. M., Green, R., Bates, L. K., & Kiely, R. C. (2009). Overcoming the challenges of post-disaster planning in New Orleans: Lessons from the ACORN housing/university collaborative. *Journal of Planning Education and Research, 28*(3), 391–400. https://doi.org/10.1177/0739456X08327259.

Reed, A. (2006, September 18). Undone by neoliberalism. *The Nation*. Retrieved from https://www.thenation.com/article/undone-neoliberalism/

Seed, R. G., & Bea, R. B. (2006, May 23). *Investigation of the performance of the New Orleans flood protection system in Hurricane Katrina on August 29, 2005* (Report No. UCB-CCRM of the Independent Levee Investigative Team). Berkeley, CA: University of California-Berkeley.

Shallat, T. (2006). Holding Louisiana. *Technology & Culture, 47*(1), 102–107. https://www.jstor.org/stable/40061011.

Spain, D. (1979). Race relations and residential segregation in New Orleans: Two centuries of paradox. *Annals of American Academy of Political & Social Science, 441*, 82–96. https://www.jstor.org/stable/1043295.

Susman, P., O'Keefe, P., & Wisner, B. (1983). Global disasters: A radical interpretation. In K. T. Hewitt (Ed.), *Interpretations of calamity* (pp. 263–283). Boston, MA: Allen-Unwin.

Tulane School of Architecture. (2007, December). The "Big Four" public housing projects are demolished post-Katrina. *New Orleans Preservation Timeline Project*. Retrieved from https://www.architecture.tulane.edu

US American Community Survey. (2016).
US Census. (2000; 2010).
Von Hoffman, A. (2000). A study in contradictions: The origins and legacy of the Housing Act of 1949. *Housing Policy Debate, 11*(2), 299–325. https://doi.org/10.1080/10511482.2000.9521370.
Wacquant, L. (2010). Crafting the neoliberal state: Workfare, prisonfare, and social insecurity. *Sociological Forum, 25*(2), 197–222. https://doi.org/10.1111/j.1573-7861.2010.01173.x.
Weber, R. (2002). Extracting value from the City: neoliberalism and urban redevelopment. *Antipode, 34*(3), 519–540.
Webster, R. A. (2016a, March 22). New Orleans public housing remade after Katrina. Is it working? *Times-Picayune*/Nola.com. Retrieved from https://www.nola.com/katrina/index.ssf/2015/08/new_orleans_public_housing_dem.html
Webster, R. A. (2016b, August 29). The poisoned promises of Agriculture Street. *Times-Picayune*/Nola.com. Retrieved from https://www.nola.com/politics/index.ssf/2015/04/agriculture_street_landfill_ne.html
Whitehall, G., & Johnson, C. (2011). Making citizens in Magnaville: Katrina refugees and neoliberal self-governance. In C. Johnson (Ed.), *The neoliberal deluge: Hurricane Katrina, late capitalism, and the remaking of New Orleans* (pp. 60–84). Minneapolis, MN: University of Minnesota Press.
Woods, C. (2017). *Development drowned and reborn: The blues and bourbon restorations in post-Katrina New Orleans* (Vol. 35). Athens, GA: University of Georgia Press.
Zapata, M. A., & Bates, L. K. (2015). Equity planning revisited. *Journal of Planning Education and Research, 35*(3), 245–248. https://doi.org/10.1177/0739456X15589967.

# Chapter 9
# Reimagining Housing: Affordability Crisis and Its Role in Disaster Resilience and Recovery

**Andreanecia M. Morris and Lucas Diaz**

## 9.1 Introduction

From a housing perspective, the New Orleans levee failures brought about by Hurricane Katrina and the weakness of the levees can be perceived as a boon and a catastrophe, simultaneously. While it is well documented that the flooding devastation after the storm destroyed a significant portion of New Orleans homes (Pistrika and Jonkman 2010), the same devastation also created new opportunities for a potential real estate renaissance. Unfortunately, with opportunities in real estate development have come the risks of continued status quo in housing inequalities, rearticulated through similar historically entrenched patterns and practices (Germany 2007), ensnaring policymakers and decision-makers at all levels of governance in a familiar game.

For residents and developers in positions to benefit, the post-Katrina New Orleans housing environment has been a boon that has brought new attention and renewed energy to the real estate market (Williams 2018). Some of the factors that have contributed to increased real estate market prices over the past 12 years include Katrina-generated citywide increases in federally funded investments for infrastructure (from roads to local parks and school buildings), coinciding with increased interest in New Orleans after the storm as a retirement and leisure destination for the wealthy, and an affordable place for newcomer entrepreneurs and millennials. These Katrina outcomes benefitted well-positioned real estate owners up to the most recent market slowdown (Thompson 2018). This renewed sense of energy in the area, coupled with a strategically inadequate distribution of tax

A. M. Morris (✉)
Housing NOLA/Greater New Orleans Housing Alliance, New Orleans, LA, USA
e-mail: amorris@housingnola.org

L. Diaz
ABD Department of Sociology, Tulane University, New Orleans, LA, USA
e-mail: ldiaz5@tulane.edu



S. Laska (ed.), *Louisiana's Response to Extreme Weather*,
Extreme Weather and Society, https://doi.org/10.1007/978-3-030-27205-0_9

incentives doled out to new development projects, also helped create sorely needed new housing stock in the city (Gotham 2015).

Unfortunately, this uptick in development activity and real estate market price growth has done little to alleviate New Orleans' need for housing that is affordable to the city's local workforce. Instead, a housing affordability shortage, created largely by the floodwaters that flowed through the levee breaches, continued to grow without adequate attention from decision-makers, advancing into the crisis the city now faces. In our estimation, the resulting influx of new, wealthy residents, the growth of luxury downtown living units, and the growth of speculative real estate development within popular or trendy neighborhoods have not only addressed the city's growing housing affordability crisis, these factors have exacerbated it, aided in no small part by local and state legislators and housing policy officials who have continued to misunderstand or misdiagnose the reach and impact of this issue. The resulting lack of coordinated, focused attention on the City of New Orleans' growing housing affordability issue at decision-making tables, we argue, poses a risk to the city's ability to develop the essential resilience needed for the type of exceptional recovery required if the city is to keep future disasters from becoming catastrophes.

This chapter provides a praxis-oriented interdisciplinary exploration of how post-Katrina's housing boom has contributed to a growing housing affordability crisis and how advocates and service providers have tackled the challenges and opportunities to provide affordable, safe housing for the local workforce. After a brief review of theoretical literature supporting data, analysis, and interpretation, we provide a snapshot of the elements constituting the housing crisis and then explore how the creation and efforts of the Greater New Orleans Housing Alliance (GNOHA) have led to the growing power of this group to inform and influence housing decision-making in the city. We provide a set of recommendations for groups and localities interested in addressing housing affordability issues, as well as recommendations for further interdisciplinary research, and conclude that ensuring housing affordability for low- to moderate-income families helps build the essential resilience families need to face future disasters.

## 9.2 From Theory to Praxis

This chapter relies on a combination of sociological and political science theories that enable the authors to locate their interpretations within a well-established tradition of disaster resilience and recovery and urban and social movement scholarship. Additionally, we rely on our own praxis as social justice advocates, as well as the praxis of others in power-building work, to provide both data and interpretation, making use of qualitative methodologies in the social sciences.

We first frame our discussion through a disaster resilience lens, using concepts that focus on human-oriented abilities to rebound well from disasters (McNicoll 1996; Tierney 1999; Laska 2012). In particular, we work from Laska's conceptualization of essential resilience as a necessary component of exceptional recovery (2012), which we utilize to connect the role housing affordability plays in low- to moderate-income families' abilities to build the type of resilience needed to help mitigate exposure to future disasters. Essential resilience here corresponds to families' abilities to rebound well from future disasters, which we posit is directly related to housing affordability.

Relying on concepts from the "urban growth machine" theory, we assume the role of economic growth and its corresponding logics of a market-centric approach to urban social problems and policies (Molotch 1976; Molotch and Logan 1983). According to Molotch and Logan, the potential negative consequences of a given economic agenda in an urban setting become secondary, or may not be considered at all, in decisions focused on ensuring economic growth (1983). Economic growth logic influences key stakeholders and power players in an urban setting in such a way as to cause them to advance business elites' economic interests at the expense of, or without adequately weighing, potential or actual negative social impacts on vulnerable populations (1983). As Tierney points out, urban growth logics points toward local power relations focused on market-oriented approaches, which apply in disaster planning and response, further exacerbating historical socioeconomic inequities when left unchecked (Tierney 2010). This understanding helps us recognize the role of power dynamics in exploring how housing-related policy and program decisions play out in a given urban setting, which leads us to further consider the roles of different players within a given field and how they operate within that field (Fligstein and McAdam 2012).

Fligstein and McAdam advance decades of social movement theories to arrive at a model that accounts for individual (micro) and regional/national (macro) factors interacting with each other in a given field for the purposes of advancing social, political, economic, or even cultural agendas (2012). The power dynamics, political interactions, and efforts undertaken to advance any group's agenda can be better explained and understood when connecting individual and group actions to the environment in which these actions take place (the field). Finally, in our exploration of GNOHA's growth, actions, and evolution, we highlight implications and potential steps for others to consider, which rely on our application of concepts from regime politics theory (Stone 1989, 1998, 2006). According to Stone, regimes in urban spaces succeed in advancing agendas when cooperation across group interests coalesces into actionable items that the majority of power brokers (across civic, business, and government) agree to support (1989, 1998, 2006).

Finally, we apply reflexive (Pensoneau-Conway and Toyosaki 2011) and institutional ethnography (Campbell 1998) in order to account for and enable the lived field expertise of the authors as data for analysis and interpretation. This methodology allows us to locate our praxis expertise as interpretable data, both as it relates to our experiences as individuals working in the housing universe and as it relates to our experiences within the institutions through which the work is conducted, as well as the field in which it takes place.

Our approach to describing this case and its analysis provided in this chapter is limited in its scope, requiring that we make explicit some of the most salient of these. First, the interpretations in this chapter are not generalizable to a broader population. Our insight is limited to the specific field of housing in New Orleans, and any lessons our interpretations may provide will be mostly of use to practitioners rather than to scholars. Second, this chapter's aim is to promote further discussion and lines of inquiry rather than to fully test a hypothesis. Thus, it is intended to provide potential insights for practitioners rather than theoretically connect causal factors to effects. As such, we encourage both praxis and academic readers to explore this chapter with an eye toward strengthening and advancing both future praxis and scholarly work related to housing affordability and its relationship to disaster resilience. It is a critical topic, we believe, after a disaster and especially in a catastrophe such as was Hurricane Katrina.

## 9.3 A Housing Affordability Crisis in the Shadow of Katrina

New Orleans is renowned for its culture and tourist attractions. More important to New Orleans locally are its neighborhoods and people. Residents perceive themselves through the lens of their neighborhoods first. To retain what we believe is our most precious asset, our residents, efforts must be made to ensure that everyone has an affordable place to call home. From first responders to hospitality employees, teachers, child care workers, and culture bearers, all New Orleanians deserve access to safe, affordable places to live. Unfortunately, today more and more of our working-class residents who work in the heart of the city find it difficult to attain and keep affordable housing that keeps them close to their jobs.

Before we highlight some of the key figures outlining the contours of today's affordable housing crisis, we must first define the term affordable housing. There are two definitions, according to the US Department of Housing and Urban Development (HUD), for affordable housing—one for renters and one for homeowners. Affordable housing for renters is housing in which households spend less than 30% of their gross monthly income on housing costs, which includes rent and utility payments. For homeowners, affordable housing is defined as housing in which households spend less than 30% of their gross monthly income on mortgage, utility, property taxes, and insurance payments. Families who spend more than this 30% threshold are said to be cost-burdened, and families who spend more the 50% of gross earnings are said to be severely cost-burdened.[1] Families who are housing cost-burdened at any level struggle to meet other financial obligations that are part

[1] Understanding the concept of cost burden is critical to making the connection to building a resilient New Orleans because we have already learned from Hurricane Katrina that it is not enough that people are able to have a roof over their heads. Being able to make the rent, or make the mortgage, or make the electricity bill (if there is outright ownership) without being able to save and having money for transportation, medical expenses, and/or food do little to build up the essential resiliency needed by low- to moderate-income families. Cost-burdened families are exactly those families who are unable to do that because their incomes are too low, and their rental/ mortgage payments are too high, and the region provides little recourse for this imbalance to shift in a way that enables these families to find relief in the future.

of life, such as health insurance, food, and transportation, to name a few. In addition, cost-burdened families struggle to save funds for rainy day emergencies, such as the type of funds needed for hurricane evacuations or other emergency needs after a disaster when weekly paychecks disappear.

It is important to note that many advocates and policy experts in housing are familiar with and make use of this definition in the work they do to advance affordable housing policies.[2] However, most people (and often elected officials not familiar with the housing universe), particularly if they hail from middle- to upper-income communities where potential affordable developments are targeted, perceive affordable housing as a proxy for substandard or slum housing. Discussed more fully below, work to advance affordable housing strategies in New Orleans has required giving considerable attention to this misunderstanding with resident leaders, as well as elected and appointed decision-makers, in order to ensure their support rather than their opposition when enlisting them to help advance housing affordability measures and policies.

New Orleans has changed dramatically from 2000 to 2015, experiencing substantial shifts in population, households, income, and housing principally from the catastrophic jolt of Katrina. Some of the significant changes are as follows: the population decreased by 28%, and households decreased by 21%, with the average size of households dropping slightly by 6% from 2.48 people per household in 2000 to 2.33 in 2013; the African American population has declined 34% (112,315 African American residents) since 2000, principally after Katrina in 2005. In 2013, 60% of the city's population was African American, down from 67% before Katrina (GNOHA 2015). Additionally, the number of people living alone has increased by 6%, and the number of non-family households has increased by 7%; meanwhile, the city's poverty rate remains incredibly high at 28% compared to 15% nationally, an overall increase of 2% since 2000 (GNOHA 2015). Despite this increase in poverty, the proportion of high-income households increased dramatically in the city, while the proportion of very low-income households also rose slightly. Housing costs have risen dramatically for both renters and homeowners when compared between pre-Katrina and post-Katrina levels, with home values increasing 54% and rents increasing 50%, while homeownership rates have remained relatively unchanged, decreasing from 46% to 45%, still well below the national average of over 60% (GNOHA 2015).

This post-Katrina housing picture of decline in many indicators combines poorly with employment statistics. The majority of jobs in New Orleans' key economic sectors, tourism and medicine, pay wages below the citywide area family median

[2] In addition to the affordable housing definition, HUD designates area median family income (AMFI) as the baseline family income measurement used for determining affordability for a given geographic unit. Using AMFI, HUD designates families as follows: earning 50%–80% of AMFI in a given geography as low-income (LI) families; earning below 50% of AMFI as very low-income (VLI) families; and earning no more than 30% of AMFI (or families who fall below the federal poverty level) as extremely low-income (ELI) families. Using HUD's affordable housing definition and its area median income categories, we can present a snapshot of the New Orleans housing crisis.

income of $33,600 (for one person household), and the housing affordability crisis starts to become apparent (GNOHA 2015). Low monthly incomes and high costs of living, including housing costs, put a strain on working New Orleanians' and their families' abilities to find and secure safe, affordable housing. Consider New Orleans' musicians, culture bearers, and tourism workers who are key to attracting tourism to the city and who do not earn wages that match the cost of living in New Orleans. According to Ramsey (2013), in 2012, New Orleans musicians reported earning an average of $17,800 a year, but 39% of those surveyed reported making less than $10,000 per year. The average income for musicians has not increased significantly since 2008, while housing prices have steadily increased.

Similarly, 33,801 people work in the Accommodation and Food Services industry in New Orleans, and these housekeepers, bartenders, fast-food employees, and hotel desk clerks earn less than $23,000 a year, on average (Committee on Financial Services 2015). For these workers, the most they can afford to pay monthly as renters or homeowners is $575 (including utilities, taxes, and insurance). Housing prices, however, have skyrocketed between 2000 and 2015. In 2015, at the time of the GNOHA's first-ever housing plan, a one-bedroom, market-rate apartment priced at $767 per month (2015), well beyond hospitality workers' ability to cover this cost without worrying if they'll have enough to cover other living expenses.

Based on the information analyzed in the HousingNOLA plan, GNOHA anticipates a need for 33,600 new or rehabbed housing units over the next 10 years, with the highest homeowner/buyer and renter demands among low- to moderate-income families (2015).[3] This means that a family of two with a combined household income of $44,575 can afford approximately a $125,000 mortgage (assuming $0 down and excellent credit). In the neighborhoods increasingly becoming desirable by either newcomer permanent residents, real estate flippers, or short-term rental operators, median sales prices were $131,000 in 2018, already beyond what the average New Orleans family can afford without experiencing financial strain (Williams 2018).

The result of these changes is an evolving New Orleans. Though the total population number still remains below pre-Katrina levels, New Orleans was one of the fastest-growing cities in America during the 10 years after the storm. It remains one of the most attractive cities for millennials (Larino 2017). Evidence shows that the growing population is not just the result of returning residents but also an influx of new residents (Larino 2017). The changing demographics contributing to high housing demand, particularly with regard to newcomer, higher-income residents, create additional challenges to housing affordability, often compounding preexisting housing inequities around the city. These changes contribute to a large portion of New Orleans families who increasingly find themselves unable to afford homes that were affordable before Katrina. Working-class families, in particular, who experience severe cost burdens, face the toughest challenges. When these families purchase or rent housing today, the bulk of their gross income is consumed by

[3] The highest homeownership demand is by families earning between $29,717 and $44,575 for two- and three-bedroom homes, and the highest renter demand is by families earning less than $11,143 and more than $37,146, for one- and two-bedroom units (GNOHA 2015).

housing costs, making them so cost-burdened that there is little room left financially to build the essential resilience that can mean the difference between poor and successful recoveries.

Before Katrina, New Orleans was a relatively insular city, a city where many people were "from here," deeply rooted in their neighborhoods, traditions, history, and unique culture, with multiple generations living in the same neighborhoods. As the city continues to change, partly as a continued effect of Katrina-related rebuilding efforts, many long-time residents are concerned about the economic effects newcomers are having on the city—especially regarding housing affordability.

When analyzed through a racialization lens (Omi and Winant 1994; Mills 1997), we find that African American households in New Orleans disproportionately pay more of their income toward housing costs than other racial and/or ethnic groups (GNOHA 2015). African Americans also comprise the majority of workers highlighted in the low-paying industries discussed above, making them one of the most cost-burdened groups in the city. New Orleans is now almost evenly divided between homeowners (47%) and renters (53%), but renters disproportionately pay more of their income toward housing costs.[4] Racialized inequities show up within this population, as well, in which more than 60% of renting African Americans are cost-burdened in New Orleans compared to just 45% of whites (GNOHA 2015). These differences along racial lines are not new. On the contrary, racialized housing inequities is a well-documented social problem that has been traced to social, governance, and economic policies orchestrated by white elites and government agencies throughout the twentieth century (Massey 1993, Anderson and Massey 2001; Gotham 2000). Unfortunately, recent and current policies and practices have continued to fail to adequately account for these historical realities, continuing to exacerbate low- to moderate-income African American communities' abilities to build the essential resilience needed to contribute to the possibility of an exception New Orleans recovery from the next disaster.

Elected and appointed officials have been purveyors of historically racialized housing inequalities, whether unwittingly or otherwise, throughout the past 12 years. Consider the federally funded $9 billion homeowner assistance program, labeled "the Road Home" program, managed by the then-newly formed state agency, the Louisiana Recovery Authority, specifically for the purpose of helping residents rebuild their damaged homes. From its inception, the program had design flaws that disproportionately affected working-class African American New Orleanians. Those who have not returned are often from poorer or minority neighborhoods in New Orleans, and their lack of resources was compounded by such design flaws. For example, to determine the award amount for rebuilding damaged homes, the Road Home program used either the pre-storm market value of a home or the cost to repair it, whichever of the two was lower. Two houses with the same square

[4] In 2013, 58% (46,433) of households spent more than one-third of their income toward housing costs, and 37% (29,271) paid more than half of their income toward housing costs. New Orleans ranks *second in the nation* for the percentage of renters paying more than half of their income on housing, as described in the HousingNOLA report.

footage and the same damage, but in different neighborhoods—for example, one in Lakeview, an affluent area of the city, and one in Gentilly, a mixed-income area of more modest homes—would have received dramatically different Road Home awards. Since many black homeowners lived in neighborhoods with lower market values, this resulted in awards much less than the cost to repair their homes. Other program rules, such as a 30% penalty for lack of flood insurance, affected homeowners with more sporadic incomes who were more likely to let policies lapse when budgets tightened. Duplication of benefits rules also reduced Road Home awards; housing assistance from FEMA was used to reduce Road Home awards, even if rebuilding was not physically possible when the FEMA assistance was distributed. Failure to account for the racialized disparities in locations of properties and household incomes, for example, at state policy-making levels resulted in a racially biased program that required court-mandated changes (Fletcher 2011; Beam 2012; Gotham 2014a).[5]

Other challenges included the tremendous burden of documentation. Many homeowners and renters lost important documents during the floods caused by the levee breaches after Katrina. Compiling enough proof to satisfy program requirements was a challenge that caused delays throughout the process. Even for homeowners who received a (smaller than anticipated) grant and were able to perform the construction work themselves, additional costs prevented homes from being fully repaired. Working on a house a little bit at a time leaves many elements exposed, creating mold from humidity, or unexpected water leaks from rain. Copper theft, especially in neighborhoods with few residents, meant that some owners had to replumb their homes or buy one air conditioner after another. Commuting costs were high in the years immediately after the storm, with many residents staying in affordable rentals outside the New Orleans area and returning to the city on weekends to work on their homes (Sheehan 2015). The barriers to recovery were numerous and compounding.

With rising housing prices and stagnant wages, it is becoming evident that the New Orleans housing market is increasingly unable to provide a sufficient supply of quality, affordable units. Further exacerbating these disparities in New Orleans is the recent private real estate sector boom, which has ushered in a process of gentrification[6] that has spread beyond the limited geographic borders of trendy neighborhoods. Affluent individuals and families, as well as real estate flippers and short-term rental investors, have been purchasing and rehabilitating housing in

[5] It is important to note here the state's role in contributing to and advancing racialized inequitable approvals of Road Home funds that further hindered low- to moderate-income African American families to recover. Important in this is the almost full decade of families of limited resources attempting to rebuild without the same support given to middle-class or more affluent families until the lawsuit wins but also the state's descaling practices by using private partners to administer the program (see Gotham 2014b).

[6] We rely on the predominant usage of the term, which incorporates the idea of lower-income population displacement through either built environment upgrading, economic upgrading, or social upgrading, or a combination of any of these, in a given neighborhood (see Lees et al. 2010; Smith and Williams 1986).

traditionally working-class African American neighborhoods all around the city, such as Hollygrove, Leonidas, Gentilly, the Upper ninth Ward, Black Pearl, and segments of the seventh Ward, pushing housing prices to levels well beyond the reach of the working folks who are being forced to look for affordable housing further out from the metro center where they have historically lived and worked.

Cities facing similar dilemmas have explored several potential solutions, from attempts to increase income, increase housing supply, or increase subsidies for housing. This issue is even more critical for New Orleans, with an economy that depends heavily on tourism. New Orleans' working-class families simply have not experienced earnings growth capable of meeting growing housing costs. With no sign of wages dramatically increasing in the very near future, it is critical that New Orleans creates housing that is affordable for the workers who support the backbone of the New Orleans economy. Doing this, however, is no easy task, as the housing universe is a field in which numerous competing interests from civic, business, and government spheres, all operating often at cross-purposes between local and state levels, do not seem to fully grasp what a wicked problem is collectively faced.

## 9.4 Building a Broad-Based Regime from the Ground Up

In 2007, just 2 years after Katrina, a group of nonprofit community development leaders came together to strengthen their abilities to help each other during the early years of recovery and rebuilding efforts in New Orleans. Over time, this group coalesced into the Greater New Orleans Housing Alliance (GNOHA), an organization of housing advocates, developers, educators, and public office decision-makers who collectively perceived a significant gap in housing affordability leadership and coordination in the City of New Orleans and the State of Louisiana. Andreanecia Morris, serving as lead organizer of the group, assumed a leadership role and in 2012 led the formal incorporation of the coalition into an advocacy and lobbying organization.

GNOHA's evolution (from monthly learning and information sharing gatherings in 2007 to monthly political influence and housing affordability policy discussions in 2012), rose out of a deepening understanding that the overly complex housing field, rife with competing power players, strong personalities, shifting political alliances, and fluctuating economic forces, needed a steady, mission-oriented[7] leadership voice. However, desiring to have or build such a voice does not guarantee attainment of the power to influence anything. While we cannot say that today GNOHA is a representative example of the type of political regime required to advance a particular social justice agenda, we do believe that GNOHA can be

[7]The word mission used here is to be understood as an antithesis to a profit agenda. That is, mission-oriented work pursues the mission of the organized group (in our case, housing affordability for GNOHA) as its goal and not profit-making for a particular group or business.

viewed as an example of how steps can be taken toward building such a regime.[8] The following brief overview of GNOHA's work and ongoing challenges highlights the essential work required to build the type of political power needed to advance housing affordability strategies, especially after an extreme weather catastrophe such as Katrina posed.

### *9.4.1 A Lesson from the Battle for Public Housing*

In 2007, the New Orleans City Council voted to allow the Housing Authority of New Orleans (HANO) to approve the demolition of "the Big Four," the four largest public housing developments in the city, to the displeasure and chagrin of an unhappy and boisterous crowd in attendance (Nossiter and Eaton 2007). Advocates who had looked to the City Council to stay the tide of a decades-long HUD initiative known as HOPE VI (mixed-income housing) felt betrayed by their locally elected officials, anticipating that the City Council would use its power to grant demolition permits to convince HANO to turn a ship around that had sailed some 10 years prior. The advocates' logic was simple: they wanted affordable housing immediately for the displaced low-income families of New Orleans who had yet to return by 2007, due to the severe shortages in housing stock in the years immediately after Katrina. Advocates relied on the moral high-ground demand to elected and appointed officials to "do the right thing," having little to no power as a collective group to influence any of the decision-makers.

Herein lies the key point that the remainder of this section highlights. Many advocates on the ground doing the work to bring affordable housing to their communities fail to appreciate the complexity of power brokers in a given urban setting, specifically how these power brokers interact and how market-oriented logics overwhelm socially oriented logics. At the time of the 2007 City Council decision to approve demolition of the "Big Four" public housing developments, GNOHA was little more than an initial set of welcoming meetings in which nonprofit housing developers, service providers, and advocates were getting to know each other and primarily reacted to experiences and changes in the field. While some members of the then-nascent group participated in activities designed to influence council members, GNOHA at that time was not in a position to participate meaningfully as an organized collective of organizations. The specter of the "Big Four" battle, however, may have provided the impetus to the group to begin to think in a more proactive manner.

[8] In applying regime politics theory, we recognize that GNOHA has initiated the process of building such a regime in New Orleans, but in its current iteration in 2018, this remains a work in progress that the organization hopes to achieve within the coming years. The expected outcome of building such a regime is more robust, sustainable policies that deeply engender housing affordability opportunities for low- to moderate-income families in the city.

### *9.4.2 The Start of Something Different*

In 2008, with nearly a full year of regular convenings and no clear direction yet as to which direction these gatherings would lead, participants of the meetings began to push for a strategic direction to the group. Some highlighted the potential power of a unified voice of housing professionals, pointing to how such a voice could help the efforts of the individual organizations assembled. Others pointed to the direct issues housing organizations were dealing with in helping residents navigate the Road Home program. Knowledgeable with first-hand experience, the group openly discussed its combined frustrations with the program's design and the lack of a strong enough voice to bring these issues to the attention of Road Home administrators at the State Capitol.

Conversations changed from reactive information sharing to participants asking for government officials to be invited to the meetings. This simple request may seem innocuous, but for advocate groups, this can be a highly polarizing line, with those in favor of including government officials pitted against those not in favor of such action, often on the grounds that members of government are not trustworthy partners. At the heart of this polarization lies a key philosophical position about how to effect social justice change as an advocate, stated simply as either working with government officials or working against them. Not everyone at the table was of the same opinion, but after some deliberation, the group agreed that it was worthwhile inviting government officials to meetings if only to educate the group on the most current thinking about housing policies and programs. With this agreement, GNOHA opened the door toward acceptance of, and ultimately adherence to, key foundational frameworks in its ongoing work to this day, research deeply, build broad relationships across the field, and work with government officials.

Between 2008 and 2009, GNOHA shifted its meetings from learning about participating members' organizational challenges to learning from government officials at different levels of governance about their housing work. The now better-organized group spoke to city, state, and federal officials across agencies, inviting them to speak at monthly meetings, ensuring ongoing learning and communication about housing policies and programs being considered and/or implemented. Through this initial effort, GNOHA began to build relationships with officials and learned about opportunities to provide input, give feedback, and present our ideas. The coalition members learned, for example, that there were consistent gaps in knowledge, with government officials lacking sufficient data and/or policy ideas that could help inform their work and local advocates and practitioners lacking clarity about decision-making parameters, timetables, and opportunities. During these learning opportunities, GNOHA leaders realized that if they could conduct the research, it could provide data-driven solutions to the same government officials with whom the members were beginning to build relationships.

The next step in development as a coalition evolved from this realization that the group could engage in policy analysis and recommendation, doing so in partnership with government officials, wherever possible. By the end of 2009, GNOHA evolved again, embarking on policy research, analysis, and writing, as well as agreeing to include government officials as members of the coalition (a decision that holds still in 2019). However, through 2015, GNOHA remained a small organization, despite having an average 40–50 annual member roster of organizations that included housing professional advocates, service providers, nonprofit developers, and government officials.[9] Compared to the Industrial Areas Foundation (IAF) model of community organizing, which has more than eight decades of successful organizing across the United States, the group was certainly not broad in its membership (Chambers 2003).

### 9.4.3 *Toward a Broad-Based Regime*[10]

Before moving further, it is important to clarify what we mean by the phrase *broad-based regime*. Beginning with the word *regime*, we want to emphasize the theoretical understanding of this word, borrowed from and framed upon Stone's conceptualization (1989, 1998, 2006), which refers to an organized coalition of power brokers across sectors in an urban setting who/that have agreed to pursue a shared vision through a set of programmatic and/or legislative policies. We cannot say that GNOHA and its sister organization HousingNOLA[11] in 2019 constitute a regime in the housing field, but it is the direction in which the organizations are intentionally heading. Without meaningful participation across business, civic, and governance sectors at local and state levels in the intentional pursuit of affordable housing solutions, GNOHA will not have the requisite power to see ideas reach adoption and implementation. Building such a regime requires dialoguing, wherever possible (we note that sometimes it is not possible), across ideological lines, across political partisan lines, and certainly across advocate-government lines, in an effort to bridge understanding and build the needed unified voice that enables solutions that benefit cost-burdened families who are quickly running out of affordable housing options in New Orleans. We conceptualize a broad-based regime as involving not only the power brokers mentioned above but also the volunteer public involvement of everyday residents.

[9] In 2015, it had one full-time staff member and two fellows. The 2-year GNOHA Road Home Liaison Group contract with the State of Louisiana added three full-time contractors from 2015 to 2017. The volunteer Board of Governors oversees policy and advocacy efforts and the chair serves as president managing the day-to-day activities.

[10] The concept builds on the IAF idea of a broad base and the regime politics theory of a powerful political regime base by combining the two to incorporate four distinct groups: everyday residents, political decision-makers, business leaders, and professional nonprofit leaders. Everyday residents comprise all folks interested in being civically involved but who are not professional nonprofit staff.

[11] The HousingNOLA plan initiated by GNOHA also gave rise to the creation of a 501(c)3 organization that could focus on the civic involvement and education of everyday residents on housing issues. This new organization, created in 2015, is named HousingNOLA (sharing the title of the report also mentioned here).

Through the years, GNOHA has had some successes in this building effort and, of course, many failures. One highly significant, if only temporary, recent success involves Governor John Bel Edwards and Louisiana Senate Bill 462. Louisiana Senate Bill 462, from the 2018 Legislative Session, targets municipalities attempting to create affordable housing units by legislating that such units be included in new, market-rate developments. This strategy is generally known as inclusionary zoning. The bill bans mandatory inclusionary zoning policies outright at the municipal level, effectively eradicating a municipality's ability to act locally in dealing with housing affordability issues (Louisiana Senate Bill 462). GNOHA members and HousingNOLA partners were aware of the bill and actively worked with other legislators, aides to Governor Edwards, and other channels to keep the bill from passing, as GNOHA deemed it severely harmful to its ability to promote one of the key strategic tools in the housing affordability toolbox. Thanks, in part (GNOHA members believe), to these efforts, Governor Edwards vetoed the bill, which had made it out of the Senate floor (Edwards 2018). However, this veto depends on a city making use of an inclusionary zoning policy before the 2019 Louisiana legislative session, with a stipulation that the governor would sign a similar bill in 2019 if it reached his desk and no city has shown a need for such a policy. While GNOHA rightly views the 2018 veto as a small win for affordable housing advocates, the governor's caveat serves as a reminder that the power to influence government officials that GNOHA's members have cultivated since 2008 remains limited in scope and reach. The work is far from done.

Of course, GNOHA members went to work locally, holding conversations with the Mayor of New Orleans, as well as New Orleans City Council officials, the City Planning Commission, and respective staff for each. Through this work, GNOHA recently experienced a partial victory when on November 13, 2018, New Orleans City Planning Commission approved a set of recommendations for zoning changes that included inclusionary zoning incentives for future multi-housing development (Evans 2018). As a follow-up to this approval, on January 24, 2019, the New Orleans City Council passed a motion to draft new ordinances. This vote rejected recommendations from the City Planning Commission that would have not included mandatory inclusionary zoning as the centerpiece of the Smart Housing Mix. While the new policy is being drafted, the Council requires data from an as-yet unfinished feasibility study that was being conducted at the time in order to determine the most viable option to consider next (Litten 2019). This brings New Orleans closer to a mandatory inclusionary zoning policy to the City of New Orleans, and if we are able to accomplish a city council vote before a similar bill to 2018's Senate Bill 462 passes through committee, house, and senate floors in 2019, then GNOHA will have accomplished a significant piece of its multi-pronged solutions recommendations (GNOHA 2015).

GNOHA's recommendation on inclusionary zoning comes directly out of our 2015 HousingNOLA plan, in which the coalition articulates a 10-year strategy and implementation plan for creating more affordable housing for New Orleans residents (GNOHA 2015). However, this plan could easily remain a shelf ornament were it not for the intentional work GNOHA began in 2012 to engage non-housing professionals in our efforts. Without a base that includes residents from all walks of life, GNOHA would simply be a coalition of housing professionals in the business, government, and civic sectors. While a feat in and of itself to bring these disparate professional groups together, GNOHA members agreed early on that inclusion of residents needed to be a priority, which began in earnest after incorporating as a 501 (c) 4 in 2012.

After agreeing to create a housing plan, GNOHA intentionally set out to build a broad base, hosting a housing summit in 2014, which served as the launching point for a set of community meetings held throughout the city, in which GNOHA members and residents codeveloped a housing plan. In 2017, GNOHA and HousingNOLA launched a multi-year campaign, titled *Put Housing First*, designed to secure 80,000 pledges over the next 10 years, and in 2018 implemented a housing affordability awareness march in order to bring attention of the issue to a broader audience. These efforts combined have the intention of informing and including residents who are not housing professionals in the effort to build a unified voice for housing affordability.

Today GNOHA continues to explore and expand ways of incorporating more residents, as the base can never be broad enough. Members of GNOHA are certain that had the group not intentionally included residents in this work, the type of actions taken by the New Orleans City Council in the last 3 months may not have occurred at all.

### 9.4.4 Work on the Buy-In

Ten years after convening with housing-oriented nonprofits, GNOHA has developed into a significant voice on issues of housing affordability. However, this work is far from over, and continues to present endless challenges. As mentioned above, GNOHA is moving toward building an affordable housing regime, but it is not there yet. One major reason for this is that the very term *affordable housing* fails to elicit a unified response. There are still many constituents, be they residents, government officials, or private developers, who scoff at the idea that New Orleans needs affordable housing, or who believe that affordable housing is a dirty word, or who simply fail to understand the basic concept that families cannot afford basic, decent, and safe housing because of external social forces. Not everyone, unfortunately, has bought in to this issue as requiring attention, despite a 2018 survey of mayors across the United States recognizing insufficient living-wage jobs and high housing costs as the top two obstacles to social mobility (Initiative on Cities 2019). We would add that, in our estimation, these are the top two obstacles to building essential resilience to disasters: housing cost burden and lack of living wages.

Building buy-in requires intentional work that targets not just housing advocates and providers, but also the general public, the business sector, government officials, and other non-housing advocates and service providers. As housing advocates, we have a clear definition for the term *affordable housing* that we believe everyone can understand, but it is not universally understood, much less accepted. In order to explore the types of solutions that can support cost-burdened families at local levels, affordable housing as a real issue needs to be better understood by cities so that discussions can incorporate the disconnects among cost-burdened families, low wages, and poor transportation, for example. GNOHA's efforts through the *Put Housing First* campaign, the march in 2018, and ongoing participation on local radio and television talk shows help promote this understanding in New Orleans, which we believe contributes to mitigating against traditional skepticisms.

## 9.5 Recommendations

The following set of recommendations for advancing housing affordability in cities is directly copied from the 2015 HousingNOLA Plan created by GNOHA, the Mayor's Office of Community Development and the Foundation for Louisiana. The plan can be downloaded directly from www.housingnola.org/main/plans. The recommendations fall under five priorities, which are as follows.

1. Preserve existing supply and expand the total supply of affordable rental and homeownership opportunities throughout your city. A recommended strategy includes creating a working partnership among city officials, other government officials, housing advocates, and housing providers to increase create/update housing stock, with time-limited targets in 5-year increments to evaluate and measure.
2. Prevent future displacement through development activities and continued study and policy review. A recommended strategy includes creating standard zoning polices and working with key stakeholders to develop workforce housing strategies and other creative ideas that assist cost-burdened families dealing with the pressures of higher costs brought about by gentrification. For example, in New Orleans, through work with the Assessor's Office, a referendum passed in the 2018 local elections to allow homeowners in gentrifying neighborhoods to ease into their increased taxes over time rather than immediately upon new assessments reflecting rapid neighborhood changes.
3. Enforce and promote fair housing policies throughout your city. This work, while not new in the affordable housing toolkit, remains underutilized. It requires comprehensive, coordinated work across sectors to improve fair housing practices in renting, selling, buying, safety, and so on. Rather than leaving this issue up to chance, cities are encouraged to actively and intentionally work with advocates and businesses to engage proactively.
4. Encourage sustainable design and infrastructure for all residents in your city. Working collaboratively with private, civic, and government leaders, study your city's market, and determine feasible incentives to consider for local or state legislative sessions.

5. Increase accessibility for all, including residents with special needs in your city. Housing affordability is not an issue for just one group of people, as it affects individuals and families across all walks of life. Ensure that housing affordability policies intentionally include and align with other city policies aimed at addressing the issues faced by special needs population.

## 9.6 Discussion

This chapter provides a brief snapshot of the complexity behind understanding housing affordability factors (both historical and current), how these factors interact at the local New Orleans level, and how advocates, along with business and government leaders, can work together with residents to build regimes with enough power to implement policies that address affordable housing inequities. As such, we hope the chapter can inform future research into the roles and relationships of and among different players in local housing fields and how these interact and build regimes capable of implementing desired policies aimed at creating housing affordability. With mayors across the country recognizing the connection between wages and housing affordability as vital to the economic vitality of their residents, we hope more cross-disciplinary channels are explored by practitioners who advocate for affordable housing and living wages throughout the United States, particularly as it relates to considering housing affordability as a key strategy for building extreme weather essential resilience among cities' residents.

## 9.7 Conclusion

A true broad-based political regime that (1) is cognizant of how housing cost-burdens keep a significant portion of the population from building the required essential resilience to respond to disasters and (2) makes housing affordability for low- to moderate-income families its goals has the potential to counter the traditionally and historically market-oriented and racialized approaches to housing that has plagued most cities and enhanced its residents' vulnerabilities. We see a key policy opportunity not just for New Orleans, but also for other cities, to install policies that tackle the growing housing affordability crisis in such a way that it can enable the thousands and hundreds of thousands of working families in all cities to be able to afford a safe, decent home that does not cost-burden them in any fashion. Accomplishing such a feat would enable families to focus on building the essential resilience they need to deal with future disasters without such disasters becoming a catastrophic event for them and the cities in which they live. If housing affordability isn't intentionally addressed post disaster, Katrina has shown us that traditional market forces left unchecked will simply push vunerable working families to the fringes of society by exacerbating and even expanding socioeconomic inequalities, which in turn contribute in converting disasters for these families into catastrophes.

## 9.8 Postscript

Our brief snapshot in this chapter provides a glimpse into the growing cost burden of working-class families in New Orleans, and within this group we want to specifically close with a brief mention of the impact gentrification is having on our culture bearers. Culture bearers include musicians, artists, Black Masking Indians, Social Aid and Pleasure Clubs, and a host of other creative individuals who contribute to the cultural experience packaged and sold around the world as being uniquely New Orleans. The majority of our culture bearers hail from working-class families who before Katrina lived in the same neighborhoods across the generations. Housing before the storm was deeply affordable, allowing culture bearers who earned less than $20,000 a year to remain home and continue to keep New Orleans culture alive. With housing prices going up, these same families and individuals are increasingly finding it difficult to stay in the neighborhoods they historically called home, straining their abilities to continue to create the cultural texture so many tourists enjoy when they visit. If we are not successful at creating the affordable housing stock these folks need, we indeed run the risk of losing what is "New Orleans."

## References

Anderson, E., & Massey, D. S. (2001). *Problem of the century: Racial stratification in the United States*. New York: Russell Sage Foundation.

Beam, J. (2012, January 5). Attorneys big winners in Road Home lawsuit. *The Hayride*. Retrieved from https://thehayride.com/2012/01/attorneys-big-winners-in-road-home-lawsuit/

Campbell, M. (1998). Institutional ethnography and experience as data. *Qualitative Sociology, 21*(1), 55–73. https://doi.org/10.1023/A:1022171325924.

Chambers, E. T. (2003). *Roots for radicals: Organizing for power, action, and justice*. New York/London: Continuum.

Committee on Financial Services. (2015). New Orleans: Ten years after the storm, Subcommittee on Housing and Insurance. Retrieved from https://www.govinfo.gov/content/pkg/CHRG-114hhrg99779/html/CHRG-114hhrg99779.htm.

Edwards, J. B. (2018, May 26). Veto Senate Bill 462 letter [Louisiana Office of the Governor correspondence]. Retrieved from http://gov.louisiana.gov/assets/docs/BillsSigned/Alario-Ltr2018-05-26-veto-SB-462.pdf

Evans, B. (2018, November 13). New Orleans steps closer to affordable housing requirement for developers. *Nola.Com*. Retrieved from https://www.nola.com/politics/2018/11/new-orleans-steps-closer-to-affordable-housing-requirement-for-developers.html

Fletcher, M. A. (2011, July 6). HUD to pay $62 million to La. homeowners to settle Road Home lawsuit. *Washington Post*. Retrieved from https://www.washingtonpost.com/business/economy/hud-to-pay-62-million-to-la-homeowners-to-settle-road-home-lawsuit/2011/07/06/gIQAtsFN1H_story.html

Fligstein, N., & McAdam, D. (2012). *A theory of fields*. New York: Oxford University Press.

Germany, K. B. (2007). The politics of poverty and history: Racial inequality and the long prelude to Katrina. *Journal of American History, 94*(3), 743–751. https://doi.org/10.2307/25095135.

GNOHA. (2015). *HousingNOLA final report and ten year strategy and implementation plan for a more equitable New Orleans (housing analysis and recommendations)*. New Orleans: Greater New Orleans Housing Alliance. Retrieved from http://flux.modiphy.com/files/view/14208.

Gotham, K. F. (2000). Racialization and the state: The Housing Act of 1934 and the creation of the Federal Housing Administration. *Sociological Perspectives, 43*(2), 291. https://doi.org/10.2307/1389798.

Gotham, K. F. (2014a). Reinforcing inequalities: The impact of the CDBG program on post Katrina rebuilding. *Housing Policy Debate, 24*(1), 192–212. https://doi.org/10.1080/10511482.2013.840666.

Gotham, K. F. (2014b). Racialization and rescaling: Post-Katrina rebuilding and the Louisiana Road Home Program. *International Journal of Urban and Regional Research, 38*(3), 773–790.

Gotham, K. F. (2015). Limitations, legacies, and lessons: Post-Katrina rebuilding in retrospect and prospect. *American Behavioral Scientist, 59*(10), 1314–1326. https://doi.org/10.1177/0002764215591186.

Initiative on Cities. (2019). *Menino survey of mayors* (p. 38). Boston: Boston University. Retrieved from http://www.surveyofmayors.com/reports/Menino-Survey-of-Mayors-2018-FinalReport.pdf.

Larino, J. (2017, June 8). New Orleans is a top-growing city for millennials: Report. *NOLA.Com*. Retrieved from http://www.nola.com/business/index.ssf/2017/06/millennials_moving_new_orleans.html

Laska, S. (2012). Dimensions of reality: Essential resiliency, exceptional recovery and scale. *International Journal of Critical Infrastructures, 8*(1), 47–62.

Lees, L., Slater, T., & Wyle, E. (Eds.). (2010). *The gentrification reader*. New York: Routledge.

Litten, K. (2019, January 25). New Orleans steps carefully toward affordable housing requirements. *Nola.Com*. Retrieved from https://www.nola.com/politics/2019/01/on-affordable-housing-new-orleans-city-council-takes-early-steps-knowing-more-work-lies-ahead.html

*Louisiana Senate Bill 462*, Baton Rouge, Louisiana.

Massey, D. S. (1993). *American apartheid: Segregation and the making of the underclass*. Cambridge: Harvard University Press.

McNicoll, G. (1996). At risk: Natural hazards, people's vulnerability, and disasters. *Population & Development Review, 22*(1), 169–170. https://doi.org/10.2307/2137699.

Mills, C. W. (1997). *The racial contract*. Ithaca: Cornell University Press.

Molotch, H. (1976). The city as a growth machine: Toward a political economy of place. *American Journal of Sociology, 82*(2), 309–332. https://doi.org/10.1086/226311.

Molotch, H., & Logan, J. (1983). Tensions in the growth machine: Overcoming resistance to value-free development. *Social Problems, 31*, 483–499. https://doi.org/10.2307/800236.

Nossiter, A., & Eaton, L. (2007, December 21). New Orleans council votes for demolition of housing. *The New York Times*. Retrieved from https://www.nytimes.com/2007/12/21/us/nationalspecial/21orleans.html

Omi, M., & Winant, H. (1994). *Racial formation in the United States: From the 1960s to the 1990s* (2nd ed.). New York: Routledge.

Pensoneau-Conway, S. L., & Toyosaki, S. (2011). Automethodology: Tracing a home for praxis oriented ethnography. *International Journal of Qualitative Methods, 10*(4), 378–399. https://doi.org/10.1177/160940691101000406.

Pistrika, A. K., & Jonkman, S. N. (2010). Damage to residential buildings due to flooding of New Orleans after Hurricane Katrina. *Natural Hazards, 54*(2), 413–434.

Ramsey, J. (2013, May 1). Sweet home: 2012 state of the New Orleans music community report. *OffBeat Magazine*. Retrieved from http://www.offbeat.com/news/sweet-home-new-orleans-releases-2012-state-new-orleans-music-community-report/

Sheehan, M. A. (2015, October 1). Detours on the Road Home. *Shelterforce*. Retrieved from https://shelterforce.org/2015/10/01/detours_on_the_road_home/

Smith, N., & Williams, P. (1986). *Gentrification of the city*. New York: Allen and Unwin.

Stone, C. N. (1989). *Regime politics: Governing Atlanta, 1946–1988*. Lawrence: University Press of Kansas.

Stone, C. N. (1998). Regime analysis and the study of urban politics, a rejoinder. *Journal of Urban Affairs, 20*(3), 249–260. https://doi.org/10.1111/j.1467-9906.1998.tb00421.x.

Stone, C. N. (2006). Power, reform, and urban regime analysis. *City & Community, 5*(1), 23–38. https://doi.org/10.1111/j.1540-6040.2006.00151.x.

Thompson, R. (2018, December 2). Despite rising prices, one segment of New Orleans' housing market has softened: $1 million-plus homes. *The Advocate*. Retrieved from https://www.theadvocate.com/new_orleans/news/business/article_ef0ff0d6-e442-11e8-84d10fd4bb524a04.html

Tierney, K. J. (1999). Toward a critical sociology of risk. *Sociological Forum, 14*(2), 215–242. https://doi.org/10.1023/A:1021414628203.

Tierney, K. (2010). Growth machine politics and the social production of risk. *Contemporary Sociology, 39*(6), 660–663. https://doi.org/10.1177/0094306110386715b.

Williams, J. (2018, June 17). Growth in New Orleans home prices may force out longtime residents, study says. *The Advocate*. Retrieved from https://www.theadvocate.com/new_orleans/news/article_9777cdb8-6e7b-11e8-b76a 575e44aca6f8.html

# Part V
# Types/Locations of Communities and Their Responses to Extreme Weather: Suburban/Mid State

# Chapter 10
# The 2016 Unexpected Mid-State Louisiana Flood: With Special Focus on the Different Rescue and Recovery Responses It Engendered

**Michelle Annette Meyer, Brant Mitchell, Shannon Van Zandt, and Stuart Nolan**

## 10.1 Introduction

When the Great Louisiana Flood occurred in August 2016, it was the worst disaster to affect the USA since Hurricane Sandy in 2012 (Yan and Flores 2016). However, it was largely overlooked by those outside of the area due to a lack of a "name" that is common to tropical storms (Scott 2016). The unprecedented amount of rain resulted damage or destruction of more than 90,000 homes (including 28,000 rental units), more than 40,000 homes without electricity, emergency sheltering of 11,699 residents, rescue of more than 28,000 people, and 13 drowning deaths during the 4-day disaster (GOHSEP 2016; Rhoden 2016; Terrell 2016). The USA saw this extreme flooding eclipsed by Hurricane Harvey and the rest of the devastating 2017 hurricane season barely a year later.

The Louisiana flooding in 2016, though quickly overshadowed, set the stage for a variety of innovations in both disaster response and recovery – making Louisiana the test bed, once again, for disaster activities, organizations, programs, and policies. In this chapter, we discuss this flood event to showcase some of the adaptation strategies Louisianans began that provide insights for addressing large-scale

M. A. Meyer (✉)
Hazard Reduction and Recovery Center, Department of Landscape Architecture and Urban Planning, Texas A&M University, College Station, TX, USA
e-mail: mmeyer@arch.tamu.edu

B. Mitchell · S. Nolan
Stephenson Disaster Management Institute, Louisiana State University, Baton Rouge, LA, USA
e-mail: bmitch9@lsu.edu; snolan7@lsu.edu

S. Van Zandt
Department of Landscape Architecture and Urban Planning, Texas A&M University, College Station, TX, USA
e-mail: svanzandt@tamu.edu



S. Laska (ed.), *Louisiana's Response to Extreme Weather*,
Extreme Weather and Society, https://doi.org/10.1007/978-3-030-27205-0_10

flooding and disaster-related needs across the country. We review two particular innovations that grew dramatically in 2016. The first is the exponential rise of civilian, water-based rescue teams. The second is multiple efforts at affordable housing recovery, particularly for renters, which is often an overlooked aspect of overall disaster recovery.

## 10.2 The Setting for a Perfect Storm

The August flooding in southern Louisiana set numerous records for rainfall in the region and extended across 26 parishes (which are county-equivalents) (GOHSEP 2016; Di Liberto 2016). The areas in and around Greater Baton Rouge (population approximately 829,000) tend to have a relatively flat topography and rely on three rivers, the Amite River, the Comite River, and the Tangipahoa River, to transport excess rainwater into Lake Pontchartrain to the south. Hundreds of tributaries and bayous also drain surface water from neighborhoods to these rivers. During the first 36 hours of the rain commencing on the afternoon of 11 August 2016, the rivers quickly filled and reached major flood stage, which is summarized by the National Weather Service as potential catastrophic flooding with the possibility to harm life and property (NWS 2017). Before the event was over, four rivers (Amite, Comite, Tangipahoa, and Tickfaw rivers) shattered previously recorded high water levels by as much as 6 feet at 11 different locations (Table 10.1) (Di Liberto 2016).

Precipitation finally diminished about 2 days later. Throughout the Greater Baton Rouge and Lafayette areas, localities recorded 10–20 inches of rain over a 48-hour period (Table 10.2). Some of the hardest hit areas – North Baton Rouge neighborhoods and South Lafayette –experienced 20–30 inches of rain, while the hardest hit parish, Livingston, recorded rainfalls reaching 32 inches. According to the National Weather Service's Hydrometeorological Design Study Center, these 48-hour rain-

**Table 10.1** River gauge records set during the August 2016 flood event organized by day the record was set

| Gauge | Old record (year) | New record (year) | Difference |
|---|---|---|---|
| Comite – Olive Branch | 23.37 feet (1961) | 26.96 feet (8/13) | +3.59 feet |
| Amite – Darlington | 22.05 feet (1990) | 22.54 feet (8/13) | +0.49 feet |
| Amite – Magnolia | 51.91 feet (1977) | 58.56 feet (8/13) | +6.65 feet |
| Tangipahoa – Amite | 24.73 feet (2016) | 26.28 feet (8/13) | +1.55 feet |
| Tickfaw – Holden | 21.04 feet (1983) | 22.16 feet (8/13) | +1.12 feet |
| Comite – Joor Rd | 30.99 feet (2001) | 34.22 feet (8/14) | +1.23 feet |
| Amite – Denham Spring | 41.50 feet (1983) | 46.20 feet (8/14) | +4.7 feet |
| Amite – Bayou Manchac | 18.85 feet (1983) | 21.50 feet (8/14) | +2.65 feet |
| Tangipahoa – Robert | 27.10 feet (1921) | 27.33 feet (8/14) | +0.23 feet |
| Amite – Port Vincent | 14.65 feet (1983) | 17.50 feet (8/15) | +2.85 feet |
| Amite – French Settlement | 7.40 feet (1977) | 9.21 feet (8/16) | +1.81 feet |

**Table 10.2** Rainfall amounts observed over a 72-hour period (National Weather Service 2019)

| 72-hour rain totals | |
|---|---|
| Livingston | 25.5 inches |
| Norwood (East Feliciana Parish) | 22.0 inches |
| Zachary (East Baton Rouge Parish) | 26.1 inches |
| Baker (East Baton Rouge Parish) | 21.2 inches |

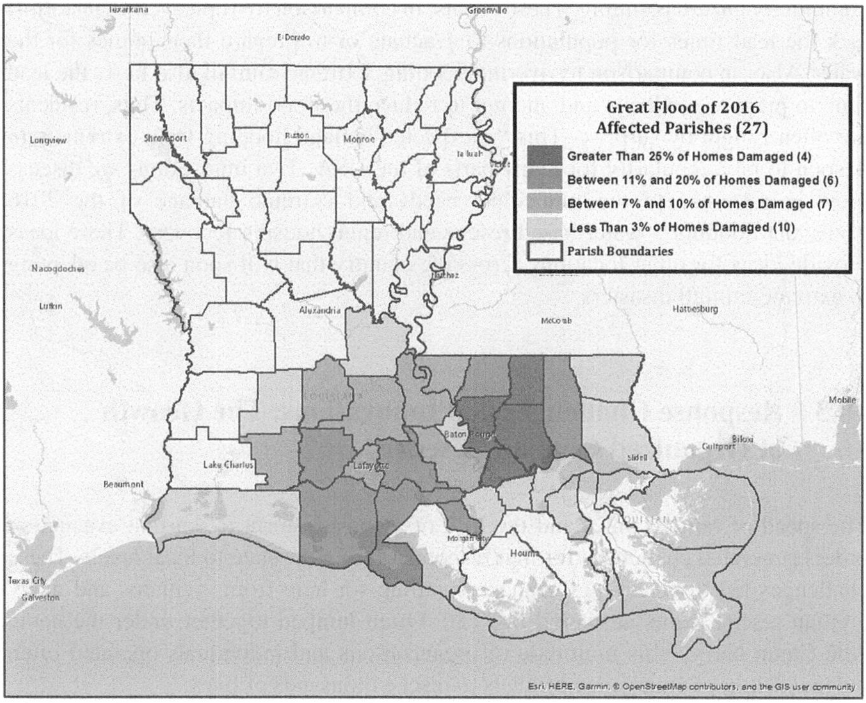


**Fig. 10.1** Map of Louisiana showing the percentage of homes damaged in the 2016 floods by parish. (Adapted from the Stephenson Disaster Management Institute and Louisiana State University)

fall totals in Louisiana exceeded a 0.2% annual probability (GOHSEP 2016). Overall, the impacted area was roughly 180 miles in length and 80 miles widespread across inland South Louisiana. Figure 10.1 shows the affected areas and the percent of homes affected in each parish.

This flooding is not unique to Louisiana. Predictions across the country with a changing climate indicate that many more inland areas will see increasing number of days with extreme rainfall (Di Liberto 2017). The Fourth National Climate Assessment (Reidmiller et al. 2017) describes increasing likelihood of extreme rainfall events across nearly the entire USA, which when coupled with decaying and damaged stormwater and flood-related infrastructure will produce higher risks of flooding throughout the country. Extreme rainfall in Southeast USA, specifically, is expected to double or triple the current average frequency. The total rainfall witnessed in Louisiana in the 2016 floods was barely less than the totals witnessed a year later during Hurricane Harvey, in which the highest recorded rainfall was 48 inches and many Texas coastal areas experienced between 20 and 40 inches of rain in just a few days (Di Liberto 2017).

Extreme rainfall events produce challenges for emergency management and community hazard planning. These events, in comparison to tropical coastal storms, lack the lead times for populations to evacuate or to prepare their homes for the water. Also, in comparison to riverine flooding, extreme rainfall also lacks the lead time to prepare sandbags and attempt to reduce the flood impacts. Thus, residents are often caught by surprise. This "unexpected" inland flooding from extreme rain is soon to be a regularity for many parts of the USA. Two innovations we discuss grew in response to the immediate needs and extreme damage of the 2016 Louisiana flooding – water-based rescue and rental housing recovery. These ideas provide ideas for other locations across the country that will soon also be adapting to extreme rainfall disasters.

## 10.3 Response Challenges and Innovations: The Growth of Organized Civilian Rescuers

The speed of rainfall onset and the lack of a coastal storm to warrant evacuation orders generated challenges for first responders from the state to local levels. These challenges led to emergency managers calling for help from civilians, and many civilian rescue teams answered this call. Often lumped together under the name "the Cajun Navy," this multitude of organizations and individuals operated often independently of emergency officials to rescue thousands of lives.

Response operations are normally a very centralized effort coordinated by state leaders from the State Emergency Operations Center (SEOC). At the time the precipitation began to accelerate on 11 August 2016, the Louisiana Governor's Office of Homeland Security and Emergency Preparedness (GOHSEP) had in place its Crisis Action Team consisting of members of its Operations Staff actively monitoring ongoing events as they were being reported by local officials. State and local officials began lifesaving actions as early as 06:00 Friday morning on 12 August 2016. However, due to the rapidly deteriorating conditions and the immediate nature of the requests from local officials for lifesaving equipment and personnel, the Louisiana Department of Wildlife and Fisheries agents, Louisiana National Guardsmen, Louisiana State Police troopers, and State Fire Marshal agents were immediately pressed into service to augment local first responders without the formalities of the usual memorandums of understanding or standard operating protocol for support requests (GOHSEP 2016). At the end of the day, 237 human rescues had been reported in 6 different parishes. Among the lifesaving actions taken this day included the full evacuation of 145 nursing home residents in the city of Denham Springs as well as the evacuation of a housing authority that included 19 individuals with access and functional needs and 11 caretakers in the city of New Iberia (GOHSEP 2016). While the exact number of people who were rescued is impossible to be determined, the State After Action Report (AAR) states that the SEOC alone coordinated the rescue of more than 15,000 people and 2200 pets with the Louisiana National Guard and their high-water vehicles responsible for the majority of those rescues.

Further recognizing the escalating need, the state called for help from civilians, a call that would foreshadow the same type of call for help by Texas jurisdictions during Hurricane Harvey (McCausland et al. 2017). The AAR acknowledged that more than 13,000 citizens were rescued by local first responders who were aided by volunteers equipped with their own flat-bottomed boats. These volunteers have since become more organized as part of a growing "Cajun Navy Movement," which includes several different nonprofit and for-profit organizations, some using the Cajun Navy moniker, who travel across the country to do water-based rescues (Meyer et al. 2018). These organizations and other civilian rescue groups have since responded to Hurricanes Harvey, Irma, and Maria in 2017 and Florence and Michael in 2018, along with other smaller-scale flooding in Texas.

Helping behavior in disasters, like that of civilian rescuers, takes a variety of forms, ranging along a continuum from spontaneous and informal efforts to more organized emergent group activity to, finally, formal organizational response (National Research Council 2006). Disaster research has long shown that survivors are most likely to be rescued and offered aid first by other survivors in close proximity (Noji 1997; Aldrich 2012; Kendra and Wachtendorf 2016). This individual or small group rescue scenario can be enlarged to emergent spontaneous groups that quickly organize to address a specific need without a formal, bureaucratic organization. They begin to perform many different types of activities quickly following a disaster impact, including providing emergency assistance, identifying and registering survivors, and engaging in cleanup and community restoration efforts (Forrest 1978; Dynes 1974). Although classic and contemporary scholarship has described spontaneous volunteer behavior in disaster settings (e.g., Barton 1969; Fritz 1961; Steffen and Fothergill 2009) as well as formal and informal organizational response (e.g., Dynes 1974; Kendra and Wachtendorf 2003; Aguirre et al. 1995; Clarke and Short 1993; Perry and Lindell 2003), as Tierney et al. (2001: 113) noted, "we know relatively little about spontaneous volunteers and even less about the other patterns of volunteer behavior" such as loosely organized, volunteer rescue efforts like that which occurred in 2016 and since.

The "Cajun Navy" is the popular moniker of outdoorsmen (and women) who began as groups of civilians mostly from southern Louisiana who used their personal flat-bottomed boats to conduct volunteer rescues. Some of these groups claim origins during Hurricane Katrina, while others began or diverged during the Baton Rouge floods in 2016 and then in storms in 2017 and 2018 (Ancelet et al. 2013). While neighbor-to-neighbor rescues during disasters are not uncommon, the Cajun

Navy Movement has quickly grown from emergent spontaneous response to an extending network of more or less structured organizations that coordinate both established and spontaneous volunteers during emergency rescue and response. These groups claim origins in Hurricane Katrina where they boated through New Orleans calling out for those in need of rescue. With the growth of social media giants Facebook and Twitter and the introduction of Zello (the walkie-talkie, two-way communication application for smartphones, zello.com), these groups were able to change their rescue tactics during the 2016 flooding to identify those in need of rescue and then coordinate with their members to get a volunteer to that person (Personal Interview 2018; Raja 2017).

This emergent and volunteer behavior is known to occur in all types of natural disasters, but the use of social media and the rise of smartphone applications allowed the Cajun Navy Movement to coalesce and spread, making it an innovation in civilian response but also a challenge to official emergency management practices. Emergency management commonly uses a command and control format in which an Emergency Operations Center, led by emergency managers, police, fire, and other essential government personnel, directs government and private resources (Dynes 1983; Schneider 1992; Siegel 1985). Emergency managers and their protocols follow strict lines of authority and maintain standard operating procedures and predetermined divisions of labor (Schneider 1992).

The Incident Command System (ICS) is the official model for command and control in emergency management and represents paramilitary, top-down, centralized rigid structure of authority in disaster response. There are 12 essential features that define how ICS is implemented, 2 of these essential features are accountability and dispatch/deployment. Accountability requires that all deployed resources must check in when they arrive to a disaster, and resource tracking is required by operations to ensure all known assets conducting operations are accounted for by command to avoid duplication of effort, among other concerns. In addition, dispatch/deployment explicitly states that personnel and equipment should not self-deploy to an event. Any self-deploying personnel and equipment to disaster areas may cause unintended stress on the established and formal command and control structure unless properly coordinated in advance.

In these systems, spontaneous behavior, including normal volunteer helping behavior, can be viewed as problematic and is prevented or, at minimum, controlled. The civilian rescuers represent a direct challenge to these traditional emergency management operations. Volunteers commonly self-deployed and often did not check in with local officials (Personal Interview 2017). For example, the officials in some areas of Texas refused to allow them into their communities so the boat rescuers circumvented the main entries to participate in rescues anyway. State and local lawmakers both encouraged rescue groups, via providing gasoline to support their efforts (Hilburn 2017), and also expressed concerns particularly around issues of liability and risk (Ballard 2018). This led Louisiana as a state, and local jurisdictions, to consider legal adaptations that could facilitate, coordinate, or, conversely, control these civilian efforts (Crisp 2018; Grueskin 2017). As rescue groups rise and make use of social media and new technologies, jurisdictions across the country will need to prepare for this coordinated influx of volunteers. And as more powerful unpredicted extreme rainfall storms increase in number, the challenges will have even more of an urgency to be addressed.

## 10.4 Housing Recovery Challenges and Policy Innovations

After rescuing ended and civilian and formal emergency responders returned home, residents across the state began the usually long recovery process. Gutted but unrepaired homes abutted properties with FEMA trailers and repairs underway throughout parts of East Baton Rouge Parish in December 2017, roughly 16 months following the flood. Other parts of the parish looked as though a 500-year flood event never happened, with homes completely rebuilt. Recovery, especially housing recovery, is the most understudied and misunderstood stage of a disaster (Rubin 2009). Recovery is uneven across neighborhoods and within neighborhoods, which may seem random, but research shows some predictable patterns (Hamideh and Rongerude 2018; Pais and Elliott 2008; Phillips et al. 2010). Affordable housing especially affordable rental housing (that which costs no more than 30% of a household's income, including utilities (HUD 2018)) is often damaged the most during disasters, making existing affordable housing crises much worse (Tulane School of Architecture 2007). Yet, little research or recovery programs address housing affordability issues or housing tenure after floods (Lee and Zandt 2018).

Hurricane Katrina showed that social stratification across class and race greatly affected return and recovery of housing. Those who returned to their previous homes the soonest following Hurricane Katrina were predominantly white, older, better educated, and homeowners and sustained less damage to their property compared to those who were displaced for longer periods of time or permanently displaced (Fussell et al. 2010). Even with equivalent amounts of housing damage, wealthier individuals, especially homeowners with insurance, returned more quickly. Higher socioeconomic status entails the financial resources to rebuild and also the cultural knowledge to maneuver the complicated US disaster aid process and political and symbolic capital to garner rebuilding assistance (Finch et al. 2010).

These disparate housing recovery outcomes from Hurricane Katrina are not unique. Social vulnerability to disasters describes how "social inequalities and historic patterns of social relations" generate differential disaster impacts and recovery trajectories (Phillips and Fordham 2010, p. 4). When disaster impacts are filtered through the US social structure of race, class, gender, nationality, and disability, for example, they create "multiple and highly unequal processes of resettlement" (Fussell and Elliott 2009, p. 389). Evidence from numerous disasters across the US and the world shows that social vulnerability affects population return and housing recovery following disasters (Thomas et al. 2013; Peacock et al. 1997, 2014; Van Zandt et al. 2012).

Social vulnerability highlights how preexisting inequalities and existing social patterns, such as the current housing affordability crisis and declining

**Table 10.3** Housing types before the flood (American Community Survey 2016, 5-year estimates)[a]

| | Total housing units | Percent single-family detached units | Percent boat/ RV/van units | Percent multifamily units | Percent mobile homes |
|---|---|---|---|---|---|
| United States | 134,054,899 | 61.6 | 0.1 | 32.0 | 6.3 |
| All 22 parishes | 775,544 | ***69.1*** | ***0.2*** | 10.2 | ***20.5*** |
| Acadia Parish | 25,867 | ***73.4*** | ***0.2*** | 8.0 | ***18.3*** |
| Ascension Parish | 44,127 | ***72.1*** | ***0.4*** | 7.2 | ***20.4*** |
| East Baton Rouge Parish | 191,397 | ***63.6*** | 0 | ***33.3*** | 3.1 |
| East Feliciana Parish | 8177 | ***67.8*** | 0.1 | 4.0 | ***28.1*** |
| Iberia Parish | 30,077 | ***66.4*** | 0.1 | 11.3 | ***22.2*** |
| Iberville Parish | 13,009 | ***71.3*** | ***0.7*** | 7.7 | ***20.3*** |
| Jefferson Davis Parish | 13,596 | ***75.3*** | 0 | 6.8 | ***17.9*** |
| Lafayette Parish | 97,847 | ***65*** | 0.1 | 23.8 | ***11.1*** |
| Livingston Parish | 53,673 | ***66.3*** | ***0.4*** | 7.0 | ***26.3*** |
| Pointe Coupee Parish | 11,298 | ***72*** | 0.1 | 6.1 | ***21.7*** |
| St. Helena Parish | 5157 | 58.3 | 0 | 3.6 | ***38.1*** |
| St. James Parish | 8702 | ***75.4*** | 0 | 9.0 | ***15.6*** |
| St. Landry Parish | 36,172 | ***68.9*** | 0.1 | 9.3 | ***21.7*** |
| St. Martin Parish | 22,536 | ***66.3*** | ***0.2*** | 6.3 | ***27.2*** |
| St. Tammany Parish | 98,916 | ***77.8*** | 0.1 | 14.2 | ***7.8*** |
| Tangipahoa Parish | 52,513 | ***62.1*** | 0.1 | 15.6 | ***22.2*** |
| Vermilion Parish | 25,869 | ***71*** | ***0.6*** | 6.2 | ***22.2*** |
| Washington Parish | 21,284 | ***70.4*** | ***0.2*** | 5.7 | ***23.7*** |
| West Baton Rouge Parish | 10,078 | ***70.5*** | 0 | 9.0 | ***20.6*** |
| West Feliciana Parish | 5249 | ***69*** | ***0.2*** | 9.5 | ***21.3*** |

[a]Bold italics indicate greater than US value

homeownership rates, create disparate disaster effects across population groups. The population that often recovers the quickest from disasters – homeowners with insurance – is declining nationwide. The Great Recession starting in 2008 began a decline in US homeownership to a 50-year low (Rohe 2017). Consequently, renting is rising across all income groups, to nearly 40% of the US population (Table 10.3). The corresponding increase in renters quickly outpaced the amount of available rental housing and drove up rental costs. In 2011, for example, there was a shortfall of 4.8 million rental units that would be affordable to persons making less than about $19,000 (Fernald 2013). Recent analysis by Freddie Mac (a federal mortgage agency) shows that newly constructed rental units, for example, are serving a greater proportion of higher-income renters than they were in 2010 (Freddie Mac 2017), even though nearly half of all current renters make less than $30,000 a year (Fernald 2013).

The affordable housing crisis is worsened by disasters in several ways. First, affordable housing, both owned and rented, often receives the most damage in disasters because it is usually located in more hazardous areas, such as floodplains, is of lower quality that doesn't withstand disaster impacts, and often lacks the mitigation upgrades to prevent disaster damage (Fothergill and Peek 2004; Peacock et al. 2018). Housing that is affordable to the lowest-income renters, furthermore, is often more than 50 years old and more likely to be of inadequate quality according to today's building standards (Fernald 2013).

Second, tenants and landlords alike have fewer incentives to undertake mitigation that would prevent disaster damage (Burby et al. 2003). Renters are dependent upon their landlords' permission to reoccupy their previous housing, which heightens the risk of displacement (McCarthy et al. 2001; Burby et al. 2003; Morrow 1999; Fussell and Harris, 2014). Landlords may not rebuild or may rebuild their properties to higher market rates (Comerio 1998). Zhang and Peacock (2009) and Peacock et al. (2014) found that rental housing came back the slowest following Hurricanes Andrew in Florida (1992) and Ike in Texas (2008).

Third, rental rates and housing costs rise due to reduced supply of affordable housing. In Baton Rouge, for example, fair market rent for a one-bedroom apartment increased from $728 to $789 between 2016 and 2018 (Grueskin 2018a; Grueskin 2018b). Disasters ignite "recovery machines" in which pro-economic growth coalitions take advantage of recovery funding to rebuild neighborhoods with higher priced housing and amenities. These efforts make it more difficult for lower-income survivors to acquire post-disaster housing (Elliott and Pais 2010; Gotham and Greenberg 2014). Political will and local funding to support fair and affordable housing and counter the recovery machine are often lacking (Weil 2009). Galveston, Texas, following Hurricane Ike, and New Orleans, following Hurricane Katrina, both made changes to their public housing structures and availability, which significantly reduced the number of affordable units (Tulane School of Architecture 2007; Walters 2018). Four large housing projects in New Orleans, for example, were replaced by mixed-income housing therefore reducing the number of fully subsidized units (Henrici et al. 2010). Evidence from longitudinal research following Hurricane Katrina showed that low-income African American mothers living in subsidized public housing were the least likely to return to their same housing, followed by renters (Fussell and Harris 2014). In East Baton Rouge Parish, 753 families using Section 8 public housing vouchers were flooded in the August 2016 floods. About 42% of those were unable to locate another unit to use their voucher in the 6 months after the flood (Jones 2017). Renters, beyond having lower income on average, also are less likely than homeowners to have various financial investments, such as retirement accounts, life insurance, stocks, certificates of deposit, or savings bonds that can be useful in crises to fund new housing options (Fernald 2013).

Fourth, and importantly, recovery programs through the government or nonprofits are overwhelmingly targeted at owner-occupied housing (Comerio 1997; GAO 2010). Louisiana, for example, allowed eligible *homeowners of any income level* to receive 100% reimbursement for repairs, providing an additional $110 million to homeowners from their Community Development Block Grant – Disaster Recovery (CDBG-DR) provided by the US Department of Housing and Urban Development (HUD) (Grueskin 2018b). Nonprofits that also address recovery housing often limit their programs to homeowners. For example, the Housing First Alliance of the Capital Area is a collaboration of about 30 local nonprofits that formed following the 2016 Louisiana Floods to address housing issues. They concentrated on restoring single-family homes first (Gallo 2017a; Gallo 2017b). Case studies have shown that governments lack clear policy strategies to address renters or rental housing issues in contrast to homeowner programs, and rental programs implemented are often ad hoc (Mukherji 2015). Government options attempted include rental subsidies or vouchers to renters to find their own housing elsewhere, temporary housing, subsidized public housing, economic incentives to rental property owners to rebuild, and homeownership programs. All these programs are often slow to begin following disaster. The "Road Home" program following Hurricane Katrina that supported owners of rental property to rebuild began *2 years* after the disaster (GAO 2010). Some programs even have unintended (or intended) consequences of spatially isolating low-income populations from others, as occurred in Japan following the 1995 earthquake (Hirayama 2000), or increasing racial and economic segregation, such as following Hurricane Katrina. Furthermore, these programs often do not address the long-term, exacerbated issue of the lack of affordable housing. As noted by Gotham (2008), many market-centered recovery programs lack coherency and sustainability and may intensify existing housing issues.

Providing or developing affordable housing, especially rental housing, is a growing post-disaster challenge for jurisdictions large and small across the country. This housing issue corresponds with a variety of other social differences that lead to heightened vulnerability to disaster (Lee and Zandt 2018). Renters, for example, are more likely than owners to be younger, unmarried, and racial minorities (due in part to discriminatory mortgage lending practices) and have lower overall incomes. Renters also are less likely to have social capital connections important to disaster recovery, have lower place attachment, and, importantly, are less politically engaged to demand attention to post-disaster needs (Aldrich and Meyer 2015; Lee and Zandt 2018).

Louisiana attempted new and expanded options following the 2016 floods to address the affordable housing concerns, including rentals. State officials specifically developed programs to address affordable housing, targeting rental housing and manufactured homes. To understand the need for affordable housing, Table 10.3 depicts total housing units and housing types for the US and for the 22 parishes that received individual assistance from FEMA (i.e., locations where households could apply for direct support from FEMA for housing). These parishes had higher percentages than the US average of single-family detached homes, mobile homes, and boat/RV/van housing. Nine of the 22 parishes receiving individual assistance from FEMA had higher rates of nontraditional housing such as boat, RV, or vans than the national rate. Multifamily units are less common in these parishes than in the nation as a whole, except for East Baton Rouge Parish, where 1/3 of housing units were in multifamily structures. The most striking statistic is the high rates of mobile home occupancy in the affected parishes. All parishes except East Baton Rouge exceeded the national rate of mobile home occupancy, a common type of affordable housing that is understudied, but extremely vulnerable, in disaster. The percent of all housing that was mobile homes ranged from 3.1% in East Baton Rouge Parish to 38.1% in St. Helena Parish, with an average of 1/5 of all housing in these parishes being mobile homes.

The 2016 flooding impacted over 28,000 rental households, of which 17,000 were very low income (Louisiana Office of the Governor 2016). The amount of rental housing needs across the affected parishes varied. Table 10.4 shows home ownership and housing costs compared to the US averages. Only East Baton Rouge Parish had a higher percent of renters than the US average, with 41% of the parish

**Table 10.4** Home ownership and housing costs before the flood (American Community Survey 2016, 5-year estimates)[a]

| | | | | Percent whose monthly costs exceed 35% of monthly income | | |
|---|---|---|---|---|---|---|
| | Percent renter-occupied | Percent owner-occupied | Percent owner-occupied without a mortgage | Owner-occupied without a mortgage | Owner-occupied with a mortgage | Renters |
| United States | 36.4 | 63.6 | 35.9 | 11.1 | 23.3 | 42 |
| All parishes | 26.14 | ***73.86*** | ***50.06*** | 7.62 | 20.0 | 40.1 |
| Acadia Parish | 28.9 | ***71.1*** | ***57.4*** | 6.4 | 13.2 | 37.5 |
| Ascension Parish | 19.4 | ***80.6*** | 35 | 5.5 | 14.3 | 34.5 |
| East Baton Rouge Parish | ***41*** | 59 | ***36.2*** | 7.6 | 20.6 | ***45.7*** |
| East Feliciana Parish | 18.5 | ***81.5*** | ***54.6*** | ***11.2*** | 18.5 | 31.2 |
| Iberia Parish | 30.2 | ***69.8*** | ***53.2*** | 5.9 | 18.6 | 40.4 |
| Iberville Parish | 24.2 | ***75.8*** | ***53.6*** | 5.2 | 22.2 | ***44.4*** |
| Jefferson Davis Parish | 28.9 | ***71.1*** | ***56.4*** | 6 | 22 | 43.1 |
| Lafayette Parish | 35.1 | ***64.9*** | ***40.2*** | 8.7 | 19 | 41 |
| Livingston Parish | 20.4 | ***79.6*** | ***40.2*** | 7.6 | 15.5 | 32.8 |
| Pointe Coupee Parish | 24 | ***76*** | ***53.6*** | 9.9 | 21.8 | 34.4 |
| St. Helena Parish | 18 | ***82*** | ***67.4*** | ***11.6*** | ***26.8*** | 41.8 |
| St. James Parish | 22.9 | ***77.1*** | ***53.5*** | 8.3 | 18.6 | 41 |
| St. Landry Parish | 30.5 | ***69.5*** | ***57.6*** | 9.5 | ***24.7*** | ***53.6*** |
| St. Martin Parish | 20.7 | ***79.3*** | ***53.1*** | 4.7 | 22.6 | 38.3 |
| St. Tammany Parish | 22.9 | ***77.1*** | 35.2 | 8.1 | 22.5 | 41.8 |
| Tangipahoa Parish | 32.2 | ***67.8*** | ***44.4*** | 9.2 | 21.6 | ***48.4*** |
| Vermilion Parish | 24.7 | ***75.3*** | ***56.4*** | 7.3 | 12.7 | 35.1 |
| Washington Parish | 29 | ***71*** | ***57.9*** | 8.5 | ***30.9*** | ***45.3*** |
| West Baton Rouge Parish | 26.9 | ***73.1*** | ***46.4*** | 3.4 | 17.6 | ***43.5*** |
| West Feliciana Parish | 24.4 | ***75.6*** | ***48.9*** | 7.8 | 17 | 27.9 |

[a]Bold italics indicate greater than US value

population renting their housing. Owner-occupied housing ranged from 59% in East Baton Rouge Parish to 82% in St. Helena Parish. Owning a home without a mortgage was much more common in these parishes than the US, with an average of 50% of parish homeowners living without a mortgage. Mortgages are important to understanding flood disaster recovery specifically because a mortgage requires insurance and flood insurance as a condition of the home loan if that loan is federally sponsored and located in a floodplain (FEMA 2018). Persons without a mortgage are able to let their homeowner's and flood insurance lapse, making them more at risk of being unable to rebuild on their own.

Renting is seeing a resurgence across the country across income categories, but in most parts of the country, it is still predominantly undertaken by low-income households (Fernald 2013). The affordability standard is 30% or less of income spent on housing costs (HUD 2018). Comparing housing costs across renters, homeowners with a mortgage and homeowners without a mortgage, a much larger percent of renters nationwide spend over 35% of their monthly income on housing costs (Fernald 2013). Five of the affected parishes had a higher percent of their renters with cost burdens than the national average.

The Louisiana Housing Corporation began with two programs targeted at owners of rental properties affected by the floods of 2016, which were funded through CDBG-DR from HUD. Some local governments have similar programs (Hardy 2017). Awardees for both programs must meet various affordability requirements (http://restore.la.gov/program-detailstimeline/). These programs include both loans (which may be entirely forgivable) and grants, depending on the applicant qualifications (Gallo 2017b). Landlords must qualify for the state program and have a bank that is willing to extend construction financing. The programs following Hurricane Katrina faced trouble due to the financial crisis of 2008, during which banks were unwilling to provide loans to landlords to qualify for the rebuilding support (Gallo 2017b). The state established agreements with three banks before the program began to counter some of these issues. To counter other problems experienced with the "Road Home" program following Hurricane Katrina, the flood recovery programs for the 2016 floods are run by the Louisiana Housing Corporation (a state agency created in 2011) rather than by contractors like the "Road Home" program (Gallo 2017b).

One program, the Multifamily Gap Program, offered zero-interest loan gap financing to multi-housing (at least 20 units) of both affordable and market rate housing (http://restore.la.gov/multifamily-gap-program/). The program started with $38.25 million for developers or housing authorities (Gallo 2017b). Agencies involved in public housing or affordable housing could receive up to $40,000 per flooded unit, and market rate rental owners could receive up to $65,000 per flooded unit. All who accepted the funds must maintain the affordability requirements for at least 5 years after renovation. This program, as of September 2018, had distributed funds totaling $1.5 million including two public housing authorities and two market rate complexes that were transitioning to affordable housing (Grueskin 2017).

A second program targeted landlords of smaller multifamily housing with seven or fewer units. The Neighborhood Landlord Rental Program provided financial assistance in taking out a loan for rebuild, repair, or new construction. If applicants comply fully, the "loan" may be completely forgiven. The program aimed to support recovery of 1,200–1,500 rental units across the state with original allocation of $36 million (Gallo 2017a; Gallo 2017b). The program specifically required landlords to keep the properties affordable for 5 years following renovation (Gallo 2017b). Nonprofits that build affordable housing can apply to rebuild flooded housing or build new affordable housing. For-profit landlords could only apply to repair flooded housing (Gallo 2017b). This program, as of September 2018, had allocated almost $36 million covering 340 units.

A third program was added in 2018 called The Piggyback 2018 program. The multifamily program received less interest than expected, and it was assumed to be so because large multiunit facilities may have had insurance. The five million dollars left in this program was transferred to the more popular Neighborhood Landlord Program. A larger amount ($17.7 million) was transferred to the Piggyback program for developers who are already using low-income housing credits to build mixed-income complexes (Grueskin 2017). This program requires more than half of the development be "affordable" to those making 80% or less of the area's median income and remain at below market rate rents for 35 years (much longer than the other two programs). Five percent of the units must be affordable to those with chronic health conditions or very low incomes. It was expected to create 500 affordable units by the definition of 80% or less of the area median income. Priority goes to parishes that were flooded in 2016 and have high rent costs relative to income. East Baton Rouge Parish, which has the most rental properties of the affected parishes, meets both requirements. The Louisiana Housing Corporation had 1804 units in their tax credit-financed portfolio damaged by the August 2016 floods. Of those, about half (974) were repaired in the first 16 months of post-flood. As of September 2018, 870 units were to begin construction in early 2019.

Another program, the Baton Rouge Rebuilds Program, provides forgivable loans for repair and reconstruction of rental housing also funded by HUD. Its three main goals are (Louisiana Housing Corporation 2017):

1. Eliminate blight and stabilize neighborhoods impacted by the floods
2. Repair damaged rental housing stock that will be made available at affordable rental rates for low-income households
3. Increase the available rental stock in flood-damaged East Baton Rouge

The program is available to owners of rental property located in the city of Baton Rouge or unincorporated East Baton Rouge Parish at the time of the flood, who are in good standing with various housing programs. Priority was given to low- to moderate-income applicants who are under 120% of area median income. One-person households making less than $57,120 and four-person households making less than $81,480 would receive priority, for example. This program targets the specific types of rental housing common to the area. Site-built, modular, *and* manufactured housing were all eligible, a unique aspect to this program. Also, the applications were aimed for smaller rental housing developments, specifically seven or fewer units. The rental properties can be located in Special Flood Hazard Areas or not, which addresses the extreme impacts of this disaster.

The above programs targeted owners of rental properties. As noted, these programs are slowly building or repairing some affordable housing units, many taking over 2 years to begin construction. What do renters do in the meantime? Some programs aimed to support displaced renters find or afford housing. Rebuild Livingston, for example, is a nonprofit supporting displaced renters through a Rapid Re-Housing program funded also by HUD's CDBG (Grueskin 2018a; Grueskin 2018b). The targeted population was specifically renters still living in FEMA-manufactured homes. Rebuild Livingston managed rental vouchers and case management to help those households find new rental housing. The program, though, already had a waiting list in early 2018 even as 180 renter households remained in FEMA-supplied housing. The main issue remains that there is not enough affordable housing for renters to use their rental voucher. Renters face tough decisions of taking higher cost units with the help of the voucher, but knowing that when the voucher ends in a year, they will have to move again. As of January 2018, 653 households received these vouchers, and 417 used the vouchers (Grueskin 2018a; Grueskin 2018b). By September 2018, 1217 households had applied, 151 had completed the program, and 661 were in leased housing. The rest were waiting to find affordable homes. Without affordable units, voucher programs may not be useful or may result in temporary housing of displaced low-income renters in unaffordable units.

## 10.5 Moving the US Forward

The 2016 Louisiana Floods were the worst US disaster since Hurricane Sandy in 2012. Then, they were quickly eclipsed by the more damaging Hurricanes Harvey, Irma, and Maria, along with wildfires in California. What Louisiana knows well is that changing weather patterns result in more "unexpected" events and more "worst ever" events. The intense, short-duration storms that happened in Louisiana are increasing in frequency. Their effects are occurring in areas outside of the traditional 100-year floodplains and in areas with ill-equipped drainage systems for these extremes. More flood damage has and will likely occur outside of historic floodplains given the atmospheric dynamics – thus we need more analyses of flood and rescue needs along with social vulnerability, especially housing tenure, and hazard mitigation planning in areas that may face these similar impacts.

Following the catastrophic flooding, some indicated that you cannot plan for this extreme of an event. And, furthermore, that such an unusual event should not affect the floodplain designations or building practices. FEMA actually does not change flood maps because of these less frequent events. Residents of Central Louisiana, for example, won a case against FEMA to have about 2000 homes removed from the high-risk flood zone in August 2016 *just before the catastrophic flooding*. A majority of those homes were underwater a few weeks later. As Central Councilman Moak quoted to the local paper, "This is an ungodly, extenuating circumstance, and I pray that it doesn't happen again, but I still believe the studies are correct; I do. They weren't arbitrarily set. They were mapped out and reapproved by FEMA"

(Allen 2016). The city and residents, even many of those flooded in 2016, wanted to rebuild without changing their practices or increasing mitigation mechanisms. Others though continue to push for greater flood mitigation standards and innovative practices. Moving forward, disasters across the country are going to raise similar important questions about risk and planning and bring more people to the table to discuss what should be done and how do we help, including in civilian rescue and rental housing.

## References

Aguirre, B. E., Wenger, D. E., Glass, T. A., Diaz-Murillo, M., & Vigo, G. (1995). The social organization of search and rescue: evidence from the Guadalajara gasoline explosion. *International Journal of Mass Emergencies and Disasters, 13*(1), 67–92.

Aldrich, D. P. (2012). *Building resilience: Social capital in post-disaster recovery*. Chicago: University of Chicago Press.

Aldrich, D. P., & Meyer, M. A. (2015). Social capital and community resilience. *American Behavioral Scientist, 59*(2), 254–269. https://doi.org/10.1177/0002764214550299.

Allen, R. (2016, September 3). Many Central homes dropped from 'high risk' zones, allowed to lose insurance weeks before flood. *The Advocate*. Retrieved from http://www.theadvocate.com/baton_rouge/news/article_c6278d8e-6576-11e6-a301173b12fac153.html

American Community Survey. (2016). 2012–2016 ACS 5-year Estimates. Washington, DC: Census Bureau. Retrieved from https://www.census.gov/programs-surveys/acs/data/summary-file.2016.html.

Ancelet, B. J., Gaudet, M., & Lindahl, C. (2013). *Second line rescue: Improvised responses to Katrina and Rita*. Jackson: University Press of Mississippi.

Ballard, M. (2018, April 6). Cajun Navy, other do-gooders' rescue efforts would be regularized in bill going to Louisiana House. *The Advocate*. Retrieved from https://www.theadvocate.com/baton_rouge/news/politics/legislature/article_7041d5f4-39ba11e8-ad22-d7335739f834.html

Barton, A. H. (1969). *Communities in disaster: A sociological analysis of collective stress situations* (Vol. 721). Garden City: Doubleday.

Burby, R. J., Steinberg, L. J., & Basolo, V. (2003). The tenure trap: The vulnerability of renters to joint natural and technological disasters. *Urban Affairs Review, 39*(1), 32–58. https://doi.org/10.1177/1078087403253053.

Clarke, L., & Short, J. F., Jr. (1993). Social organization and risk: Some current controversies. *Annual Review of Sociology, 19*(1), 375–399.

Comerio, M. C. (1997). Housing issues after disasters. *Journal of Contingencies and Crisis Management, 5*(3), 166–178. https://doi.org/10.1111/1468-5973.00052.

Comerio, M. C. (1998). Disaster hits home: New policy for urban housing recovery. Oakland, CA: Univ of California Press.

Crisp, E. (2018, May 11). Legislation inspired by "Cajun Navy" to coordinate volunteer rescues nears final passage. *The Advocate*. Retrieved from https://www.theadvocate.com/baton_rouge/news/politics/legislature/article_17e6a262-55311 1e8-8982-4f17c307d35f.html

Di Liberto, T. (2016, August 19). August 2016 extreme rain and floods along the Gulf Coast. *National Oceanic & Atmospheric Administration*. Retrieved from https://www.climate.gov/news-features/event-tracker/august-2016-extreme-rain-and-floods-along-gulf-coast

Di Liberto, T. (2017, September 18). Reviewing Hurricane Harvey's catastrophic rain and flooding. *National Oceanic & Atmospheric Administration*. Retrieved from https://www.climate.gov/news-features/event-tracker/reviewing-hurricane-harveys-catastrophic-rain-and-flooding

Dynes, R. R. (1974). *Organized behavior in disaster*. Columbus: Disaster Research Center, Ohio State University.

Dynes, R. R. (1983). Problems in emergency planning. *Energy, 8*(8), 633–660.

Elliott, J. R., & Pais, J. (2010). When nature pushes back: Environmental impact and the spatial redistribution of socially vulnerable populations. *Social Science Quarterly, 91*(5), 1187. https://doi.org/10.1111/j.1540-6237.2010.00727.x.

Federal Emergency Management Agency (FEMA) (2018). The National Flood Insurance Program. Retrieved from https://www.fema.gov/national-flood-insurance-program

Fernald, M. (2013). *America's rental housing: Evolving markets and needs. Resource document.* Cambridge, MA: Harvard University. http://www.jchs.harvard.edu/sites/jchs.harvard.edu/files/jchs_americas_rental_housing_2013_10.pdf.

Finch, C., Emrich, C. T., & Cutter, S. L. (2010). Disaster disparities and differential recovery in New Orleans. *Population and Environment, 31*(4).. https://www.jstor.org/stable/40587588), 179.

Forrest, T. R. (1978). Group emergence in disasters. In E. L. Quarantelli (Ed.), *Disasters: theory and research* (pp. 105–125). London: Sage.

Fothergill, A., & Peek, L. (2004). Poverty and disasters in the United States: A review of recent sociological findings. *Natural Hazards, 32*(1), 89–110.

Freddie Mac (2017). Multifamily in focus: Rental affordability is worsening. Retrieved from http://www.freddiemac.com/multifamily/pdf/rental_affordability_worsening.pdf

Fritz, C. E. (1961). Disaster. In R. K. Merton & R. A. Nisbet (Eds.), *Contemporary social problems* (pp. 651–694). New York: Harcourt, Brace, and World.

Fussell, E., & Elliott, J. R. (2009). Introduction: Social organization of demographic responses to disaster: studying population—Environment interactions in the case of Hurricane Katrina. *Organization & Environment, 22*(4), 379–394.

Fussell, E., & Harris, E. (2014). Homeownership and housing displacement after Hurricane Katrina among low-income African-American mothers in New Orleans. *Social Science Quarterly, 95*(4), 1086–1100. https://doi.org/10.1111/ssqu.12114.

Fussell, E., Sastry, N., & Vanlandingham, M. (2010). Race, socioeconomic status, and return migration to New Orleans after Hurricane Katrina. *Population and Environment, 31*(1–3), 20–42.

Gallo, A. (2017a, June 17). Here's how Louisiana programs aim to rebuild thousands of flooded rentals, get owners "over the goal line." *The Advocate*. Retrieved from https://www.theadvocate.com/louisiana_flood_2016/article_8d7bead0-52c8-11e7-99ffcb4057db2000.html

Gallo, A. (2017b, October 24). $150,000 grant to help Baton Rouge nonprofits build affordable rentals. *The Advocate*. Retrieved from https://www.theadvocate.com/baton_rouge/news/article_0d7e448c-b8e3-11e7-9b2ac39c9eb20e23.html

GAO (2010). *Disaster assistance: federal assistance for permanent housing primarily benefited homeowners; opportunities exist to better target rental housing needs (GAO 10-17).* Washington, DC: U.S. Government Accountability Office.

Gotham, K. F. (2008). From 9/11 to 8/29: Post-disaster recovery and rebuilding in New York and New Orleans. *Social Forces, 87*(2), 1039. https://doi.org/10.1353/sof.0.0131.

Gotham, K. F., & Greenberg, M. (2014). *Crisis cities: Disaster and redevelopment in New York and New Orleans*. New York: Oxford University Press.

Governor's Office of Homeland Security and Emergency Preparedness (GOHSEP) (2016). State of Louisiana Flooding August 2016 After Action Report/Improvement Plan. Baton Rouge, LA: GOHSEP.

Grueskin, C. (2017, December 17). New "piggyback" funding program targets rebuilding of mixed income rentals in flooded parishes. *The Advocate*. Retrieved from https://www.theadvocate.com/baton_rouge/news/article_e0faa738-e1c4-11e7-9f15132ff3a097b6.html

Grueskin, C. (2018a, February 5). A harsh reality: Program to house flooded renters living in FEMA trailers runs into lack of housing. *The Advocate*. Retrieved from https://www.theadvocate.com/baton_rouge/news/communities/livingston_tangipahoa/article_ccc573a-078b-11e8-a69f-ef38c5208b9a.html

Grueskin, C. (2018b, July 13). Restore Louisiana Task Force votes to give homeowners full reimbursement for flood repairs. *The Advocate*. Retrieved from https://www.theadvocate.com/baton_rouge/news/article_7c944026-86c7-11e8-9cc843bb05594f02.html

Hamideh, S., & Rongerude, J. (2018). Social vulnerability and participation in disaster recovery decisions: public housing in Galveston after Hurricane Ike. *Natural Hazards, 93*(3), 1629–1648. https://doi.org/10.1007/s11069-018-3371-3.

Hardy, S. (2017, July 26). Baton Rouge begins distribution of flood funds to help landlords and the homeless. *The Advocate*. Retrieved from https://www.theadvocate.com/baton_rouge/news/article_e0d69642-7246-11e7-85e22381f48760df.html

Henrici, J. M., Helmuth, A. S., Fernandes, R. (2010). Mounting losses: Women and public housing after Hurricane Katrina (Publication No. D491). Retrieved from https://iwpr.org/wpcontent/uploads/wpallimport/files/iwpr-export/publications/D491.pdf

Hilburn, G. (2017, August 29). Governor Edwards refuels Cajun Navy at border. *News Star*. Retrieved from https://www.thenewsstar.com/story/news/2017/08/29/gov-edwards-refuels-cajun-navy-border/613245001/

Hirayama, Y. (2000). Collapse and reconstruction: Housing recovery policy in Kobe after the Hanshin Great earthquake. *Housing Studies, 15*(1), 111–128. https://doi.org/10.1080/02673030082504.

Jones, T. L. (2017, February 12). Renters continue to struggle finding affordable rental units in Baton Rouge area after flood. *The Advocate*. Retrieved from https://www.theadvocate.com/louisiana_flood_2016/article_6653239a-dce1-11e6-ad4a63b860ee4b50.html

Kendra, J. M., & Wachtendorf, T. (2003). Elements of resilience after the World Trade Center disaster: Reconstituting New York City's emergency operations centre. *Disasters, 27*(1), 37–53. https://doi.org/10.1111/1467-7717.00218.

Kendra, J. M., & Wachtendorf, T. (2016). *American Dunkirk: The Waterborne Evacuation of Manhattan on 9/11*. Philadelphia: Temple University Press.

Lee, J. Y., & Zandt, S. V. (2018). Housing tenure and social vulnerability to disasters: A review of the evidence. *Journal of Planning Literature, 34*, 156. https://doi.org/10.1177/0885412218812080.

Louisiana Housing Corporation (2017). Baton Rouge Rebuilds. Retrieved from http://www.lhc.la.gov/page/baton-rouge-rebuilds

Louisiana Office of the Governor (2016*). 2016 State of Louisiana historic flooding – Rebuilding and recovery update*. Retrieved from http://gov.louisiana.gov/assets/docs/RestoreLA/SupportingDocs/Meeting-11-18-16/LACongressional-Flood-Update-11-16-16.pdf

McCarthy, G., Van Zandt, S., & Rohe, W. (2001). *The economic benefits and costs of homeownership: A critical assessment of the research. (Working paper No. 01-02)*. Arlington: Research Institute for Housing America.

McCausland, P., Gosk, S., & Helsel, P. (2017, August 27). Flooded Houston braces for more rain as Harvey rescues continue. *NBC News*. Retrieved from https://www.nbcnews.com/storyline/hurricane-harvey/flooded-houston-braces-more-rain-rescues-continue-n796476

Meyer, M., Mitchell, B., & Dombroski, S. (2018). Working with the Cajun Navy: Optimizing the use of volunteers in disaster response. *National Science Foundation*. Retrieved from https://www.nsf.gov/discoveries/disc_summ.jsp?cntn_id=296509&org=NSF&from=news

Morrow, B. H. (1999). Identifying and mapping community vulnerability. *Disasters, 23*(1), 1. https://doi-org.ezproxy.library.tamu.edu/10.1111/1467-7717.00102.

Mukherji, A. (2015). From tenants to homeowners: Housing renters after disaster in Bhuj, India. *Housing Studies, 30*(7), 1135–1157. https://doi.org/10.1080/02673037.2015.1008423.

National Research Council. (2006). *Facing hazards and disasters: Understanding human dimensions*. Washington, DC: National Academies Press.

National Weather Service (NWS) (2017). NWS Directives (NWSI) 10-950 (Hydro definitions and general terminology). Retrieved from http://www.nws.noaa.gov/directives/sym/pd01009050curr.pdf

National Weather Service (NWA) (2019). August 2016 Record Flood Summary Page. Retrieved from https://www.weather.gov/lix/August2016flood

Noji, E. K. (1997). *The public health consequences of disasters*. New York: Oxford University Press.

Pais, J., & Elliott, J. (2008). Place as recovery machines: Vulnerability and neighborhood change after major hurricanes. *Social Forces, 86*(4), 1415–1452.

Peacock, W. G., Morrow, B. H., & Gladwin, H. (Eds.). (1997). *Hurricane Andrew: Ethnicity, gender and the sociology of disasters*. New York: Routledge.

Peacock, W. G., Van Zandt, S., Zhang, Y., & Highfield, W. E. (2014). Inequities in long-term housing recovery after disasters. *Journal of the American Planning Association, 80*(4), 356–371.

Peacock, W. G., Dash, N., Zhang, Y., & Van Zandt, S. (2018). Post-disaster sheltering, temporary housing and permanent housing recovery. In *Handbook of disaster research* (pp. 569–594). Springer.

Perry, R. W., & Lindell, M. K. (2003). Preparedness for emergency response: Guidelines for the emergency planning process. *Disasters, 27*(4), 336–350.

Phillips, B. D., Thomas, D. S. K., Fothergill, A., & Blinn-Pike, L. (Eds.). (2010). *Social vulnerability to disasters*. Boca Raton: CRC Press.

Phillips, B. D., & Fordham, M. (2010). Introduction. In B. D. Phillips, D. S. K. Thomas, A. Fothergill, & L. Blinn-Pike (Eds.), Social Vulnerability to Disasters (pp. 1-26). Boca Raton, FL: CRC Press.

Raja, T. (2017, August 27). A makeshift navy struggles to respond to Hurricane Harvey. *The New Yorker*. Retrieved from https://www.newyorker.com/news/news-desk/houston-a-makeshift-navy-struggles-to-respond-to-hurricane-harvey

Reidmiller, D. R., Avery, C. W., Easterling, D. R., Kunkel, K. E., Lewis, K. L. M., Maycock, T. K., & Stewart, B. C. (Eds.) (2017). Impacts, risks, and adaptation in the United States: fourth national climate assessment, *Vol.* 2. US Global Change Research Program. doi:https://doi.org/10.7930/NCA4.2018.

Rhoden, R. (2016, August 19). Where they died: Interactive map shows Louisiana flooding victims. *NOLA.com/The Times-Picayune*. Retrieved from https://www.nola.com/weather/index.ssf/2016/08/la_flood_victims_black_and_whi.html

Rohe, W. M. (2017). Tackling the housing affordability crisis. *Housing Policy Debate, 27*(3), 490–494. https://doi.org/10.1080/10511482.2017.1298214.

Rubin, C. B. (2009). Long term recovery from disasters – The neglected component of emergency management. *Journal of Homeland Security and Emergency Management, 6*(1).

Schneider, S. (1992). Governmental response to disasters: The conflict between bureaucratic procedures and emergent norms. *Public Administration Review, 52*(2), 135–145. https://www.jstor.org/stable/976467.

Scott, M. (2016, August 16). National media fiddle as Louisiana drowns. *NOLA.com/The Times Picayune*. Retrieved from https://www.nola.com/weather/index.ssf/2016/08/national_media_louisiana_flood.html

Siegel, G. B. (1985). Human resource development for emergency management. *Public Administration Review, 45*, 107–117.

Steffen, S. L., & Fothergill, A. (2009). 9/11 volunteerism: A pathway to personal healing and community engagement. *The Social Science Journal, 46*(1), 29–46.

Terrell, D. (2016). The economic impact of the August 2016 Floods on the State of Louisiana. *Louisiana Economic Development*. Retrieved from http://gov.louisiana.gov/assets/docs/RestoreLA/SupportingDocs/Meeting-9-28-16/2016-August-Flood-Economic-Impact-Report_09-01-16.pdf

Thomas, D. S. K., Phillips, B., Lovekamp, W. E., & Fothergill, A. (Eds.). (2013). *Social vulnerability to disasters* (2nd ed.). Boca Raton: CRC Press.

Tierney, K. J., Lindell, M. K., & Perry, R. W. (2001). *Facing the unexpected: Disaster preparedness and response in the United States*. Washington, DC: Joseph Henry Press.

Tulane School of Architecture (2007). The "Big Four" public housing projects are demolished post-Katrina. *New Orleans Preservation Timeline Project*. Retrieved from http://architecture.tulane.edu/preservation-project/timeline-entry/1427

U.S. Department of Housing & Urban Development (HUD) (2018). *Affordable Housing*. Retrieved from https://www.hud.gov/program_offices/comm_planning/affordablehousing/

Van Zandt, S., Peacock, W. G., Henry, D. W., Grover, H., Highfield, W. E., & Brody, S. D. (2012). Mapping social vulnerability to enhance housing and neighborhood resilience. *Housing Policy Debate, 22*(1), 29–55.

Walters, E. (2018, April 16). "It's our form of apartheid": How Galveston stalled public housing reconstruction in the 10 years after Ike. The Texas Tribune. Retrieved from https://www.texastribune.org/2018/04/16/galveston-public-affordable-housing-hurricane-ike/

Weil, H. J. (2009). Finding housing: Discrimination and exploitation of Latinos in the post Katrina rental market. *Organization & Environment, 22*(4), 491–502. https://doi.org/10.1177/1086026609347194.

Yan, H. & Flores, R. (2016, August 19). Louisiana flood: Worst US disaster since Hurricane Sandy, Red Cross says. *CNN*. Retrieved from https://www.cnn.com/2016/08/18/us/louisianaflooding/index.html

Zhang, Y., & Peacock, W. G. (2009). Planning for housing recovery? Lessons learned from Hurricane Andrew. *Journal of the American Planning Association, 76*(1), 5–24. https://doi.org/10.1080/01944360903294556.

# Part VI
# Types/Locations of Communities and Their Responses to Extreme Weather: Rural

# Chapter 11
# Challenges of Post-Disaster Recovery in Rural Areas

**Alessandra Jerolleman**

## 11.1 Introduction

Although disaster losses frequently occur in rural and agricultural areas, a significant majority of the existing disaster research has focused on urban areas and coasts, often overlooking rural populations and communities (Cutter et al. 2016; Tierney 2013). Our research-based understanding of the recovery of housing post disasters in rural areas is even more limited, again with much of the current scholarship focused on urban areas and cities (Ganapati et al. 2013). Furthermore, the majority of the limited studies that have taken place in rural communities have focused on environmental or technological disasters, such as mining-related incidents, and not on more frequently occurring events such as disaster losses from flooding (Scott et al. 2012).

Rising disaster losses and increasing frequency of events across the United States coupled with a current political climate that does not result in a national consensus demand more local responsibility for disaster recovery, when less federal aid is offered as a result, making rural disaster studies a particularly pressing issue. Even if recommendations to address climate change are taken, communities will continue to experience increasing impacts and will be expected to take on a greater percentage of the burden for disaster recovery (Coppola 2016). Research has shown that disaster impacts can best be mediated at the local level, where the most effective risk reduction measures can be undertaken and the most effective policies enacted. Thus, there is a "silver lining" to shift to more local disaster recovery and adaptation attention. Unfortunately, achieving success in risk reduction is far more challenging for communities that lack sufficient resources to ensure the success of these measures and much less have the resources to fund their own adaptation programs (Haddow 2016a, b).

A. Jerolleman (✉)
Department of Emergency Management, Jacksonville State University,
Jacksonville, AL, USA



S. Laska (ed.), *Louisiana's Response to Extreme Weather*,
Extreme Weather and Society, https://doi.org/10.1007/978-3-030-27205-0_11

Even in larger urban areas, resources are often scarce at the local level, where the government is facing pressures to balance budgets and cut expenditures while also experiencing reductions in federal support. The limited tax base of more rural communities makes these pressures even more acute, exacerbating many of these impacts and lessening the community's ability to invest in resilience. These challenges can result in underinvestment in preparedness and hazard mitigation, at a time when those investments are most needed (Skertich et al. 2012). This creates a magnifying effect on what is already a classic example of a "wicked policy problem" (Aldrich and Meyer 2015). The term "wicked problem" denotes those policy challenges for which solutions are not definitive; there is no means to fully test a solution; there is no ability to learn from trial and error because the consequences of attempts at solutions are high; there is no clear set of potential solutions; the problem can be considered a symptom of other problems; multiple explanations exist for the problem and may be contradictory; and the government is liable for the consequences of the actions generated (Rittel et al. 1973).

Crises and catastrophes often have no clear technical or policy solution, a characteristic of wicked problems. They also involve multiple stakeholders and present ripple effects. In other words, a wicked problem is without a clear solution, in a context that further limits the available responses to the issue.

This chapter will begin by summarizing the current literature on post-disaster recovery and resilience in rural areas, followed by presenting a case example from Northern Louisiana,[1] and concludes with some recommendations for strengthening the resilience of rural areas. In order to begin this discussion, it is important to define just what is meant by *rural*. There are several definitions that might be considered, all of which are premised upon being outside of an area defined as urban. The US Census has utilized various definitions over time, with the earliest definition being simply places outside of cities and towns with populations of under 2500 persons (Ratcliffe et al. 2016). More recent definitions also consider density, but the basic definition remains that of areas outside of urban areas. According to the US Census, rural populations declined from 54.4% of the overall US population in 1910 to 19.3% in 2010 (Ratcliffe et al. 2016). This chapter will use the term rural in the broader sense, simply taken to mean areas that are not defined as urban in character. A more specific definition is not feasible given the wide range of definitions utilized within the limited literature and the failure of much of that literature to clearly delineate the bounds of rurality. What is clear, however, is that many rural communities are losing population and in some cases economic viability, a challenge to their ability to adapt to the increasingly frequent and diverse disaster events.

The following map (Fig. 11.1) illustrates the location of rural census tracts across Louisiana.

[1] The abbreviated case study included in this chapter is based upon an ongoing recovery effort. As a result, there is limited data available regarding final outcomes of the recovery process. Over the span of 2017–2018, the time frame during which this chapter was being written, the impacted communities have continued to seek ways to work with the federal and state government to continue their recovery and to improve their adaptation. I would like to thank Olivia Porter for her assistance in locating media and reports about the community's progress.

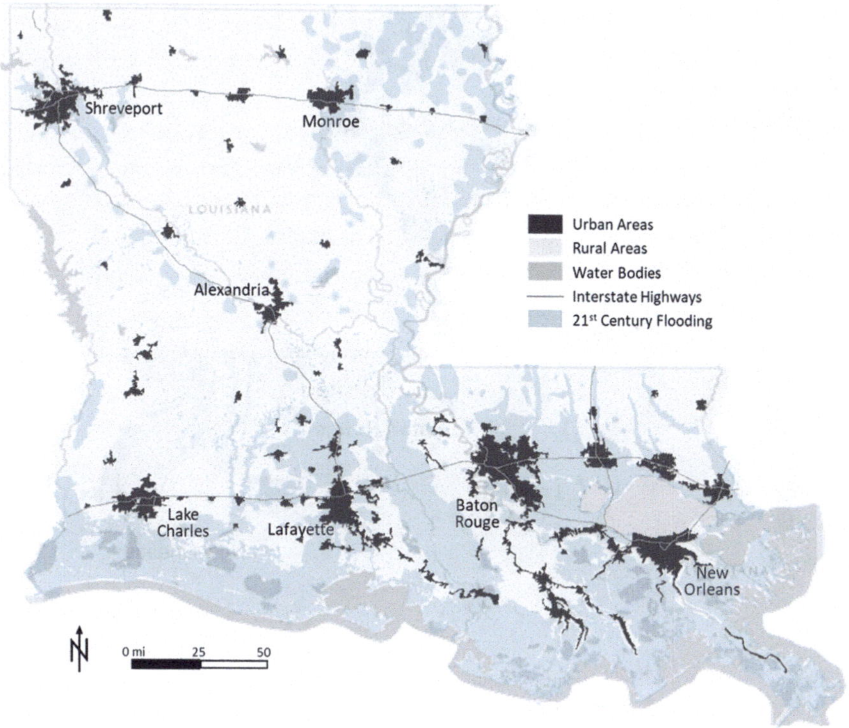


**Fig. 11.1** Map showing the distribution of rural and urban areas in Louisiana. Both urban areas (>50,000 people) and urban clusters (2500–50,000 people) are defined by the 2000 US Census. (Data retrieved from the 2000 US Census Summary File 3)

The state of Louisiana received a disaster declaration on March 13, 2016, DR-4263, based upon several days of flooding that had primarily impacted the northern portions of the state. The flooding began in the northwestern portion of the state, sweeping from Shreveport down through Central Louisiana and along the I-20 Corridor, causing extensive damage across both smaller urban areas and rural areas.

The following map (Fig. 11.2) illustrates the extent of the flooding.

## 11.2 Literature Review

### *11.2.1 Is There a Difference Between Rural and Urban Areas?*

The current literature broadly presents two differing views of the inherent differences between urban and rural areas when it comes to disaster vulnerability and recovery. Some researchers describe rural communities as having more limited capabilities than urban areas, significantly lacking in resources, unable to fully benefit from the resources made available after a disaster, failing to deliver basic

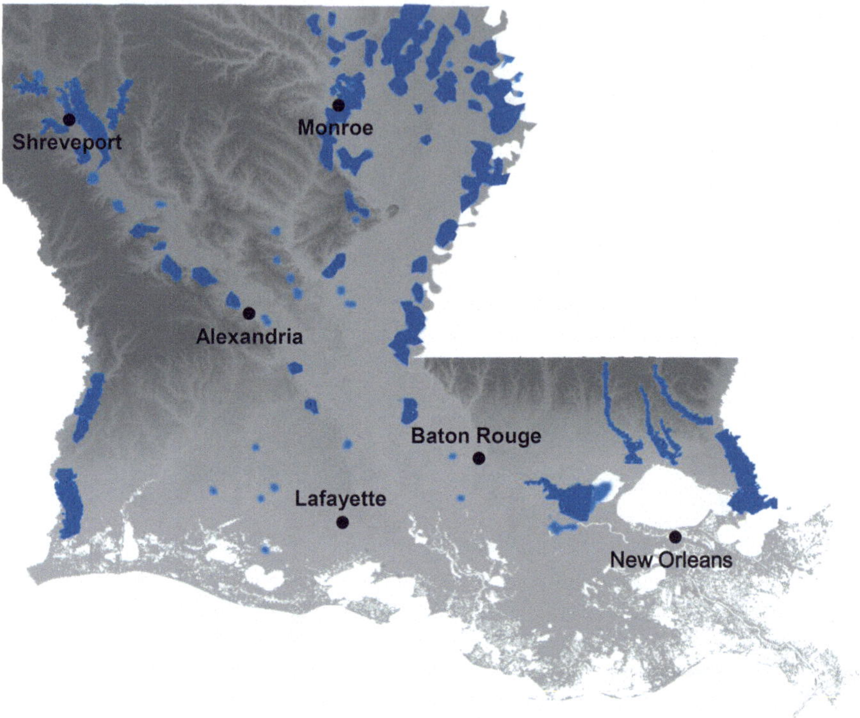


**Fig. 11.2** Louisiana flooding in March 2016. (Flood map from Fig. 1.1)

services day-to-day, and quickly losing social support systems in the face of systemic pressures such as shrinking local economies (Downey 2016; Mogle 2017; Tierney 2013; Doherty 2004). This particular view of rural communities is succinctly stated by Tierney, who describes rural communities as "under resourced places in which the capacity to anticipate, cope, and adapt has been seriously compromised (Tierney 2013, p. xv)."

Seen through this lens, larger jurisdictions, such as those in urban areas, have an inherent advantage in terms of larger budgets and staff capacity for building code enforcement. These larger and more resourced jurisdictions can also engage in other resilience promoting actions such as having a "rainy day" fund or investing in resilient infrastructure (May 2013). Communities that are able to maintain a rainy day fund, for example, can begin work without waiting for federal dollars to be made available and can afford to wait several months for reimbursements (Landy 2008). They are also better able to participate in more effective emergency management networks, characterized by crossing sectors and organizations, which have been found by much of the research to improve recovery outcomes (Demiroz et al. 2013). In rural areas, particularly those further for urban centers, these networks, when they exist, have a tendency to be more centralized and include fewer actors (Demiroz et al. 2013).

The second view presented by the research is that rural communities can, in some cases, have a greater sense of self-reliance, stronger social bonds, a defined sense of community, and access to natural resources (Cutter et al. 2016). In other words, rurality can confer certain advantages when it comes to resilience. Cutter, Ash, and Emrich used an index titled the Baseline Resilience Indicators for Communities (BRIC), a model that looks at six different capitals (social, economic, community, institutional, infrastructural, and environmental), in order to analyze whether the rural nature of a community had an impact on resilience. These different capitals align with many of the key elements discussed in the resilience literature, and with several areas in which distinctions are drawn between larger and smaller, or urban and rural communities.

In their research Cutter, Ash, and Emrich found a great deal of variation from region to region and rural community to rural community. However, they did find that community capital variables were generally more prevalent in rural communities, while economic resilience variables were more prevalent in urban areas. Such findings support the assertion that rural communities often possess greater levels of social capital but are also facing extensive challenges relative to their economic base which detract from the benefits of the greater social capital.

However, an interesting finding of the study was the extent to which there was regional variation. Across the United States, the research found pockets of lower economic resilience along the Arkansas, Mississippi, and Louisiana border, as well as increased housing vulnerability due to a prevalence of mobile home building stock in the south. It is worth noting that North Louisiana is included in one of the regions of the country where a greater correlation between rurality and vulnerability was found. Cutter, Ash, and Emrich speculated that the increased vulnerability in certain areas had to do with a history of economic disparities along racial lines and thus the weaker social capital that results from the subcommunity economic differences (2016). The role of racial and economic disparity has been explored by other researchers in the context of disaster vulnerability and will be discussed further below. Given prior experience of the author working within North Louisiana, as well as upon informal conversations she has had with local rural leaders, both of these factors can be said to be present in the flooded areas of North Louisiana.

Another area in which rural communities are at times described as having an advantage is in the complexity, or lack thereof, of local governance structures. In other words, although larger governments with more departments may be more able to provide manpower for disaster recovery, a more centralized and flexible governance structure can present certain advantages. Rural communities may often have a less complex local government landscape to navigate in disaster recovery, but they are also far more cash strapped, have a less diversified economic base, and are less able to maintain a rainy day fund (Caruson and MacManus 2011; Waugh 2013). This advantage may be eliminated by the concurrent difficulties that arise in interacting with disaster recovery mechanisms that require a certain amount of cash reserves and staff capacity.

Recent research into disaster recovery by Mogle found that recovery committees can be more difficult to establish in communities that lack a lot of manpower and are already reeling from the resources required to navigate complex governmental recovery processes (2017). These committees are a key component of the national framework for disaster recovery and are expected to mirror state and federal frameworks, serving as the key mechanism for receipt of technical assistance. A recovery committee was established in North Louisiana following the 2016 flooding and was negatively impacted by the lack of personnel available to serve the necessary functions. Under the National Disaster Recovery Framework, there are six Recovery Support Functions (RSFs), each focused on a different aspect of recovery, such as economic recovery and housing. These RSFs each require staffing and local engagement in order to successfully participate in a recovery process.

Mogle also found that recovery committees have a very steep learning curve before they can become effective, a challenge exacerbated by limited pre-disaster capabilities (2017). Jurisdictions that have been able to invest in pre-event planning, relationship building, and training around recovery have an inherent advantage. Unfortunately, this is not feasible in communities that are already stretched far too thin. This lack of institutional capacity can have negative impacts on families and households, particularly in the context of a federalist structure that leaves the primary responsibility for emergency management at the local level (Mogle 2017). In fact, comparative research into international disaster recovery has found that recovery in a democracy is negatively impacted by variations in institutional quality, leading to more disparate outcomes than might be found under a more centralized government structure (Persson and Povitkina 2017).

Returning to a broader comparison of urban and rural areas, Caruson and MacManus argued that urban areas might be more vulnerable due to having a wider range of vulnerable populations, infrastructure with greater vulnerabilities, and other limitations (2011). However, even this research recognized some advantages for urban areas, such as greater eligibility for grants and better access to financial resources. These advantages may be more readily available to rural communities that are in sufficient proximity to urban areas so as to participate in regional partnerships or benefit from collaboration and partnerships. In fact, Waugh found that rural populations on the fringe of urban areas could access more assistance than those located at a greater geographic distance, observing empirically the impacts of distance from the urban core that had been reported anecdotally in North Louisiana (2013).

Similarly, Brody and Gunn, looking at community resilience through the lens of development along the Gulf Coast, found that rural jurisdictions contained more pervious surfaces, an indicator of reduced flood risk correlating to the environmental capital included in the Baseline Resilience Indicators for Communities index, but were also experiencing a greater percent loss of wetlands (2013). This finding could also be explained by the fact that the urban areas had already destroyed much of the previous wetland cover which exacerbated their flood risk, while the rural areas were perhaps embarking down the same path. It is interesting to note that although limited development may confer some advantages to rural areas, the lack of building codes may mean that the development, when it does occur, will have even greater detrimental impacts. This is particularly concerning in terms of disaster recovery, because research in rural areas following disasters have found a greater increase in housing growth in rural areas as opposed to suburban. This is sometimes a function of post-disaster migration away from urban areas, such as what occurred after Hurricane Katrina. This type of migration often drives the most vulnerable away due to rising housing costs as disasters often have a detrimental impact on the affordability of housing (Ganapati et al. 2013). Ganapati found that growth in rural counties after Hurricane Katrina was impacted by domestic migration and resulted in an increase in the percentage of mobile homes. One unfortunate downside to efforts to increase resilience through building codes can be the inability of former residents to afford to rebuild at the new improved standards. This can drive gentrification of urban areas and push the urban poor out to the suburbs or further out to rural communities, both to lower-lying areas and in flood plains. Areas inland from New Orleans received urban residents, some from the iconic Lower Ninth Ward, after Hurricane Katrina only to have these resettlers experience flooding during Hurricane Isaac in 2012 and the August 2016 storm.

Yet another framing of rural resilience involves utilizing the lens of adaptive capacity, built upon the assumption that rural communities are more likely to have a culture of adaptation to challenges and of living with the environment. Cox and Hameln, looking at adaptive capacity in rural Canada, described resilience as the foundation of rural life (2015). In their analysis, rural resilience is treated largely as an offshoot of a broader ethos of adaptation. However, they did find that many rural communities lacked the civic infrastructure to fully participate in resilience planning and that this type of planning effort might not be realistic for smaller rural communities, even when efforts are made to make the process accessible and resources are provided. This finding speaks to the need to adopt a wide range of resilience strategies that can take into account the unique nature of different communities, including the presence of civic infrastructure and governance structures, many qualities which are assets if appreciated and supported in ways possibly different from urban. It is not just offering what is offered to the urban communities but investing the effort to determine what approaches would be supportive of the adaptive skills the rural areas already have as well as the different challenges they experience.

In the communities with which Cox and Hameln interacted, they found governance issues, economic issues, and limitations in capacity that limited the ability to engage for resilience. These limitations, although in the Canadian context, are strikingly similar to the findings from other researchers in US rural jurisdictions. These similarities perhaps open the door for lessons from international contexts and not simply from the United States, arguing that the challenges faced by agricultural and rural communities across the world may share certain similarities and provide valuable lessons.

This chapter will argue that rural communities are in fact facing different challenges in terms of their ability to increase disaster resilience, practice risk reduction, and successfully recover following a disaster. These challenges are exacerbated by the increasing media and policy focus on cities, as the leaders in the climate resilience movement, with much less attention paid to the successes and challenges in rural America (Haddow 2016a, b). While it is certainly true that cities have been innovative and influential in the climate adaptation movement, it must also be recognized that urban leaders often have greater control over the physical infrastructure, stronger social connections (particularly those that garner resources), and access to more financial resources (Haddow 2016a, b). Furthermore, while direct property losses are much worse in urban areas, due to the concentration of population and the financial value of the built environment, relative impacts are far worse in rural communities (Cutter et al. 2016). This results in a challenge for drawing comparisons, as most damage data is considered in terms of the number of units or the population impacted, with little consideration for the percent impacts upon a particular community. These impacts can be assumed to be magnified in a close-knit, smaller community. However, as the literature has shown, rural communities do have strengths upon which to focus when building resilience, and these should not be discounted in favor of a simplistic assumption that rural jurisdictions are at a significant disadvantage in terms of disaster recovery and resilience.

### *11.2.2 Challenges Faced by Rural Communities*

One of the primary challenges faced by rural communities, even before a disaster, is a more limited ability to not only provide basic services, such as health care and access to broadband, but also extensive limitations in their ability to take action to reduce risk through hazard mitigation (May 2013, 2016). Rural areas in the United States often lack zoning and building codes or lack the capacity to fully enforce existing codes (May 2013; Schwab 2016). At a more fundamental level, rural communities have been struggling to maintain even basic government services in light of economic conditions and changing rural environments and economies, such as the loss of jobs and displacement of small farms (Doherty 2004; Waugh 2013). Rural populations have been declining, with a drop of nearly 200,000 people between 2010 and 2016, due to factors such as outmigration of young adults, fewer births, and an aging population. At the same time, job growth in rural counties following 2011 was substantially lower than in urban counties and the poverty rates much higher (USDA 2017). This has led to the disappearance of support and social systems, at the same time that necessary services such as health services are becoming less available (Doherty 2004).

These challenges have been described as a slowly growing crisis, even outside of a more traditional natural disaster. According to Doherty, even among communities with high levels of social capital and cohesion, a crisis can disrupt the sense of balance and negate the benefits of social capital (2004). This can be particularly problematic when there are competing narratives regarding the origins of the crisis, such as when there are questions around responsibility and impacts and when the detrimental effects are concentrated on particular segments of the population (Aronoff and Gunter 1992). At times the benefits, such as those from economic development, can be concentrated solely in particular segments of the population while exacerbating vulnerabilities of other segments. Questions of responsibility, particularly around industrial accidents or hazardous materials incidents, can have detrimental effects on rural communities' social capital and on the ability to recover.

The preexisting capacity deficits, coupled with the long-term impacts of sustained crises, leave rural communities unable to fully take advantage of the resources that become available following a disaster (Downey 2016). Bolin and Bolton found this to a be a problem as far back as the late 1980s, finding that rural disaster victims tended to receive aid from less government sources and less total aid (1986). For example, one rural community in Texas struggled to recover following a tornado due to the lack of nonprofit organizations and infrastructure needed to both accept and distribute funds and donations to the victims (Mogle 2017). As with previous examples, the experience of this community illustrates the value of adaptive recovery models that do not demand as much inherent staff and organizational capacity at the local level.

One study of local governments in Pennsylvania, primarily rural governments, by Skertich, Johnson, and Comfort, found four key contributing factors related to the demands and constraints placed upon the provision of public safety and public health services, despite the legal requirements that these services be provided. These contributing factors are (1) the actual legal requirements for the provision of these services, (2) reductions in available economic resources as a result of fiscal stress, (3) cutbacks in personnel as a result of efforts to balance municipal budgets, and (4) increasing demand for services as infrastructure ages and increasing numbers of vulnerable people are living in regions (Skertich et al. 2012). Limitations in levels of household and individual preparedness were also found to have an impact on the success of public safety and public health efforts. As the research shows, communities are facing legal requirements to provide services which they lack the resources to provide while also facing extensive pressures to further cut budgets and staffing. These challenges are difficult to navigate without the added strain of a disaster on aging infrastructure and a vulnerable population.

In their study, Skertich, Johnson, and Comfort found that local governments had turned to both consolidation and privatization as a solution to the complex demands and constraints described above. This approach, which also included utilizing volunteers and placing personnel in dual roles, served to mask the actual limitations that the communities faced. These limitations become glaringly obvious when a disruption occurs. Communities in Pennsylvania were found to have utilized one of three key strategies: (1) increased reliance on technology, (2) reduction of engagement with community participation and preparedness efforts, and (3) increased reliance on larger organizations such as regional or neighboring governments. These three strategies all carry different implications for disaster recovery. The second and third strategies, in particular, can be seen to have very negative impacts on the ability to recover, as in the case of lessened preparedness and reduced community participation that might also negatively impact social capital and civic engagement but, also in case of the third strategy, should the larger organizations have impacts that exceed their own abilities and therefore be less able to render assistance to the smaller jurisdictions that have also been impacted. Overreliance on partners who may not always be available can result in further decreasing local capabilities. The third strategy, in particular, might be less viable for rural areas that are further removed from larger urban centers.

Revisiting the broader question of whether there is a tangible difference between urban and rural communities, it is clear that the resource constraints described above impact all communities in some way. However, as shown by the current literature, they may have particularly devastating impacts on rural communities where the demands placed upon local government far exceeded its ability to perform. Cutbacks in personnel and other resources are likely to have a direct impact on resilience, as well as on the ability to recover following a natural disaster. Simply stated, when resources are strained during day-to-day operations, there is no capacity to absorb the added demands created by a disaster. Planning is another area in which these constraints have a disproportionate impact.

## *11.2.3 Constraints to Planning*

Another challenge that rural jurisdictions face is limitations in their ability to plan successfully for hazard mitigation and for post-disaster recovery. As might be expected, lower capacity for planning and program management related to hazards has been found to be a concern in rural areas (Waugh 2013). Research by Berke and Campanella found that more immediate and pressing concerns, such as those described in the preceding paragraphs, eclipsed efforts to plan for longer-term issues such as disaster recovery (2006). This inability has detrimental effects on successful recovery management, either because a plan has not been developed or due to the lack of a knowledgeable constituency, as can be created by a robust planning process, who are more likely to support sound risk reduction policies. In fact, this relates directly back to the finding that government capacity to protect its population has direct impacts on the degree of human suffering following a disaster (Persson and Povitkina 2017). If a community cannot effectively plan or prepare, then loss of life and other impacts will be greater.

Most federal assistance programs are premised on risk sharing and risk reduction, both of which require extensive preplanning and a full understanding of risk. In other words, communities that are unable to successfully engage in risk reduction through robust planning and forward looking policies face a substantial disadvantage, even in their interactions with federal programs. This is in keeping with the findings from Persson and Povitkina that disaster prevention can be considered a public good to which access is uneven (2017). They found that democratic institutions fail to protect their populations well in the face of poor planning or incompetence in public administration, both of which can arguably be said to be more prevalent in rural areas due to a lack of capacity and resources. In other words, inadequate planning processes significantly impact government's ability to protect the population.

In keeping with this concern, research into hazard mitigation plan quality found that rural communities faced significant disadvantages in the development of their plans, including an aging population base, fewer resources, greater isolation, a lack of in-house expertise, limited resources for hiring consultants, inferior housing stock, and poverty (Horney et al. 2017). These challenges impact both infrastructure and social structures, going beyond the traditional scope of a hazard mitigation planning process but clearly impacting the jurisdiction's ability to successfully implement a hazard mitigation strategy. Difficulties in not only hiring consultants but doing so effectively are particularly problematic, given how limited local resources are. Research by Mohr et al. found that smaller rural communities had the most challenges with contracting for services (2010). This is the result of several factors including the fact that smaller governments receive fewer bids, therefore losing the advantages of competition, and have a more difficult time enforcing contract provisions with limited staff or volunteers. Regional planning, as well as regional project implementation, may provide a usable alternative for some communities, but rural jurisdictions are often at a disadvantage in terms of resource allocation through regional mechanisms (Horney et al. 2017).

Program management, in particular, is negatively impacted by the limited borrowing ability and low revenues that small governments often face. However, other organizations such as faith-based organizations may play a key role in supporting local government when they are present and have sufficient capacity (Horney et al. 2017).

### *11.2.4 Social Capital*

The literature on disaster recovery and resilience indicates that social capital plays a key role in a successful recovery and is perhaps particularly vital when other resources are lacking. The suggestion that social capital might play a larger role in rural disaster recovery is an important consideration for this chapter. In order to fully discuss social capital, it is important to begin with an operational definition. This chapter will utilize definitions from Aldrich and Lalone, two leaders in the academic discourse regarding social capital and its applications to the disaster context. Aldrich defines social capital as "networks that connect individuals to each other through weak or strong ties" (2017, 358). This definition emphasizes the connectivity and relational aspect of social capital.

Lalone provides the following definition, which is similar to Aldrich's framing, but also explains some of the mechanisms through which social capital becomes beneficial: "Social capital refers to the resources of support that are embedded within social networks, and that are cemented and reinforced through relationships of trust and social norms emphasizing reciprocity and mutual assistance" (2017, p. 3). This framing of social capital as mutual assistance is one that is seen in much of the literature regarding the beneficial nature of social capital in rural disaster recovery. The literature further distinguishes between bridging and bonding capital, in which bonding capital links similar individuals, such as family members, and bridging capital links people with different backgrounds, typically through institutions or other organizations. These types of institutions and organizations are often more prevalent in urban areas than they are in more isolated rural communities. A third type of social capital, linking, will be discussed later in this chapter.

Reciprocity, a key component of social capital, has historically been a norm in many farming and mining communities across the rural United States where families supported each other with labor, food, and other resources when employment and crops fluctuated (Lalone 2012). This history of reciprocal assistance can have a tremendous impact on the ability of a community to come together following a crisis and provide support to each other. Bonding social capital, in particular, can increase the prevalence of emergent social action but can also have the unintended effect of reducing the likelihood of seeking external or formal aid, a problem when impacts exceed the capacity of local resources to fully address all needs (Aldrich and Meyer 2015).

A 2012 study of the mobilization of social capital in a rural Appalachian region in Virginia following a 2011 tornado found mobilization of labor and supplies in the response to the event. Local churches quickly came together to provide goods and shelter, families and neighbors stepped into assist with debris, and people came from across the region to provide assistance. In fact, the community response was so successful that formal shelter only remained open for 2 days and the level of community support overwhelmed the state emergency management structures, particularly by the number of volunteers and donations. Local government in neighboring jurisdictions including emergency managers also stepped in to help (LaLone 2012).

This is a clear example of a tradition of reciprocity, between neighbors and also across a broader geographic region, laying the foundation for a more successful disaster recovery. At the same time, it also serves as an example of the difficulty that formal emergency management structures have in coordinating with these existing structures. In this instance, the formal mechanisms do not appear to have impeded the informal recovery mechanisms, but that is not always the case. In some instances, there is a direct conflict between the formal and informal structures, with opportunities for utilizing local capital discarded or, at worst, detrimental impacts on existing social capital as a result of disaster recovery. A highly publicized example of the rural social capital confronting the official urban is the "Cajun Navy" – boat owners from rural Louisiana coastal areas who rushed to New Orleans to help rescue those marooned on rooftops after Hurricane Katrina only to be turned away at the parish line because they had no official role and could have been subject to liability cases. Some managed to get through the barriers to help and now they are considered an emergency group and continue to respond to events such as the recent floods in North Carolina from Hurricane Florence.

Aldrich writing about the role of social capital in disaster recovery described the majority of the post-disaster needs as collective action challenges, situations in which a collective identity and willingness to work together becomes crucial (2017). Social capital, which provides a mechanism for informal mutual aid, is a key asset for collective action challenges. Strong networks, particularly those that include bridging capital, can provide access to resources and information regarding the trustworthiness of different actors and the best means to both access and utilize resources. Rural communities in closer proximity to urban areas or in larger regions may benefit more from these types of networks. In some cases, bonding social capital can even act like informal insurance. However, by that same token, a disaster can negatively affect social cohesion due to displacement, the extent of losses, or uneven impacts – leading to the creation of a corrosive community (Aldrich and Meyer 2015).

As previously mentioned, researchers have argued that the bonds from social capital are stronger in some rural communities than they are in urban areas because of the prevalence of long-standing relationships based upon reciprocity and mutual assistance, both creating these relationships and becoming stronger as a result of them. This might result in an advantage for communities that do possess strong social capital, as access to social capital can provide resources when other resources are lacking and can correlate to civic engagement which has tangible benefits for the economy and recovery more broadly. Research following an earthquake in Tokyo found that voter turnout was a better indicator of post-disaster population growth than other economic indicators, damage levels, or population density (Aldrich and Meyer 2015).

In fact, research has found that community ties, shared goals, and other intangibles can matter just as much as political and government infrastructure in terms of successful disaster recovery (Ireni-Saban 2012). There is also a documented positive correlation between the number of nongovernmental organizations and social groups active within a community and post-disaster population recovery (Aldrich and Meyer 2015; LaLone 2012). Social capital can also have an empowering impact or even a therapeutic impact for groups that are often considered vulnerable such as women (Ganapati 2012).

In the context of a disaster, these ties can serve to connect communities to power structures and decision-makers, through linking capital – a third type of social capital that extends beyond the immediate community. Social capital can also allow groups to mobilize far more easily and to assist their members through informal assistance and insurance mechanisms. However, social capital may also have occasional negative impacts as it can prevent people from leaving disaster-impacted regions and can support the mobilization of certain groups that might utilize their capital and access to the detriment of others. This is seen in the frequent resistance to post-disaster placement of temporary housing by those neighborhoods in which temporary mobile home parks or to the placement of trailers on personal property while homeowners repair their damaged homes. This is more often the case in more affluent neighborhoods, for example, where the resistance can be based in part upon the exclusion of particular groups such as renters but also upon fear that home values will be reduced (Aldrich and Meyer 2015).

One of the most immediate and important decisions following a disaster is whether to stay or go. Strong community bonds reduce exit, an effect that may have both positive and negative impacts on disaster resilience and recovery. On the one hand, a community in which people come together with a shared commitment to recovery will have certain advantages in terms of the mobilization of resources and manpower, as well as in community reinvestment and the maintenance of a tax base. On the other hand, an unwillingness to consider leaving may have a negative impact on individual and family recovery, particularly when the community is unable to fully recover and support the recovery of housing, infrastructure, and the economy. Unfortunately, all of these factors have the potential to impact household recovery negatively.

Although social capital plays a key role in effective disaster response and recovery, researchers have found that a failed response can have a detrimental impact on community bonds (Ireni-Saban 2012). This is a concern due to the fact that as communities face an ever-growing array of threats, the potential for a failure becomes even greater and the resulting erosion of social bonds can further damage the community's ability to respond in the future. Given the more significant challenges faced by rural communities, and the value of social bonds within those communities, this is a particularly pressing concern. Furthermore, the erosion of social capital may have longer-term detrimental impacts upon the community as a whole eroding the day-to-day systems of reciprocity, which enable the success of the rural community.

### *11.2.5 Vulnerability*

Much of the research on post-disaster recovery has looked at the exacerbation of social vulnerability as a result of the disaster, as well as at the disparate impacts that result across socioeconomic and demographic characteristics. Although the majority of this research has not explicitly considered differences between rural and urban areas, it is possible to apply the lessons from this research to the rural context. One particularly relevant finding is that recovery in rural areas tends to concentrate socially vulnerable populations in a different physical displacement pattern than what is seen in urban areas (Cutter et al. 2016). This concentration of vulnerability has cascading impacts on the ability to successfully recover, much less to build resilience. Caruson and MacManus argued that true assessments of vulnerability should also consider the ability to manage events and their impacts, not just the socioeconomic and demographic characteristics of the survivors (2011). In other words, the impacts of vulnerability may be exacerbated within the rural context where there is a concentration of socially vulnerable populations coupled with a lessened ability to manage events and their impacts.

Minority racial status and lower-income social class, which have been shown empirically to correlate with higher levels of vulnerability, may also have disparate impacts in rural areas, particularly those with complex histories of racial and economic disparities, as well as ongoing tensions. Research by Highfield, Peacock, and Van Zandt looking at the impacts of Hurricane Ike on Houston found that hazard exposure, structural characteristics, and socioeconomic characteristics were all predictors of structural damage (2014). However, "…even after controlling for [all of] these factors areas with higher proportions of non-white residents and lower-valued homes received more damage than their counterparts in predominantly white areas despite being further from or outside high-risk areas" (p. 289). In other words, the increased damages and impacts to nonwhite residents and lower-valued homes were found to be the case even outside of the areas of highest risk. This finding is particularly troubling, as it shows that the greatest correlation with disaster impacts was not the characteristics of the event or the condition of the structure but the race of the impacted households. The authors speculated that these disparate impacts had to do with a history of lack of investment by local government in infrastructure and maintenance within the minority communities.

Similarly, a comparison of economic recovery in New Orleans and Gulfport, Mississippi, following Hurricane Katrina, and focusing on the effects of race and poverty, found a greater negative effect on recovery from race outside of urban areas. The research found that community heterogeneity had a positive impact on recovery and that the plight of minorities outside of urban areas did not garner the media coverage and attention that it did in urban areas (Downey 2016). This has significant implications for rural communities that are not heterogeneous and have complex histories of economic and racial disparities. In fact, the social capital of the white residents may enable their recovery at the expense of others.

Perceptions of recovery both within one's own socioeconomic group and looking across groups also appear to be impacted by race and class. A 2017 study of the recovery from the 2011 tornadoes in Joplin, Missouri, and Tuscaloosa, Alabama, found a distinct difference in the ways in which white residents and residents of color perceived the impacts of race, class, and gender on recovery (McKinzie 2017). The study found that white people reported a leveling effect, in which they felt that recovery and disaster impacts rendered community members equal, while people of color often disagreed. The researchers speculated that the history of race relations in both of the cities had an impact on the ways in which race impacted recovery perceptions across the two cities. These differing perceptions of recovery, including the assumptions that a leveling effect has taken place, may in fact contribute to the creation of a corrosive community.

Aranoff and Gunter in their research on communities that successfully avoided the creation of a corrosive community found that three key factors were crucial to avoiding the corrosive community: (1) effective prior leadership, (2) self-identification as a "survivor community," and 3) ties of residence and occupation. In other words, the perception of having social capital as well as the existence of bonding capital had a positive impact on the community's ability to avoid becoming a corrosive community (1992).

Another important consideration relative to the nexus between perceptions and vulnerability is the role that subjective perceptions of individual vulnerability play, particularly in terms of human decision-making around preparedness and risk reduction. Kusenbach and Christman found differences in the perception of risk and social inequalities between experts and individuals within the community 2013). This difference in perception has real implications for populations, such as the population they focused on – that of individuals and families living in mobile homes – because these populations can be considered vulnerable based upon their housing stock and economic characteristics but may not see themselves in that light and may not respond positively to messaging that requires them to see themselves as a vulnerable population. Rural coastal Louisiana residents who experienced the impacts of the BP oil spill resisted strongly the characterization of being vulnerable, or victims, a requirement of receiving compensation for their fisheries being "oiled" (Laska et al. 2015).

In their research of at-risk mobile home communities, Kusenbach and Christman found that some of the respondents felt a sense of agency and control that did not necessarily align with their risk-reduction efforts. In fact, many respondents had undertaken little to no preparation and had very limited resources but still considered themselves to not be vulnerable (2013). This perception may have to do with resilience characteristics that were not visible to the researchers, but they may also have to do with a lack of understanding regarding the actual level of risk. In other words, the perception of resilience might negatively impact preparedness and self-protective behaviors. The authors go on to suggest that the perception of vulnerability is itself a component of vulnerability, as a condition has to be seen as a problem in order to be responded to. This raises the question of whether or not rural communities might perceive themselves as being more vulnerable than urban. In the case of the communities impacted by the 2016 flooding, there was a definite understanding that their resilience was impacted by a lack of resources and access to power structures, both of which correlated to distance from the capital (Personal communication 2016).[2]

This also creates a nexus with considerations around social capital and perception of inherent resilience and self-sufficiency, such as those held in some rural communities. It may well be the case that focusing on a message of resilience, which is in closer alignment to individual perceptions of the community, will be more effective. This aligns with Aranoff and Gunter's finding, mentioned previously, that self-identification as a resilient community has a positive impact on resiliency.

Aranoff and Gunter identified three key strategies that communities utilized in order to navigate a lack of resources, a component of the community vulnerability described above, in the face of larger needs: (1) making do, (2) taking charge, and (3) working within the system (1992). These strategies have clear applications to disaster recovery, but making do in particular is relevant to this discussion of perceptions, as its utilization required recognition of the community's preexisting socioeconomic place and acceptance that inequities would remain. This recognition may have negative impacts on the resilience of a community, as it may impact the communities' perception of its own resilience.

[2] Personal communications with local officials and community members in late 2016.

### 11.2.6 Local and State Responsibilities: The Role of Federalism in Rural Recovery

In the context of disaster impacts that exceed a local jurisdiction's capability to successfully respond, there is often an assumption that the next level of government, state, and the federal will intercede and assist. As a result, many assume that an overwhelmed local government can simply request assistance. Although federal aid can be considered generous, it is surrounded by false expectations regarding speed and the needs it can meet (Landy 2008). It also often includes requirements for local match and administration that can be very challenging for a small local jurisdiction. In order to fully understand how to improve recovery and resiliency, it becomes imperative to truly understand the role that our federalist system plays. In the context of homeland security, there is a good bit of confusion regarding the federal and state roles, much less the distinction between local and state. Outside of warfare, public safety is a state and locally incorporated city or county responsibility when the impacted area is not incorporated. Police, fire, and public health all reside with the state itself, although local governments can be granted rights and responsibilities by the state constitution. The federal government may provide training and equipment, as well as occasional grants, but the states serve as intermediaries (Eisinger 2006).

Generally speaking, the US Constitution only recognizes the federal and state governments. The states are sovereign jurisdictions with their own constitutions that establish local government rights and responsibilities. The extent to which the state can intervene in local decision-making varies, with home-rule states enforcing strict limitations (Col 2007). The federal government tends to provide aid to the state, which then can assist localities. This can be done through block grants, grant programs, and other mechanisms (Landy 2008). The state is the primary decision-making entity. In some cases, feuding between local entities has a negative impact on assistance received as competition between entities can harm the relationship with the state.

### 11.2.7 Cooperation and Networks

Recent research in emergency management has found that effective disaster recovery, particularly in communities that are lacking in resources, requires a shift in mind-set towards interagency cooperation. Cooperation and regional networks can also be a mechanism through which local communities can support each other in meeting their responsibilities relative to public safety. However, maintaining such networks can be very difficult and require extensive, continual resource investments (Bowman and Parsons 2013).

This constitutes a paradigm shift for many government departments or the entire communities, particularly when they are used to sharing some minimal information but not sustaining consistent action-based involvement with partners (Bowman and

Parsons 2013; Skertich et al. 2012). However, integrated emergency management, which is characterized by the use of networks, is far more effective at coordinating with other entities, both locally and vertically. Resources-strapped rural communities can benefit greatly from this kind of coordination and may help communities learn how to collaborate in post-disaster recovery when the emergency response collaboration is practiced.

Research by Bowman and Parsons into what constitutes functional and effective emergency management networks found that the best partners are often other jurisdictions and state agencies (2013). They found that the most successful networks function as performance regimes with a focus on actions and goals. They also found that counties with higher capacity tend to interact less with smaller towns and jurisdictions, a missed opportunity for the smaller governments, and that counties win geographic proximity to state actors engage more fully with the state. These findings have real implications for rural communities seeking to engage in functional networks.

As this indicates, rural communities face certain disadvantages when attempting to work within network structures, including negative impacts from population instability and economic downturns (Bowman and Parsons 2013; Caruson and MacManus 2011). One study by Choi and Kim found that power, defined as the ability to get things done, is a determinant of network effectiveness (2007). The research looked at several types of power within networks: structural (formal power within the design of the network), resource, actor, cognitive, and political. Unfortunately, rural emergency managers operating within a network that includes larger partners may be lacking in all of these forms of power. Cognitive-based power, stemming from local knowledge, might be an asset in some situations. However, formal emergency management structures, including the systems for engagement with the nonprofit sector, often fail to really utilize local knowledge or to take into account community values and norms. This results in an inability to make the best use of local nonprofits and community-based groups that are best suited to serve the impacted community. Ireni-Saban identified this phenomenon following Hurricane Katrina, and it is a recurring concern across other disasters (2012). For rural communities, where informal actors are most important, the lack of appreciation for local knowledge and community values and norms is particularly problematic.

## 11.3 2016 Louisiana Flooding

The flooding in March of 2016 impacted many communities in the northern areas of Louisiana. Although there are some larger population centers that were impacted, generally speaking, Central and Northern Louisiana communities have more of an agricultural and industrial economy than Southeastern Louisiana. These communities, like much of rural America, have suffered from the economic downturn and general changes in the national and state economies. A review of existing hazard mitigation plans, such as the one for Ouachita Parish (the county upon which this case focuses), reveals a backlog of needed infrastructure work. The parish, and its incorporated jurisdictions, also struggles with blighted properties and a general lack of safe and affordable housing (Ouachita Parish 2018).

The storm-impacted communities suffered extensive damage to housing, infrastructure, agriculture, and economy – all of which were already strained. Ouachita Parish reported an estimated 9500 flood-damaged homes, with around 5400 homes reported to have been completely flooded (Robichaud 2017). FEMA approved approximately $94 million in housing and other needs assistance through its Individual Assistance Program relating to the March flooding (FEMA 2017). Infrastructure impacts throughout the northern region were severe as many access roads, highways, waterways, and railways in these rural communities were reported flooded due to record river crests (Vagell 2016). Damage to these transportation routes caused cascading impacts to the many industries dependent upon them (US Economic Development Administration 2017). Road flooding also affected rural school bus routes and caused school closures across the region (Associated Press 2016). In total, FEMA reported having provided more than $47 million to repair infrastructure and conduct emergency work in the communities affected by the spring flooding (FEMA 2017).

Impacts on the agriculture industry were also significant, partly due to the rain event having occurred just after the state's corn planting season (Gautreaux 2016). Thus, impacts to agriculture included lost revenues from damaged crops and the cost of replanting flooded fields, as well as crop yield losses, lost livestock, and costs associated with relocating surviving herds (McClure 2016). Total impacts to the agricultural industry from the March flooding event totaled over $80 million (Louisiana Office of Community Development Disaster Recovery Unit 2019).

In addition to the damage to the agricultural industry, economic impacts of the flooding in the northern region were also felt by the area's small businesses. Businesses having to evacuate during the rain event suffered lost revenue as well as damage from floodwaters. By December 2016, the Small Business Administration had approved over $15 million in Business and Economic Injury Disaster Loans to business in the region affected by the flooding (Louisiana Office of Community Development Disaster Recovery Unit 2019). Additionally, Louisiana Economic Development approved 698 loans of more than $36 million in assistance to small businesses affected by the flooding (LED 2016).

The rural nature of many of the impacted communities, coupled with the distance from the state capital and a lack of experience navigating the federal public assistance process, led to a failure to fully document damages. Local officials and volunteers reported limited assistance from the state in damage documentation and repeatedly voiced concerns that they were not equipped to fully document damages. Although local officials did request additional assistance and did work to bring political pressure to bear, the subsequent flooding of more populous areas closer to Baton Rouge further impacted their ability to access the needed technical resources (Personal communication 2016).[3]

[3] Personal communications with local officials and community members in late 2016.

The distance from the capital followed by the flooding of communities with much stronger linkages to the state government led to a perceived lack of investment in Ouachita's recovery and to the perception that other areas of the state constituted a higher priority for the state government (O'Donoghue 2016). This was reflected in challenges around damage assessment, in variations in resource allocation for studies, and in direct access to key decision-makers.

These disparities were further exacerbated when areas closer to the capital, including many suburbs, were flooded in August of 2016. The August flooding diverted attention from the remaining needs in North Louisiana and re-impacted some of the areas that had flooded in March (Grueskin 2018). Although there is little data currently available regarding the status of the recovery, anecdotal data indicates an ongoing struggle to fully recover including many of the challenges that the literature would indicate might be expected.

Efforts at the state and local level to address regional resilience to extreme weather events like the March 2016 floods are underway, yet much of this planning is still in the development stage. Following the March and August 2016 flooding, a state agency initiative to develop regional "watershed coalitions" was created by the Louisiana Resilient Recovery Initiative. In 2018, Governor Edwards issued the Executive Order JBE18–16 that formally created the Council on Watershed Management. The council is comprised of individual parish government entities, OCD, CPRA, GOHSEP, the Department of Transportation and Development, and the Department of Wildlife and Fisheries. The council is tasked with creating a floodplain management plan based on watershed data. The plan is in the development stages, with an implementation goal of March 2019 (Louisiana Watershed Initiative 2019).

Ouachita Parish has successfully worked with the state and federal government to initiate a recovery planning process. Public meetings took place in July of 2018, over 2 years after the flooding. They were focused on how the parish might become more resilient (Merritt 2018; Ouachita Strong 2018). This process is still ongoing, but local officials hope that it will lead to greater investment in the resilience of the community (Personal communication 2018)[4] . Thus, it is evident that efforts at the state and local level are underway to address future resilience to flooding events like that experienced by Northern Louisiana in March 2016. It is too soon to know how successful they will be.

## 11.4 Conclusion and Recommendations

There are three key recommendations that emerge from the research: (1) planning for community resilience should consider means to build upon existing social capital while also increasing local social capital; (2) networks and interaction between peers may be the best means to improve recovery outcomes in under-resourced

[4] Personal communications with local officials and community members in 2018.

communities; and (3) disaster recovery processes must take into account the differences between communities and actively strive to ensure equitable access to resources as well as equitable outcomes.

These recommendations support the increased resilience of all communities, not just rural, but they are particularly relevant to communities that have a shortage of capacity or financial resources.

**Recommendation #1**

Both informal social capital networks and formal policy/planning channels are critical to achieving community resilience (LaLone 2012). These channels are often overlooked by traditional emergency management actors, who focus more on command and control and do not understand the contributions that arise from social capital (LaLone 2012). Social capital can be increased through the nurturing of community relationships of reciprocity and trust (Ireni-Saban 2012). Aldrich and Meyer found that social capital can also be increased through what they termed time banking, building a history of time spent together at regular gatherings, and supported through spatial design (2015). In other words, it is simply creating spaces and reasons for individuals to spend time interacting with each other and building a "bank" of time spent together. It can also be built through face-to-face interactions, supportive policies, institutional support, and leadership programs. The Ouachita Strong Resilience Strategy planning process has sought to identify ways in which to foster community relationships but has also brought to the forefront several preexisting tensions.

Several key barriers to resilience can be overcome through the increasing of social capital. As Rivera and Settembrino observed, a community may face few environmental hazards but still suffer from social and economic pressures and problems that have a negative impact upon their resilience (2013). Key barriers to building resilience in the face of these pressures are mistrust of the government and the lack of access to resources. These barriers can be mediated through an increase in bonding and bridging social capital, as individuals foster relationships that increase their access to resources and personal connections to those within the government. Furthermore, focusing on community capacity building has been found to be more effective than focusing on reducing administrative shortcomings (Ireni-Saban 2012). Such an approach might be tied into existing planning processes or might be accomplished through partnerships and networks.

**Recommendation #2**

Research has shown that informal peer-to-peer and bottom-up interactions can be more effective at building resilience than formal planning processes and that these types of interactions can be crucial to the success of even formal processes (Brunner 2016). These interactions can take place at any time and are most effective when there is a long history of network building. The resulting relationships can provide direct assistance but can also assist impacted communities with navigating the large amounts of information that are made available following a disaster. In fact, there are multiple clearinghouses purporting to provide information vital to disaster resiliency and recovery, but the filtering of that information is largely left up to the users (Brunner 2016). When that user is already operating at diminished capacity and struggling to meet basic demands, the time necessary for successful filtering is simply not available. Peers can provide a much-needed source of input regarding what resources may be most useful, as well as what the trade-offs might be for any particular resource being considered, both during planning processes and following a disaster.

This begs the question of how these types of interactions can be promoted and whether government can work to develop empathetic relationships with community members in order to facilitate collaborative actions when disaster strikes (Ireni-Saban 2012). These types of collaborative relationships will support the creation of emergency management networks, an invaluable resource at any phase of the emergency management cycle, while also increasing social capital. A workshop in 2012, comprised of 80 practitioners in the field of climate adaptation, found that building relationships, having the right people at the table, and promoting collaboration within and across groups was the most successful at promoting the adoption of adaptation measures (Brunner and Nordgren 2016).

Additionally, the government can also make policies that show that local knowledge is valued and that focus on the community as a whole and do not just treat households and individuals as being without social context.

**Recommendation #3**

Whenever permissible through programmatic and regulatory mechanisms, recovery resources should be targeted towards building local capacity (Downey 2016). These resources should also be rendered flexible and adaptable, in order to meet distinct local needs (Brunner and Nordgren 2016).

One of the key recommendations that emerges from the research is to identify the ways in which recovery processes can become better able to meet a wider range of community needs.

There are various ways in which the state government can assist local jurisdictions that are suffering from a lack of resources. One strategy, utilized after Hurricane Katrina by the state of Mississippi, is to assist local governments with borrowing money for recovery (Landy 2008). This can help local jurisdictions to weather the immediate shortfalls and to be able to remain solvent while awaiting reimbursement. Landy goes so far as to suggest that all localities be required to have a rainy day fund but perhaps to receive assistance in creating one. Additionally, partnerships with the state government along with coordination with regional entities can help to ensure that local and regional projects do not compete against each other for similar pots of funding (Brunner and Nordgren 2016).

One means through which differing community needs might be addressed is through additional empowerment of local FEMA representatives, allowing for greater use of bureaucratic discretion and for decisions to be made at the local level. This would require moving away from the strong oversight mentality and towards a focus on improving recovery outcomes (Landy 2008). Another means of increasing local adaptability and flexibility might be to create a version of CDBG that includes fewer requirements.

## References


Aldrich, D. P. (2017). The importance of social capital in building community resilience. In W. Yan & W. Galloway (Eds.), *Rethinking resilience, adaptation and transformation in a time of change* (pp. 357–364). New York: Springer.

Aldrich, D. P., & Meyer, M. A. (2015). Social capital and community resilience. *American Behavioral Scientist, 59*(2), 254–269.

Aronoff, M., & Gunter, V. (1992). It's hard to keep a good town down: Local recovery efforts in the aftermath of toxic contamination. *Industrial Crisis Quarterly, 6*, 83–97.

Associated Press. (2016). North Louisiana flooding: Is this what we can expect? *The Times-Picayune.* Retrieved from https://www.nola.com/weather/index.ssf/2016/03/north_louisiana_flooding_is_th.html

Berke, P. R., & Campanella, T. J. (March 2006). Planning for postdisaster resiliency. *Annals AAPSS, 604*, 192–207.

Bolin, R., & Bolton, P. (1986). *Race, religion, and ethnicity in disaster recovery. Program on environment and behavioral science.* Boulder: University of Colorado.

Bowman, A. O., & Parsons, B. M. (Jan-Feb. 2013). Making connections: Performance regimes and extreme events. *Public Administration Review, 73*(1), 63–73.

Brody, S. D., & Gunn, J. R. (2013). Examining environmental factors contributing to community resilience along the Gulf of Mexico Coast. In N. Kapucu, C. V. Hawkins, & F. I. Rivera (Eds.), *Disaster resiliency: Interdisciplinary perspectives* (pp. 160–177). New York: Routledge.

Brunner, R. D. (2016). Missed opportunities: Evaluating what works. In J. Bullock, G. Haddow, K. Haddow, & D. Coppola (Eds.), *Living with climate change: How communities are surviving and thriving in a changing climate* (pp. 145–158). Boca Raton: CRC Press.

Brunner, R. D., & Nordgren, J. R. (2016). Climate adaptation as an evolutionary process: A white paper. In J. Bullock, G. Haddow, K. Haddow, & D. Coppola (Eds.), *Living with climate change: How communities are surviving and thriving in a changing climate* (pp. 134–144). Boca Raton: CRC Press.

Caruson, K., & MacManus, S. A. (2011). Gauging disaster vulnerabilities at the local level: Divergence and convergence in an "all-hazards" system. *Administration & Society, 43*(3), 346–371.

Choi, S. & Kim, B. (2007). Power and cognitive accuracy in local emergency management networks. Public Administration Review, 67, 198-209.

Col, J. (2007). Managing disasters: The role of local government. *Public Administration Review, 67, 114.*

Coppola, D. P. (2016). Community risk implications. In J. Bullock, G. Haddow, K. Haddow, & D. Coppola (Eds.), *Living with climate change: How communities are surviving and thriving in a changing climate* (pp. 35–70). Boca Raton: CRC Press.

Cox, R. S., & Hamlen, M. (2015). Community disaster resilience and the rural resilience index. *American Behavioral Scientist, 59*(2), 220–237.

Cutter, S. L., Ash, K. D., & Emrich, C. T. (2016). Urban-rural differences in disaster resilience. *Annals of the American Association of Geographers, 106*(6), 1236–1252.

Demiroz, F., Kapucu, N., & Dodson, R. (2013). Community capacity and interorganizational networks for disaster resilience: Comparison of rural and urban counties. In N. Kapucu, C. V. Hawkins, & F. I. Rivera (Eds.), *Disaster resiliency: Interdisciplinary perspectives* (pp. 334–354). New York: Routledge.

Doherty, G. W. (2004). Crisis in rural America: Critical incidents, trauma and disaster. *Traumatology, 10*(3), 145–164.

Downey, D. C. (2016). Disaster recovery in black and white: A comparison of New Orleans and Gulfport. *American Review of Public Administration, 46*(1), 51–74.

Eisinger, P. (2006). Imperfect federalism: The intergovernmental partnership for homeland security. *Public Administration Review, 66*, 537–545.

Federal Emergency Management Agency. (2017). *After 2016's spring rains, a flood of assistance in Louisiana*. (R6-17-006). Washington, DC: FEMA.
Ganapati, N. E. (2012). In good company: Why social capital matters for women during disaster recovery. *Public Administration Review, 72*(3), 419–427.
Ganapati, N. E., Cheng, S., & Ganapati, S. (2013). Resilient rural communities: Housing recovery patterns following Hurricane Katrina. In N. Kapucu, C. V. Hawkins, & F. I. Rivera (Eds.), *Disaster resiliency: Interdisciplinary perspectives* (pp. 99–120). New York: Routledge.
Gautreaux, C. (2016). 2016 was a tough year for Louisiana agriculture. Delta Farm Press. Retrieved from https://www.farmprogress.com/cotton/2016-was-tough-year-louisiana-agriculture
Grueskin, C. (2018). Despite suffering two floods in 2016, some people denied assistance from Restore LA. *The Advocate*. Retrieved from https://www.theadvocate.com/baton_rouge/news/communities/article_240fd5d2-0783-11e8-9a75-efa2d0baf96f.html
Haddow, G. (2016a). Conclusions and recommendations. In J. Bullock, G. Haddow, K. Haddow, & D. Coppola (Eds.), *Living with climate change: How communities are surviving and thriving in a changing climate*. Boca Raton: CRC Press.
Haddow, K. (2016b). Learning to survive and thrive in a changed climate. In J. Bullock, G. Haddow, K. Haddow, & D. Coppola (Eds.), *Living with climate change: How communities are surviving and thriving in a changing climate* (pp. 1–34). Boca Raton: CRC Press.
Highfield, W. E., Peacock, W. G., & Van Zandt, S. (2014). Mitigation planning: Why hazard exposure, structural vulnerability, and social vulnerability matter. *Journal of Planning Education and Research, 34*(3), 287–300.
Horney, J., Nguyen, M., Salvesen, D., Dwyer, C., Cooper, J., & Berke, P. (2017). Assessing the quality of rural hazard mitigation plans in the Southeastern United States. *Journal of Planning Education and Research, 37*(1), 56–65.
Ireni-Saban, L. (2012). Challenging disaster administration: Toward community-based disaster resilience. *Administration & Society, 45*(6), 651–673.
Kusenbach, M., & Christman, G. (2013). Understanding hurricane vulnerability: Lessons from mobile home communities. In N. Kapucu, C. V. Hawkins, & F. I. Rivera (Eds.), *Disaster resiliency: Interdisciplinary perspectives* (pp. 61–83). New York: Routledge.
LaLone, M. B. (2012). Neighbors helping neighbors: An examination of the social capital mobilization process for community resilience to environmental disasters. *Journal of Applied Social Science, 6*(2), 209–237.
Landy, M. (Dec. 2008). Mega-disasters and federalism. *Public Administration Review., 68*, S186.
Laska, S., Peterson, K., Rodrigue, C., Cosse', T., Philippe, R., Burchett, O., & Krajeski, R. (2015). 'Layering' of natural and human caused disasters in the context of anticipated climate change disasters: The coastal Louisiana experience. In M. Companion (Ed.), *The impact of disasters on livelihoods and cultural survival: Opportunities, losses and mitigation* (pp. 225–238). Boulder: University Presses of Colorado.
Louisiana Economic Development. (2016). *SBA continues flood relief assistance at limited Louisiana locations*. Baton Rouge: LED. Retrieved from https://www.opportunitylouisiana.com/led-news/news-releases/news/2016/04/12/sba-continues-flood-relief-assistance-at-limited-louisiana-locations.
Louisiana Office of Community Development Disaster Recovery Unit. (2019). *January 2, 2019 thru March 31, 2019 Performance Report*. (Grant B-16-DL-22-0001). Baton Rouge: Louisiana Division of Administration.
Louisiana Watershed Initiative. (2019). *Managing future flood risk in Louisiana through watershed-based solutions*. Baton Rouge: Louisiana Office of the Governor.
May, P. J. (2013). Public risks and disaster resilience: Rethinking public and private sector roles. In N. Kapucu, C. V. Hawkins, & F. I. Rivera (Eds.), *Disaster resiliency: Interdisciplinary perspectives* (pp. 126–145). New York: Routledge.
McClure, O. (2016). March flooding will cost Louisiana farmers at least $10 million. Louisiana State University Ag Center. Retrieved from https://www.lsuagcenter.com/profiles/lbenedict/articles/page1461012122900

McKinzie, A. (2017). A tale of two cities: Variations in perceptions of disaster recovery and the importance of intersectionality. *Sociology of Race and Ethnicity, 3*(4), 522–537.

Merritt, K. (2018). Ouachita Parish introduces strategy to help with disaster recovery. *KNOE News*. Retrieved from https://www.knoe.com/content/news/Ouachita-Parish-introduces-strategy-to-help-with-disaster-recovery-483996081.html

Mogle, D. (2017). East Texas tornadoes case study to help rural communities prepare for and deal with disasters. Tyler Morning Telegraph. Retrieved from https://www.cbs19.tv/article/news/east-texas-tornadoes-case-study-to-help-rural-communities-prepare-for-and-deal-with-disasters/501-503997125

Mohr, R., Deller, S., & Halstead, J. (2010). Alternative methods of service delivery in small and rural municipalities. *Public Administration Review, American Society of Public Administration, 70*(6), 894–905.

O'Donoghue, J. (2016). Feds to shoulder bigger share of Louisiana flood costs. *The Times-Picayune*. Retrieved from https://www.nola.com/politics/index.ssf/2016/09/louisiana_flood_federal_fundin.html.

Ouachita Parish. (2018). *Ouachita Strong Resiliency Strategy*. Ouachita Parish long-term recovery: Public meeting(s) RECAP.

Persson, T. A., & Povitkina, M. (2017). "Gimme shelter": The role of democracy and institutional quality in disaster preparedness. *Political Research Quarterly, 70*(4), 833–847.

Ratcliffe, M., Burd, C., Holder, K., & Fields, A. (2016). *Defining Rural at the U.S. Census Bureau*. Washington, DC: U.S. Census Bureau.

Rittel, H., Webber, W. J., & Melvin, M. (1973). Dilemmas in a general theory of planning. *Policy Sciences, 4*, 155–169.

Rivera, F. I., & Settembrino, M. R. (2013). Sociological insights on the role of social capital in disaster resilience. In N. Kapucu, C. V. Hawkins, & F. I. Rivera (Eds.), *Disaster resiliency: Interdisciplinary perspectives* (pp. 48–60). New York: Routledge.

Robichaud, D. (2017). Remembering the March 2016 flooding – Experts say it could happen again. *KNOE News*. Retrieved from https://www.knoe.com/content/news/Remembering-The-March-2016-Flooding-Experts-Say-It-Can-Happen-Again%2D%2D415737073.html

Schwab, J. (2016). Planning and climate change: Creating resilience in US communities. In J. Bullock, G. Haddow, K. Haddow, & D. Coppola (Eds.), *Living with climate change: How communities are surviving and thriving in a changing climate* (pp. 71–81). Boca Raton: CRC Press.

Scott, S. L., McSpirit, S., Breheny, P., & Howell, B. M. (2012). The long-term effects of a coal waste disaster on social trust in Appalachian Kentucky. *Organization & Environment, 25*(4), 402–418.

Skertich, R. L., Johnson, D. E. A., & Comfort, L. K. (2012). A bad time for disaster: Economic stress and disaster resilience. *Administration & Society, 45*(2), 145–166.

Tierney, K. (2013). Foreword. In N. Kapucu, C. V. Hawkins, & F. I. Rivera (Eds.), *Disaster resiliency: Interdisciplinary perspectives* (pp. xiii–xxvi). New York: Routledge.

U.S. Economic Development Administration. (2017). *Success story: Economic disaster recovery – The calm after the storm*. Washington, D.C.: U.S. EDA, Retrieved from https://www.eda.gov/news/blogs/2017/09/01/success.htm.

United States Department of Agriculture Economic Research Service. (2017). *Rural America at a Glance*. Washington, D.C.: U.S.D.A.

Vagell, Q. (2016). Over 26 inches of rain triggers record flooding in the south, including the Sabine River. *Weather.com*. Retrieved from https://weather.com/storms/severe/news/historic-south-flooding-march-2016/

Waugh, W. L. (2013). Management capacity and rural community resilience. In N. Kapucu, C. V. Hawkins, & F. I. Rivera (Eds.), *Disaster resiliency: Interdisciplinary perspectives* (pp. 291–307). New York: Routledge.

# Part VII
# Types/Locations of Communities and Their Responses to Extreme Weather: Coupled Coastal-Inland

# Chapter 12
# Regional Resilience: Building Adaptive Capacity and Community Well-Being Across Louisiana's Dynamic Coastal–Inland Continuum

**Traci Birch and Jeff Carney**

## 12.1 Introduction

Climate change impacts pose significant risk to both coastal and inland communities. This is particularly true in Louisiana, which has lost 1900 $mi^2$ of coastal wetlands since 1930, and another 1750 $mi^2$ are estimated at risk of loss over the next 50 years (Couvillion et al. 2017; CPRA 2012, 2017b). Coastal Louisiana is a young and dynamic landscape, built over the last 8000 years through regular Mississippi River spring floods, which left rich sediment behind as the waters receded (Couvillion et al. 2017). As the river migrated east or west searching for the path of least resistance to the Gulf of Mexico, wetlands developed along coastal edges, bayous, and estuarine landscapes (see Fig. 2.2 in Boesch). Land loss is a result of several complex factors, but the primary culprits are levees separating the deltaic plain from the sediment-rich Mississippi River, hydrological alteration from oil and gas exploration, and accelerating eustatic sea level rise (Day et al. 2007).

Land loss is profoundly changing the nature of Louisiana's social and natural environments, and diminishing many of its benefits, including storm protection, fisheries habitats, and distinct cultural practices (Costa 2018; Groves and Sharon 2013). Land loss increases community flood risk because healthy coastal wetlands, swamps, barrier islands, and ridges provide a critical buffer against slow-onset and abrupt climate change impacts. While migration away from the most vulnerable areas is ongoing, 47% of Louisiana's population still lives in the coastal zone.

T. Birch (✉)
Coastal Sustainability Studio, School of Architecture, Louisiana State University, Baton Rouge, LA, USA
e-mail: tbirch@lsu.edu

J. Carney
School of Architecture, Florida Institute for Built Environment Resilience (FIBER), University of Florida, Gainsville, FL, USA
e-mail: j.carney@ufl.edu



S. Laska (ed.), *Louisiana's Response to Extreme Weather*,
Extreme Weather and Society, https://doi.org/10.1007/978-3-030-27205-0_12

When extended to encompass coastal watersheds, more than 70% of Louisiana's population are coastal dwellers – in large part because this is home to the bulk of the state's economy. Louisiana's coastal zone is a major locus of seafood, oil and gas, maritime, and petrochemical industries for the nation – what Laska et al. (2005) refer to as "immovable industries." This includes major cities, vast suburbs, and small villages along south Louisiana's rivers and bayous – all of which rely on healthy coastal marshes and forests for flood protection. Coastal flood risks include sudden catastrophic impacts from tropical storms and hurricanes, as well as less severe but more common blue sky tidal flooding exacerbated by low topography, relative sea level rise, and land loss.

Inland communities far from the coastal edge but connected ecologically, infrastructurally, and culturally as part of the Lower Mississippi River Delta Plain also face risks from severe rain that overwhelms riverine floodplains. This is exacerbated by the increasing convergence of coastal and inland processes from a retreating coastline (Birch and Carney 2019). This was made apparent by a series of floods experienced by inland communities across the state from March to August 2016. Research across disciplines suggests these impacts will continue to increase in frequency and intensity over time (Kolker et al. 2011; Prein et al. 2016; Vermeer and Rahmstorf 2009). However, entrenched industry coupled with residents' deep place attachment developed through generational tenure and intimate knowledge of local environments requires adaptation for survival in the face of coastal change (Burley et al. 2007). Sustainability of the overall ecological, infrastructural, economic, and social systems of this and other coupled coastal–inland regions rely on local capacity for incremental adaptation and replicable design strategies, in addition to the larger-scale structural protection and ecological restoration efforts.

In the face of these challenges, the Louisiana Coastal Protection and Restoration Authority (CPRA) developed *Louisiana's Comprehensive Master Plan for a Sustainable Coast* (2017 Coastal Master Plan), defining a 50-year strategy for coastal flood risk reduction and sustainable natural landscape production. The Master Plan reflects the state's environmental complexity and climate change reality, proposing economic, ecological, and hydrological adaptation priorities to ensure long-term coastal resilience and viability. Built on a platform of infrastructural protection, hydrological and ecological restoration, and hazard mitigation, the plan proposes to reduce community and economic risk through engineered defenses, both built and natural. However, a heavy reliance on ecological and engineered solutions calls into question whether the Master Plan provides an effective framework for community-scaled resilience. This chapter highlights resilience thinking as a critical but underdeveloped element of risk reduction and protection in the state. Coordinated community planning and innovative design may hold the greatest long-term risk reduction potential against the impacts of climate change but are generally underutilized (Lyles et al. 2014). This chapter examines coastal design and planning in Louisiana, providing an overview of existing large-scale efforts with an eye to their achievements and barriers. Further, there is an investigation of current community-based frameworks addressing resilience and adaptation gaps at the architectural, neighborhood, and community scales. Finally, the authors identify points of opportunity to further integrate community planning and design into the coastal protection and restoration framework to protect the health, safety, and welfare of coastal and inland communities. While Louisiana's situation is extreme (in both the scale of the problem and the extent of the proposed solution), it provides a meaningful bellwether to planners, designers, and others for identifying both coastal vulnerability, restoration approaches, and community level impacts. As coastal hazards increase, regions from Texas to New York are contemplating large-scale restoration and protection efforts proportional to those in Louisiana, and much can be learned from this experience.

## 12.2 Literature on Resilience Thinking for Community Resilience and Adaptation

In the face of increasingly costly human and environmental disasters, resilience has risen to prominence in community planning and design scholarship related to mitigating hazards and enhancing capacity to cope with environmental change and disturbance. While a recent addition to the disciplinary lexicon, it is not a new concept. Coming from the Latin root *resi-lire*, meaning to spring or bounce back, it was first used by physical scientists to describe the stability of nonliving materials and resistance to external shocks. Through this lens, resilience measures structural elasticity and the speed by which an engineered design returns or bounces back to a previous equilibrium after a disturbance (Gunderson 2000). It is widely recognized that Holling (1973) extended the concept of resilience to include ecosystem stability and change. Rather than bouncing back to a previous state, Holling (1973, 1996) emphasized ecological complexity and the evolutionary capacity of natural systems to adapt and transform over time. Hence, this measure of resilience rejects a single state of equilibrium, measuring not just how long it takes a system to return to resume normal functions but also how much disturbance a system can take before it fundamentally changes (Davoudi 2012). Davoudi (2012, p. 301) notes what underpins both perspectives is the "belief in the existence of equilibria in systems, be it a pre-existing one to which a resilient system bounces back (engineering) or a new one to which it bounces forth (ecological)." An increased awareness of the interconnections between environment and society, resilience has since been redefined and extended to encompass ecological, socio-ecological, and economic systems (Folke 2006; Holling 2001; Walker and Salt 2006). Equilibrium resilience has been highly influential across a range of social science disciplines, including environmental psychology, economic geography, disaster studies, and environmental planning, as a way to predict or model socio-ecological change.

Resilience thinking proposes a systems approach to human–environmental relations, having evolved mainly through the application of ecological concepts to social systems (Cote and Nightingale 2012). Resilience thinking emerged out of a dissatisfaction with models of ecosystem dynamics that (a) focused too heavily on a return to previous states and (b) paid too little attention to interactive dynamics between human and biophysical systems (Berkes and Folke 1998; Berkes et al. 2000). Within this work, humans and their environment are not conceived as separate systems but rather as a single dynamic, adaptable, and interrelated socio-ecological system (SES). In this perspective, resilience is understood not as a fixed asset, but as a continually changing process as systems are confronted with disturbance and stress. This means that, for example, people might become more resilient not in spite of adversities but because of them. Emphasis on the fundamental role of adaptive capacity and transformation in the analysis of SES allows change to happen and systems to adapt rather than trying to control and avoid it (Folke et al. 2010).

Within resilience and SES thinking, there is a recognition that resilience is an important counter-narrative to conventional anthropocentric approaches such as "maximum yield" or "carrying capacity," which separate human actions from ecosystem impacts. As noted by Cote and Nightingale (2012, p. 478), resilience thinking plays an important role in shifting the focus away from the quantitative availability of resources and toward a more dynamic and forward-looking approach to human–environmental change that is defined by ecological rather than political boundaries. Innovations in resilience thinking provide genuine promise to the domain of planning for changing urban and environmental conditions. This is due mainly to the field's emerging commitment to complexity as seen through the integration of a diversity of knowledge holders (Cote and Nightingale 2012).

While expressed in new nomenclature, the resilience discourse is in fact similar to the positivist methods applied across the planning fields to address urban infrastructure (Mehmood 2016). Planning is inherently a systems-based endeavor – promoting the scientific, aesthetic, and orderly disposition of land, resources, facilities, and services with a view to securing systemic physical, economic, and social efficiencies. In particular, land-use planning and design is intended to promote more desirable social and environmental outcomes while minimizing conflicts between uses. However, within multijurisdictional regions, planning is also highly fragmented across decision-making bodies with different goals. Each has authority to determine its own land uses, development practices, and management strategies to achieve locally determined objectives. Financial solvency, for example, is a primary concern of each jurisdiction and results in radically different land-use strategies. As noted by Brody (2008, p. 21), while natural systems are "intricately connected over broad spatial and temporal scales, the land-use decision-making framework is limited to local jurisdictions and some limited input from regional planning councils. Uncoordinated local land use decisions have cumulative negative impacts on the system as a whole." Fragmented planning has often yielded undesirable results including wasteful land-use patterns, degraded air and water, loss of biodiversity, displacement of the poor, and natural disaster losses. Also noted by Brody (2008, p. 22) is that "having the ability to look at the entire ecological system, even if it extends beyond a planner's jurisdiction is a critical aspect to effectively managing ecosystems."

While the translation of resilience from ecology to social systems can be problematic, as we will discuss, it provides a framework to embrace complexity and extend planning boundaries to ecosystems rather than maintaining the politically defined units that perplex planning. The incorporation of adaptation and transformation as the rule rather than the exception undermines the assumptions of the steady state on which the linear extrapolations of planners often rely. This provides opportunities for collaboration and innovation as components of any system must be conceived to allow change to happen, rather than requiring control to avoid it (Davoudi 2012).

Another promising aspect of resilience thinking is the genuine commitment to integrating professional disciplines and nontraditional stakeholders. Resilience scholars recognize that an ecosystem of knowledge includes information and experiences from researchers and decision-makers, as well as knowledge gained through extensive personal observation shared among local resource users. The inclusion of local ecological knowledge (LEK) is key to understanding ecological and institutional change, which can in turn modify the way resources are managed, and thereby the landscape itself (Berkes and Folke 1998; Folke 2006; Cote and Nightingale 2012). LEK is a cumulative body of knowledge handed down through generations about the relationship of living beings (including humans) with one another and with their environment (Berkes and Folke 1998, p. 3). The use of LEK in coordinated land-use and community design can play a key role in augmenting ongoing scientific modeling efforts, galvanizing support for coastal restoration projects, employing traditional adaptive measures, and developing compromises given the reality of limited resources (Tompkins et al. 2008). LEK builds upon the historic human experiences and adapts to social, economic, environmental, spiritual, and political change. Grumbine (1994, 1997) notes that ecosystem management must include recognition that people are embedded within natural systems and that societal values play a dominant role in ecosystem management. Ecosystem frameworks entail understanding natural system properties and processes and understanding systems as dynamic while promoting stakeholder engagement and coordination between partners (Armitage et al. 2009).

While there is great promise, it is important to note that the application of ecological concepts to social concerns in the name of resilience problematically assumes that social and ecological dynamics are essentially similar. Recent extensions of ecological resilience to social systems have given rise to the concept of social resilience, defined as "the ability of groups or communities to cope with external stresses and disturbances as a result of social, political, and environmental change" (Adger 2000, p. 347). The rise of social resilience in theory has also given rise to critical literature due to the perceived overemphasis on the role of physical shocks and an underemphasis on history, economy, and power dynamics that perpetuate inequality (Davoudi 2012; Fainstein 2015). In ecological literature, there are no rewards or punishments, just consequences. In society, there are always rewards and punishments. Resilience building means that some will gain while others lose. Cote and Nightingale (2012, p. 475) argue that an analysis of the SES capacity to adapt and change must be framed within an understanding of cultural values, historical context, and ethical standpoints of the kinds of actors involved. It is less about "getting the rules right" and more about subjective identities, effective relationships, and lay knowledge as a better way of understanding human–environmental processes and change – the resilience of what and for whom. Planners and designers must engage in discussions of power and social difference that are central to critical social science research, because resilience planning has as much to do with who shapes the challenges as it does how to responding to them (Davoudi 2012). With regard to community development, the heavy focus on adaptive capacity is more likely to encourage small incremental change rather than significant systemic transformation, which disadvantages socially marginalized groups (Cretney 2014).

While reconceptualization of planning to embrace some sort of evolution and change broadens the potential for resilience thinking across disciplines, some argue clarity and practical relevance have suffered (Brand and Jax 2007; Cutter 2016a; Davoudi 2012; Fainstein 2015). The intent and foundation of resilience as both a state of bouncing back and jumping forward has given way to a blending of contradictory descriptive aspects that make definition, operationalization, and assessment difficult – what Markusen (2003) refers to as "fuzzy conceptualization." Cutter (2016b, p. 110) observes "such vagueness has its merits, especially in the policy world where the goals and motivations of proponents are highly variable and politicized." However, Markusen (2003) points out that fuzzy conceptualization also makes implementation challenging. Matyas and Pelling (2014, p. S1) note the ambiguity surrounding resilience means "it is a concept caught between the abstract and the operational." From the planning and design perspective, resilience frameworks call into question whether traditional planning tools (i.e., toolkits, etc.) are adequate to address complex challenges – or whether they are doomed to solving yesterday's problems (Taylor 2005). Others note that malleability creates flexibility and opportunity to foster communication between science and planning practice (Brand and Jax 2007; Davoudi 2012).

Communities, however, face challenging economic, social, and environmental changes requiring attention. There is a growing need for effective ways to support adaptation-related decision-making due to both slow-onset and rapid environmental change. Government agencies, businesses, and individuals increasingly find themselves fundamentally unprepared for meeting the challenges of climate change. Typically, local decision-making such as infrastructure construction and the types of zoning and development regulations implemented assume environmental stability. Yet many coastal and inland communities are faced with increasing uncertainty and vulnerability associated with climate change. Local governments also have core regulatory powers in the land-use, transportation, and waste sectors critical to comprehensive climate change responses (Trisolini 2010). Building flexibility, adaptability, and durability into local decision-making is key to building resilience (Beatley 2009; Godschalk 2003; Vale and Campanella 2005). The National Research Council (2012) notes enhancing community resilience and sustainability requires both "bottom-up" approaches at the individual and community level and the "top-down" strategies at the federal and state levels. Further, Beatley (2009) notes the qualities of resilient communities include hazard mitigation of the built environment, strong and diverse economic conditions, and a robust social infrastructure including strong social systems and networks. The following discusses how the State of Louisiana is attempting to build resilience into ecological and social systems through both top-down and bottom-up approaches.

## 12.3 Louisiana's Comprehensive Master Plan for a Sustainable Coast

As early as the 1970s, researchers and coastal residents recognized that flood protection and industrial interventions in the Louisiana landscape were increasing coastal land loss (Barrett 1970; Chabreck 1972; Gagliano 1973; Gagliano and Van Beek 1975). In response to land loss, Louisiana has developed a series of restoration plans since the 1990s prioritizing hydrological and ecological restoration as the centerpiece of ecosystem management (CPRA 2007, 2012, 2017b; LCWCRTF 1998; USACE 2004). Each successive plan has added new data, science, and innovation to emphasize the perilous state of the coast and the need for restoration to prevent total ecological and economic collapse (CPRA 2017b, p. ES-10). However, the dual impacts of Hurricanes Katrina and Rita in 2005 proved too much for established coastal governance in the state. After these devastating impacts, which affected coastal and inland communities alike, Louisiana consolidated the tasks of coastal restoration and hurricane protection under one agency called the CPRA. At the same time, the state redefined the coastal zone to include all lands subject to storm or tidal surge. The expanded inland boundary was determined by a range of landscape-based calculations (e.g., tidal influence, vegetation, topography) and varies from 20 to more than 100 miles inland (La. Rev. Stat. § 49:214.24 2015) (see Fig. 7.3 in Chap. 7.)

The authority to *protect* coastal property and *restore* coastal ecosystems first came together in CPRA's 2007 *Louisiana's Comprehensive Master Plan for a Sustainable Coast* (2017 Coastal Master Plan) and has been expanded in subsequent 5-year updates (2012 and 2017). The Master Plan builds upon the Multiple Lines of Defense Strategy, which proposes a coastal management structure of both natural and man-made features to hold back storm surge, maintain threatened habitats, and promote landscape resilience (Lopez 2009). Written primarily by engineers and ecologists, somewhat logically the Master Plan embraces both the "bounce-back" and "bounce-forward" resilience scenarios of its authors (Bahadur et al. 2010; Pendall et al. 2009). In particular, coastal protection measures are designed to resist storm impacts and maintain structural integrity, while enhanced coastal landscapes can adapt and transform over time to absorb climate change impacts. There are eight protection and restoration typologies (Fig. 12.1) providing a range of proposed projects phased in near- (1–10 years), mid- (11–31 years), and long-term (31–50 years) timeframes. Projects in the near-term are relatively close to engineering and

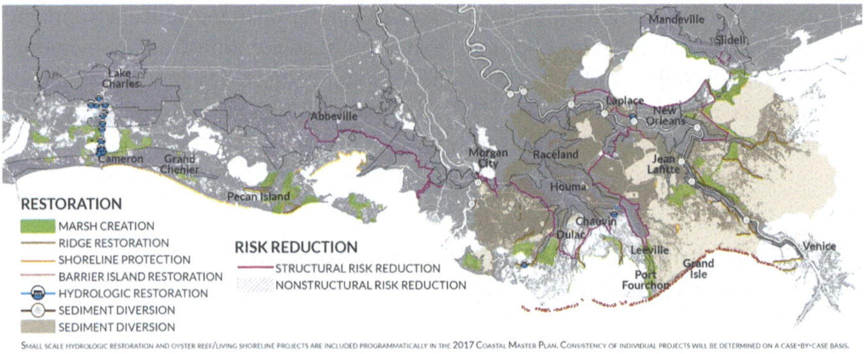


**Fig. 12.1** 2017 Master Plan proposed projects. (Image source: Louisiana Coastal Protection and Restoration Authority (CPRA))

construction, while middle- and long-term proposals are more speculative. The effort is supported by state and federal resources, but settlements from the 2010 BP oil spill provide a significant down payment of $8B on the plan's $50B price tag. Following the 2012 Master Plan, the CPRA estimates roughly $1B annually has been spent on restoration and protection measures across the state's coastal zone (CPRA 2017b). As noted by the CPRA (2017b, p.ES-2), "the master plan identifies a long-term program of construction, operations, and maintenance, and adaptive management…to be implemented as funds become available." The state recognizes, however, that even if all Master Plan projects are implemented, land will continue to erode and landscapes will change in unexpected ways due to impacts from restoration projects (CPRA 2017b). Further, risk to both coastal and inland communities will continue as coastal waters encroach and thus stronger, engineered protection measures will be needed to maintain community and economic interests in the coastal zone.

The Master Plan takes a systems approach to coastal restoration and management, broadening the scope of traditional coastal planning to consider a wide range of ecological and economic interactions within an ecologically defined region (Boesch 2006). Restoration and management units are defined by watershed and habitat boundaries and functions rather than political designations, requiring greater intergovernmental coordination to manage complex restoration scenarios. The Master Plan computationally models, synthesizes, and evaluates projects designed to protect coastal infrastructure, habitats, and settlements against natural hazards (Groves and Sharon 2013). Restoration actions are intended to reestablish coastal landscapes to protect regional ecology and major economic interests (Colten 2017). This ecosystem approach emphasizes regional biophysical ecology and hydrologic processes, but has generally not been expanded to include social and cultural dimensions (Colten 2017; Peyronnin et al. 2013). While significant scientific knowledge of landscape change is employed to project how the natural system – hydrology, flora, and fauna – will transform and adapt, the properties of human-centered systems are far less studied and modeled in a systems context. Structural protection focuses on building or enhancing physical barriers against storm surge and sea level rise to protect existing communities and infrastructure. Nonstructural hazard mitigation reduces risk to human settlement through flood proofing, elevation, and limited voluntary acquisition of at-risk structures (CPRA 2017b, p. ES-15). Yet there are few if any proposed changes to traditional community development patterns or future growth scenarios. CPRA contends this is the largest risk reduction program in the country, recommending mitigation of 26,000 structures coast-wide – with an estimated $6B price tag (CPRA 2017a, p. 18). Unlike other proposed Master Plan projects, however, community-based components are voluntary, relying on local political will and policy change for implementation. A number of problems emerge from this strategy, including a lack of planning and design capacity in underserved communities, disincentives for local communities to reduce population in the floodplain and potentially displace residents from their community, and a dearth of accurate information related to environmental and social conditions. Further, no funds budgeted for coastal protection and restoration have been awarded for nonstructural projects since 2007. And it is unclear whether BP oil spill funds will be dedicated to this effort. Using federal disaster recovery funds appears to be the only possible source.

To support the Master Plan, CPRA developed a modeling approach reflecting both the environmental complexity and climate reality of the state (Groves and Sharon 2013). Modeling efforts reflect ecological, hydrological, and geological components of the region, but there is no comparable accounting of acceptable restoration compromises given limited resources or societal disruptions that will accompany ecological restoration. Local community members often possess abundant ecological knowledge. As noted by Burley (2010), Louisiana is one of a handful of locations in the USA where most trace their familial ties generations back, and many are engaged in natural resource-based activities providing a lifetime of environmental observation and experience. However, current models of stakeholder engagement generate extensive transcripts of public opinion but are limited in terms of scope and genuine stakeholder contribution (Bethel et al. 2014). Public meetings often present technical information in a form to which laypersons do not easily relate and in an atmosphere that can limit useful input into the process – a complaint voiced by residents regarding the 2012 Coastal Master Plans (Colten 2014). As a result, public hearings and stakeholder group meetings may do little to empower the public in helping to shape decision-making. Further, there seems to be an increasing mistrust in scientific findings about resource management. Such skepticism undermines the public perception regarding the integrity of scientists and the scientific process (Ko et al. 2017).

A resilience thinking approach to Louisiana's coastal ecosystem therefore involves not only a shift in resource management practices but also a fundamental restructuring of the historical practices of community engagement, land use, and urban design (Beatley 2000; Brody 2003; Montgomery et al. 1995). The ability to overcome the politics of land use and the exclusion of local knowledge holders to achieve collaborative solutions will be the key to successful ecosystem restoration and community resilience in the coastal zone. While some states provide mandates and resources for local planning and design, the majority (including Louisiana) leave decisions of land use and community development to local authorities. The Master Plan (2017a, p. 4) fully supports local planning, recognizing that improved land use and community design "should emphasize resilience, systems thinking, community engagement, equity, implementation, and adaptation in order to meet the challenges today's communities face." However, a lack of dedicated funding underscores the disjunction between community resilience and an agency whose mandate and expertise are more closely aligned with engineering and natural sciences (2017a, p. 10). Rather, CPRA makes policy recommendations and calls upon other state agencies to develop programmatic initiatives and write community resilience policies. Despite statements of protection and restoration in the name of preserving coastal communities, the Master Plan's primary focus on ecological restoration and avoidance of community-based planning and design in many ways overlooks key elements of social resilience. Colten (2017, p. 699) notes managing Louisiana's coastal wetlands is a human endeavor, "but the responsible government bodies sometimes carry out their task as if the environmental processes they direct were detached from the local society." Without resources and technical support for resilience, there is a lost opportunity to drive elements critical to comprehensive risk reduction, such as (1) where to (and not to) develop in coastal areas based on future loss and restoration strategies, (2) need-based capital expenditures and infrastructure investment, and (3) LEK, priorities, and acceptable trade-offs.

In Louisiana, there is much debate about governmental responsibility for working with communities on issues related to resilience and retreat. In 2010, following Hurricanes Gustav and Ike, the US Department of Housing and Urban Development and the Louisiana Office of Community Development, Disaster Recovery Unit (OCD-DRU) set aside $10M in disaster recovery funds to enhance community resilience through innovative planning and design. This competitive grant program – a precursor to ensuing design competitions such as Resilient by Design and the NDRC – supported hazard mitigation and sustainability measures. Thirty impacted communities received funding for projects ranging from comprehensive planning and zoning to design-driven housing and water resource management. Notably, these funds were used to write the Greater New Orleans Urban Water Plan, a comprehensive, integrated, and sustainable water management strategy for the New Orleans region.

In 2016, Governor John Bel Edwards issued an executive order calling for all state agencies to align major infrastructure investments and coastal restoration objectives. This effort prioritizes interagency and intergovernmental coordination to support community planning, floodplain management, and hazard mitigation. Following the 2016 floods, the state recognized weaknesses in floodplain management that stopped at the coastal zone boundary, and it is in the nascent stages of expanding intergovernmental coordination across watersheds reaching hundreds of miles inland. In a state with a relatively weak history of planning, these are positive steps, but questions remain about which agencies and policies are capable of coordinating community resilience, floodplain management, and coastal restoration moving forward. This also recognizes that the Master Plan is not likely to be the place where community resilience is addressed, nor is it able to address regional issues with origins outside the coastal zone. Parishes and city governments recognize a lack of capacity and resources to implement resilience planning measures within their own boundaries and little if any ability to address systems outside their jurisdiction (Manning-Broome et al. 2015). There is a call for useful guidance on impacts and mitigation measures for both projected land loss and coastal restoration measures. There is also a belief that there will be less backlash from constituents if community design directives come from the state rather than local government agencies (Manning-Broome et al. 2015). While there is a trade-off between exercising strong state government authority and allowing greater local autonomy, Louisiana has chosen the latter in matters of coastal ecosystem restoration. In light of this, it is imperative that the state establish an effective and enforceable framework supporting land-use planning, risk reduction, and the development of best practices for regional urban and ecosystem management.

Community-based planning and design are in many ways the best tools available to address regional infrastructure coordination, coastal restoration impacts, and necessary compromises that are equitable, just, and adaptable for Louisiana communities. However, traditional land-use planning is not without critiques in its ability to simultaneously address regional environmental and community development issues. In comparison with the restoration and protection processes put forth by the CPRA, there remains a question of whether traditional methods of urban planning and design are sufficiently scalable to reduce risk coast-wide. Successful solutions achieved through architecture and urban design are time intensive and unique to specific conditions of place. Often, the desirable "bottom-up" qualities of design, without a mechanism for scalability, make large-scale implementation unfeasible. Projects underway through the NDRC and research emerging from the grant *Inland from the Coast (IFC)* provide two innovative planning and design frameworks being implemented in Louisiana that may serve as precedents for others as they tackle climate change, ecological restoration, and flood recovery issues. Within the context of the Master Plan, these planning and design methodologies provide improved information and guidance related to community-level hazard mitigation, strengthening of economic conditions, and strong social network support.

## 12.4 National Disaster Resilience Competition: Sowing the Seeds for Adaptive Planning and Community Design in Coastal Louisiana

Following devastating disasters such as Hurricanes Katrina in 2005 and Sandy in 2012, federal, state, and local emphasis has shifted to developing community risk reduction strategies through building-, community-, and regional-scaled design. Following the success of Rebuild by Design (2013) and the ongoing 100 Resilient Cities initiative, the US Department of HUD and the Rockefeller Foundation developed the NDRC to provide meaningful support for enhancing community resilience. NDRC was a two-phase competitive process awarding $1B to help communities across the USA with disaster recovery and the development of replicable resilience frameworks. The competition encouraged "American communities to consider not only the infrastructure needed to become resilient, but also the social and economic characteristics that allow communities to quickly bounce back after disruption" (HUD 2015, p. 2). The competition encouraged multidisciplinary approaches that considered equity and long-term environmental stability alongside innovative engineering and design responses to hazard mitigation. In January 2016, HUD awarded the State of Louisiana $92M to address climate change impacts in coastal communities facing sea level rise, wetlands loss, and severe hurricane damage. Of that, $41M was dedicated to adaptation planning efforts known as *Louisiana's Strategic Adaptations for Future Environments* (LA SAFE) and $48M for the relocation of the Isle de Jean Charles Band of the Biloxi–Chitimacha–Choctaw tribe (See Chap. 6, Jessee, of this book). HUD also awarded the City of New Orleans $141M for the Gentilly Resilience District – a neighborhood-scale effort to reduce flood risk, slow land subsidence, and encourage revitalization. The following provides a general overview of these efforts and how each incorporates key elements of resilience thinking.

### *12.4.1 Louisiana's Strategic Adaptations for Future Environments (LA SAFE)*

In its application for NDRC funding, the state recognized that structural and ecological proposals contained in the Master Plan are essential to long-term occupation of the coastal landscape, but it is also crucial to develop a process of locally determined design, planning, and decision-making (OCD-DRU 2015). LA SAFE introduced CPRA ecosystem management approaches into community-based design processes to create a replicable framework for developing catalytic adaptation projects. The community planning and design effort was led by Louisiana's OCD-DRU in partnership with nonprofits and design professionals. The challenge facing coastal and inland planners is providing for the movement of coastal residents incrementally as land loss continues or in larger waves of migration following storms. Out-migration will have significant impact on communities as they depopulate but will also have effects across the entire state as pressures to accommodate additional population mount (see also Chap. 7, Peterson, of this book for discussion of this latter challenge). The state, through LA SAFE, recognizes the need to migrate, though this discussion has primarily focused on moving people within the coastal zone (OCD-DRU 2015). As seas rise and homes and communities are abandoned, research shows that people are likely to migrate along existing infrastructure pathways taking them inland and away from coastal areas rather than moving incrementally a few miles at a time (Black et al. 2011; Findlay 2011). Considering the flexibility and adaptive capacity of the overall coastal region requires a systems approach that considers communities as part of the complex coastal ecosystem.

Enabling communities to move within a regional network following storms, economic disruptions, or long-term change can help the region thrive even as the environment becomes increasingly dynamic.

Between 2017 and 2018, LA SAFE engaged six parishes (counties) – Jefferson, Lafourche, Plaquemines, St. John the Baptist, St. Tammany, and Terrebonne – all recently impacted by coastal storms and facing significant land loss. In a context of tremendous Master Plan investment in coastal ecosystems, LA SAFE sought to fill the gap between increased risk and the eventual benefits of coastal protection and restoration projects (OCD-DRU 2015). The program set goals to teach people about their current and future vulnerabilities: (a) build projects to address this and (b) change the minds of development community to reimagine development patterns for the long-term (OCD-DRU 2015). By 2022, LA SAFE will culminate in the construction of six adaptation projects, each developed by community-based planning and design processes within each of the six parishes. LA SAFE adopted a codesign planning process that integrated planning expertise, science, and LEK to develop a range of adaptation strategies. Meetings were facilitated by local community members to build trust and gather LEK, and proposed projects were chosen through community consensus. Projects include a business incubator, resilient housing prototypes, and a wetland education center. While most assessments of LA SAFE are favorable, there are concerns about the larger framework. First, from the very beginning, there was not the political will to cross parish boundaries or consider fundamental shifts in development patterns that might move people away from the coastal zone (Birch and Brand 2019). When faced with the realities of climate change and projected land loss, consensus between community members and decision-makers was to focus on the narrow problems of today rather than dealing with the problems of tomorrow (OCD-DRU 2015). Given the project-based nature of the funding dictated by HUD, LA SAFE projects can neither systematically address climate impacts nor breach the imperative need for large-scale resettlement. Second, the "reshape, retrofit, or resettle" concept of adaptation was developed before a series of inland floods, calling into question whether areas below 3' BFE (Base Flood Elevation) are appropriate for densification (Fig. 12.2). The reality is that climate impacts are moving many from coastal locations further inland to urban centers such as Baton Rouge, Hammond, Lafayette, and Houston, low-lying inland communities that have flooded since 2016 due to severe rain events. As this program moves from planning to construction, additional questions have emerged about economic development priorities related to the projects, pilot project scale related to the scale of climate-related issues, and lack of overall regional coordination.

### *12.4.2 Gentilly Resilience District*

When HUD awarded NDRC funds to the City of New Orleans for the Gentilly Resilience District, it validated over a decade of innovative water resource planning supporting the city's ongoing recovery. Following Hurricane Katrina, New Orleans recognized the devastating impacts of the "levee effect," where flood risk is actually

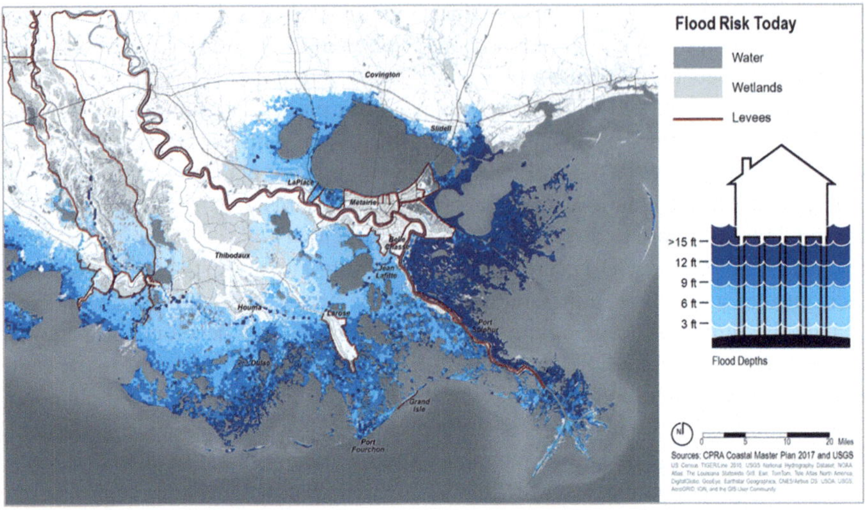


**Fig. 12.2** Current flood risk in participating LA SAFE parishes influencing reshape, retrofit, or resettle scenarios. (Image source: Louisiana Office of Community Development-Disaster Response Unit)

increased by the compaction/sinking of land behind the levees which also offer a false sense of security, despite well-documented environmental, technical, and social weaknesses (Tobin 1995). In an effort to sustain the city in spite of increased climate risk, the design community in partnership with philanthropic organizations, set out to reimagine New Orleans' management of, and relationship with, water. Beginning with the 2006 Dutch Dialogues®, teams of practitioners developed a series of ambitious water management and resilience strategies culminating in the federally funded projects *Greater New Orleans Urban Water Plan* and subsequent *Resilient NOLA*. These efforts were realized primarily through a sustained period of engagement between city government and community design professionals to develop fundable landscape-scale mitigation interventions. The Gentilly Resilience District includes multiple water management and community development projects designed to reduce flood risk, slow land subsidence, and encourage neighborhood revitalization. The centerpiece of this effort is the Mirabeau Water Garden – a 25-acre site formerly owned by the Catholic Church – redesigned to provide open space and storage for 10M gallons of neighborhood stormwater (see Fig. 12.3). In total, the seven water management projects showcase efforts to improve social well-being in some of New Orleans' most floodprone areas. Though reflective of nearly a decade of urban design and planning, the effort to present landscape-scale solutions has been criticized for lacking meaningful local community engagement in decision-making throughout the process (Anguelovski et al. 2018). Such a critique highlights the significant challenge this work presents to design practitioners. Together, these projects represent an emergent ecosystem approach embracing design and renewal projects. However, the strength of these effo1ts appears more

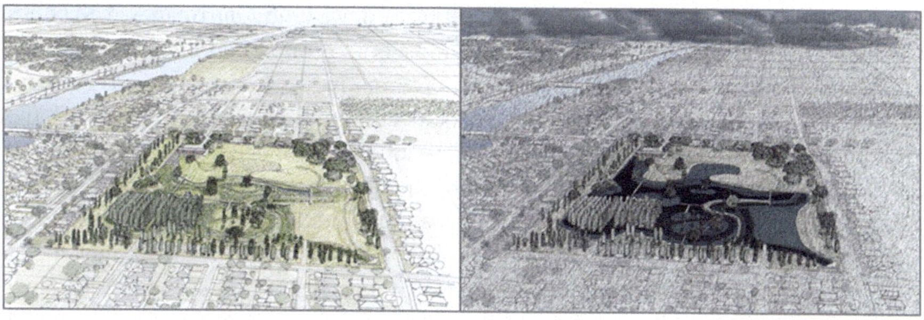

**Fig. 12.3** The Mirabeau Water Garden, currently in the design phase, in New Orleans' Gentilly neighborhood, during both dry (left) and rainy (right) conditions. (Image source: Office of Resilience and Sustainability and 100 Resilient Cities)

dependent on individual *designers* than on integrative, replicable, and scalable *design processes*. The design competition and design workshop models demonstrated through these projects have achieved significant progress in resilience planning and design. However, it remains unclear how these processes can be scaled up, repeated, and quantitatively valued for implementation through mechanisms like the Master Plan.

## 12.5 Inland from the Coast: Providing Opportunities for Coupled Coastal–Inland Resilience Thinking

In August 2016, a low-pressure system dropped 22–31" of rain in 2 days across Louisiana's capital region. Resultant flooding took 13 lives and caused damage to an estimated 145,000 structures. The region was brought to a standstill due to multi-day closures of Interstates 10 and 12, leading to transportation and economic disruption across the Gulf Coast. Touted as a "one-in-1000-year flood," this was the third such event in 2016 to hit the southeastern USA and one of nine since 2010 (NOAA NWS 2016). Climate change predictions indicate these severe precipitation events are likely to increase in frequency and intensity in the future (Prein et al. 2016). Through the Coastal Master Plan, Louisiana has demonstrated large-scale risk reduction through structural protection and ecological restoration. However, significant inland flooding places like Baton Rouge indicate that coastal restoration and protection alone are not sufficient to reduce flood risk within coupled coastal–inland systems (Birch and Carney 2019). While there has been an expansion of resiliency-based design projects, there remains a sizable gulf between demonstrated community needs and the systematic deployment of new practices at sufficient scale. Recognizing a need to bridge the gap between restoration, protection, and adaptation, the authors, through the Louisiana State University Coastal Sustainability Studio (LSU CSS), launched a multidisciplinary research effort known as *Inland from the Coast: A Multi-Scalar Approach to Regional Climate Change Responses*

(IFC). Founded in 2009 in the wake of a series of devastating hurricanes, LSU CSS is a transdisciplinary research center engaging a core group of architects, landscape architects, planners, coastal scientists, and civil engineers to research and respond to issues of resettlement, restoration, and socioeconomic sustainability. In 2017, the Gulf Research Program of the National Academies of Science, in partnership with the Robert Wood Johnson Foundation, funded IFC as part of the *Thriving Communities III* program. The three-year grant expanded the LSU CSS collaborative model to include faculty and students from geography, psychology, and social work. The expanded group also includes researchers from the University of New Orleans Department of Planning and Urban Studies, Louisiana Sea Grant's Law and Policy Program, and the Florida Institute for Built Environment Resilience (FIBER) at the University of Florida. The grant requires innovative approaches to applying research directly through practice. To achieve this, the collaboration also includes members from the local chapters of American Institute of Architects, American Planning Association, and American Society of Landscape Architects.

IFC poses three fundamental research questions to address specific needs not currently met by the Master Plan or other planning efforts: (1) How is climate change impacting or likely to impact communities connected across a coupled coastal–inland system? (2) How can greater understanding of environmental risk and community well-being increase adaptive capacity? (3) How can well-being and adaptation scholarship be incorporated into community design? The project takes a multi-scalar approach to present and future environmental conditions modeling, community well-being research, and applied community design for ongoing flood recovery and long-term resilience. Unlike the Master Plan and NDRC projects, which are relegated to the coast and near-coast inland Parish of St. John the Baptist, IFC addresses ecological, social, and infrastructural issues across the Amite River basin, which spans a region that includes coastal, transitional, and inland communities (Bilskie and Hagen 2018). Efforts are structured to develop adaptive design opportunities in flood-damaged communities facing increased coastal and riverine flooding. The project takes an iterative approach, linking university researchers with design professionals, policy makers, and community members throughout the process to (1) improve understanding of coastal–inland environmental conditions and vulnerabilities, (2) define current and future community health and well-being, and (3) develop design and planning best practices for reducing risk and increasing regional adaptive capacity.

### 12.5.1 Modeling Coupled Systems for Stormwater Management

On August 12, 2019, meteorologists began sounding the alarm that a low-pressure system would move inland from the Gulf of Mexico and deliver significant rainfall over the south central portion of the state. While tropical in nature, had this unnamed storm been a hurricane, there would have been advance warning and time to let people prepare. Instead, the storm moved inland quickly and paralyzed the Baton Rouge and Lafayette regions with historic rainfall and flooding. While considered today to be unprecedented, this was actually the second storm of 2016 to catch state and local agencies and residents off guard. In March 2016, a one-in-500-year event impacted areas slightly east of Baton Rouge – in many cases flooded the same residents twice. It is notable that many of the communities flooded in 2016 have experienced significant growth in recent years, in part due to in-migrations from vulnerable coastal communities who, along with community development professionals, presumed these areas were safe from flooding (Fig. 12.4). In reality, as many as 91% of homes damaged were not identified by FEMA as within Special Flood Hazard Areas, thus requiring flood insurance. As a result, the majority of impacted homeowners lacked flood insurance coverage. Bookended by Hurricanes Matthew in 2016 and Harvey in 2017, this storm has largely faded from national consciousness, but its impact provides evidence that migration inland from vulnerable coastal areas does not necessarily reduce overall risk (OCD-DRU 2015). The Amite River is a shallow, slow-moving river running from southern Mississippi through southeast Louisiana before emptying into Lake Pontchartrain and the Gulf of Mexico (Fig. 12.5). Over time, channelization, resource extraction, and urban development have drastically altered the capacity of the river. One of the greatest impacts to the system not previously accounted for was the construction of Interstate 12 (I-12) through the region. As the Amite River moves north–south through the region, it is intersected by I-12, greatly restricting natural flow and capacity and acting as a levee in many areas. In August 2016, this constraint of the river was apparent as extreme rainfall north of the roadway filled drainageways and was unable to

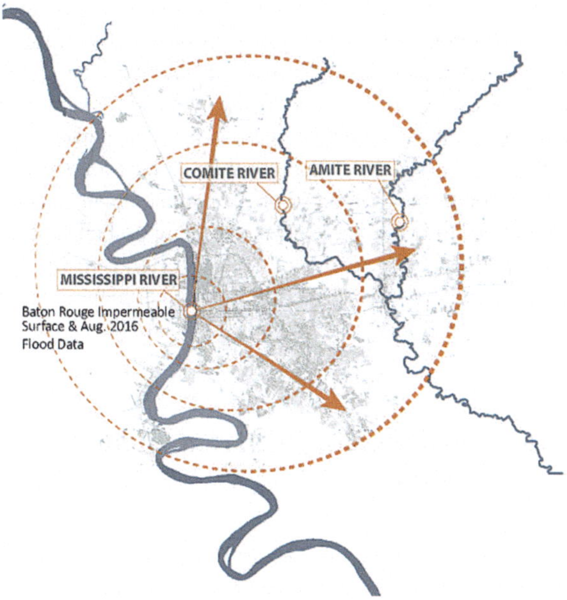


**Fig. 12.4** Suburban expansion of Baton Rouge starting from the historic core along the Mississippi River high-ground to the lowlands of the Amite River watershed. (Image source: LSU CSS)

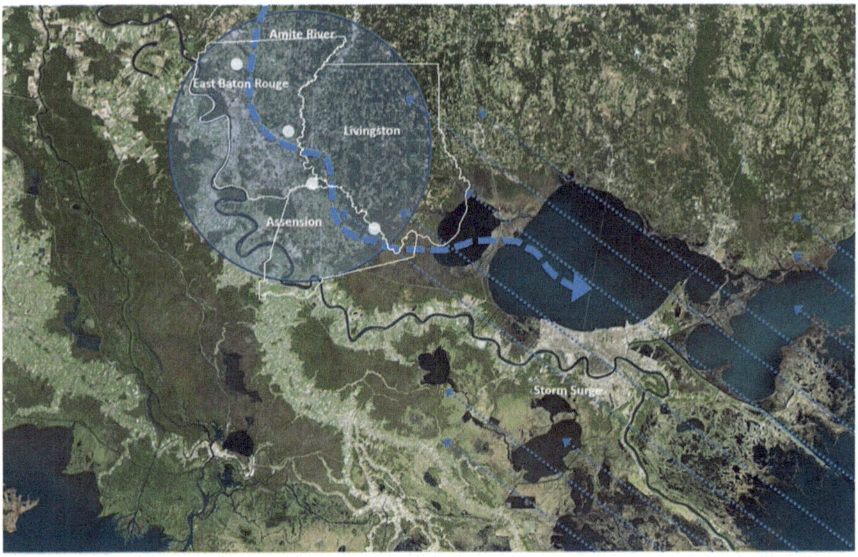


**Fig. 12.5** Three scales of the *Inland from the Coast Project*: coastal–inland flood interface, watershed–parish boundary conflicts, and local projects. East Baton Rouge, Livingston, and Ascension Parish are all within the Amite River watershed. (Adapted from LSU Coastal Sustainablity Studio)

drain, causing significant backwater flooding into communities north of the interstate. Ultimately, I-12 was overtaken in numerous spots by stormwater flows, stranding those trying to evacuate flooded communities.

To address uncertainty about current and future environmental conditions, IFC researchers from the LSU Center for Coastal Resiliency and Center for River Studies are developing a hydrodynamics model of the Amite River watershed to inform stormwater management. Concurrently, cultural geographers completed focus groups with LEK experts (e.g., floodplain managers, planners, long-time residents, elected officials) to understand how the system reacts under varied conditions. Conversations with local experts provide more than a check on the veracity of a model – they are being used to inform and improve mathematical assumptions by inserting information related to landscape obstruction, temporal change, and cumulative impacts. This approach enables researchers to model future hydrological conditions at a level of detail appropriate for design intervention at the scale of the neighborhood and building and to test these models against the experience of residents.

## 12.6 Understanding Current and Future Community Well-Being

When disasters strike, disparities in community health and resilience become glaringly apparent. In few places has this been better demonstrated than south Louisiana, where coastal land loss, human-made disasters, and extreme weather impact the region at regular intervals. The circumstances that people are born into, grow up with, and live in greatly impact well-being and overall ability to cope and adapt over time. Prolonged or repeated exposure to negative environmental impacts or crises result in physical and mental tolls to both individuals and communities. Repeated impacts from catastrophic events produce a social crisis context that disrupts all aspects of daily life, in line with what Picou et al. (2004) refer to as "corrosive communities." Building adaptive capacity to respond to disruption and become more resilient requires a framework that allows communities to define their own well-being and prioritize those environmental, cultural, and social values essential for recovery. Community well-being is a complex combination of physical, mental, emotional, political, and socioeconomic conditions contributing to one's ability to cope with the normal stresses of life (McCrea et al. 2015; Wiseman and Brasher 2008). Central to IFC is that community well-being is spatially related to flood risk. In Baton Rouge, large-scale alteration of the environment, through flood control projects and unrestricted suburban development, has resulted in significant alteration of the environment and a "muddling" of residents' understanding of environmental conditions and risk. This leads to reduced capacity among community members to respond to and effectively cope with flooding leading to reduced community resilience.

Improved modeling of the Amite River watershed allows floodplain managers to address flood risk and planners to guide development as described previously. Similarly, enhanced knowledge of specific indicators of community well-being informs and enhances future recovery and resilience needs. IFC researchers developed a Community Well-being Index weighing social, environmental, and health indicators to provide a quantifiable measure of a communities' pre- and post-disaster capacity to recover. The foundation of this index uses the methods of Burton (2015) and Cutter et al. (2010) (Fig. 12.6). However, IFC researchers determined that simply measuring social vulnerability is too narrow, missing many vulnerabilities associated with disaster risk and impacts. For example, many heavily flooded communities north of I-12 appear to be fairly resilient according to traditional social vulnerability indices due to higher incomes, education, and property values. However, these are also the areas where more than 90% of homeowners were without flood insurance, and thus income alone likely doesn't capture economic stress experienced by residents. IFC researchers compiled through literature review, geo-spatial data analysis, and survey results over 100 variables relating to well-being in East Baton Rouge (EBR), Ascension, and Livingston parishes at the census-tract level. Missing were measures related to disaster impacts (i.e., mental health concerns, financial impacts to those without flood insurance) included in the IFC index. Geo-spatial data sources measuring pre-flood baseline conditions and post-flood recovery progress include US Census counts and estimates, environmental quality measures, public health data, mental health survey results, and both personal and community-wide economic figures. Data gathered and analyzed by IFC researchers was useful in creating well-being profiles for communities that consider unique characteristics as well as commonalities across the region and in providing new evidence of risk factors threatening well-being and community resilience.

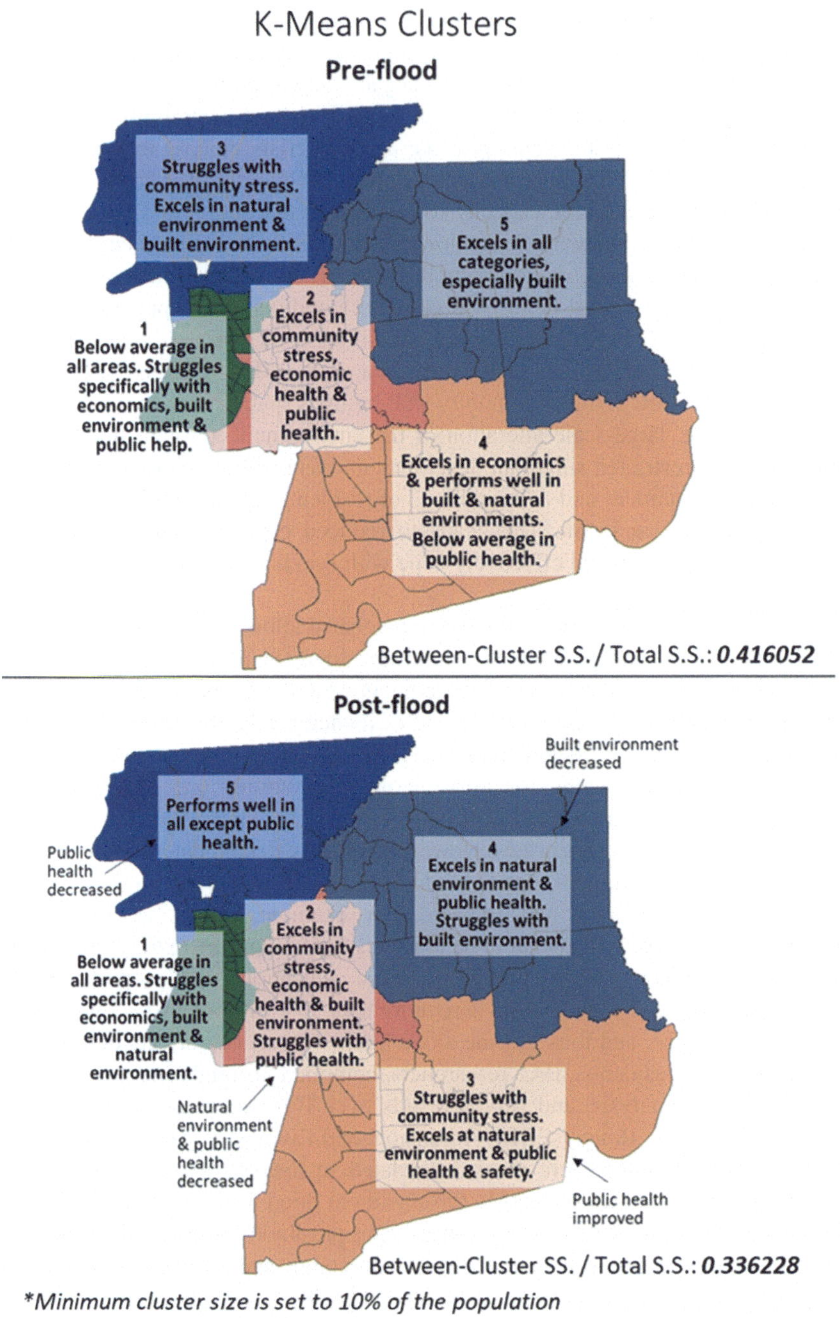


**Fig. 12.6** The resilience index builds a composite understanding of well-being both before and after the 2016 floods. (Image: LSU Coastal Sustainability Studio)

## 12.7 Design Application and Policy Framework

Connecting research and practice requires prolonged and sustained engagement of stakeholders, including vulnerable populations, with locally engaged practitioners committed to the long-term resilience of the community (Abendroth and Bell 2015). Applying new concepts and techniques (e.g., coastal–inland flood modeling, place-based well-being research, and design best practices) within communities, especially following a disaster, requires trust and community support. Design and planning professionals with knowledge of the local context and culture offer a bridge to community members through a professional lens that can measurably improve community resilience and build future professional capacity to implement projects (Nassauer and Opdam 2008; Perkes 2009). This component of IFC engages faculty and students from architecture, planning, landscape architecture, and law with design professionals affiliated with the AIA, ASLA, and APA, regional policy makers, and community members who participate in design workshops applying flood conditions and community well-being priorities at the design project level. Communities are engaged at the site, neighborhood, and city scales in specific, locally championed, design projects to identify opportunities for increasing adaptive capacity in inland communities.

Local engagements (particularly with vulnerable populations) in design activities considering community-driven definitions of well-being are important to urban resilience and successful adaptation to the impacts of climate change (Beatley 2009; Doherty and Clayton 2011). IFC public agency partners include the EBR Redevelopment Authority, the cities of Baker and Denham Springs, Ascension Parish, and the Capital Area Transit System. Working groups are organized around specific local and regional design challenges to infuse the consideration of stormwater management, public health, and social equity into community planning and design decision-making. While the projects are ongoing, outcomes will include six community-specific strategic urban design and implementation plans, each addressing an issue generalizable to other communities (e.g., green infrastructure, stream restoration, or corridor redevelopment). Results will be compiled into a regional best practices compendium – highlighting interventions at architectural, neighborhood, and community scales.

Throughout the course of the project, changes to the National Flood Insurance Program, including its proposed reauthorization (expected October 2020) and new Community Rating System Manual are being researched and evaluated for community impacts. In-depth legal research and outreach includes potential local government liability for development decisions in light of improved hydrologic modeling, policy options for translating mapping into land use directives, and issues related to takings and stormwater management. Though local governments have broad legislative authority to implement land-use decisions based on public health, safety, and welfare, the ability of local governments to regulate based on future climate conditions is less clear. The processes and partnerships developed through the design, planning, and policy work are intended to build capacity and implement change at the site, neighborhood, and community scales. Also, and perhaps most important, is the evaluation of such interventions for their scalability and collective impact on the greater region. To achieve a measurable impact on the resilience of the coupled coastal–inland system, successful projects cannot afford to be limited to individual success; they must be scalable and translate through improved governance policies coast-wide.

## 12.8 Discussion and Conclusion

In this work, we examined several planning and design efforts addressing issues of risk and resilience in Louisiana. This includes top-down and bottom-up strategies focused in varying degrees on physical hazard mitigation and building social infrastructure. Given current and future risk facing Louisiana's coastal and inland communities, and the emphasis on risk reduction and resilience at state and federal levels, we evaluated *Louisiana's Comprehensive Master Plan for a Sustainable Coast* (2017 Coastal Master Plan) as an effective framework for community resilience and how effectively other community planning and design frameworks are being employed in Louisiana to fill resilience gaps. While the Master Plan is unequivocal in its support of local planning and design, the lack of committed resources underscores the disconnect between community resilience and the actions of an agency whose structure and mandate is more closely aligned with engineered solutions. Unlike protection and restoration efforts which can be modeled to inform future risk reduction and selected by a cost/benefit ratio, the translation of architectural- and urban-scaled design into broader quantified values is more difficult. Social dynamics present sometimes contradictory responses to seemingly straightforward challenges and rational solutions. Such "wicked problems" will not be easily resolved through the technical apparatus of the Master Plan and may challenge the cost/benefit model driving Master Plan decision-making in the future. Social challenges are often not solved through such rational solutions but through education, consensus, and trust (Lazarus 2009; Rittel and Webber 1973).

The LA SAFE focus on reshaping, retrofitting, and resettlement demonstrates the state is willing to take on difficult community-based associated with climate change and land loss. The Gentilly Resilience District in New Orleans demonstrate a skillful approach to complex design challenges and an ability to follow through. However, these approaches alone are not sufficient to reduce risk in coupled coastal–inland systems (OCD-DRU 2015). As demonstrated by the Baton Rouge flood of 2016, the process of inland migration has already begun, and movement away from the coast does not automatically reduce risk. To overcome this gap, CSS's IFC research initiative has adopted a multi-scalar regional to local systems-based approach, recognizing the integrally linked nature of coastal and inland environmental processes (Birch and Carney 2019). IFC is enhancing hydrological modelling to understand the region's complex environmental flows, adding fidelity to the layers of community wellness and social complexity surrounding risk and the benefits of resilience to the overall process of adaptation. Most importantly, the project engages designers not in the final expression of scientifically based decision-making but as a partner throughout the process. Through this work, the tools, the knowledge, and the ability to demonstrate greater resilience through design is emerging (Birch and Carney 2019).

The Rockefeller Foundation (2018) defines resilience as "the capacity of individuals, communities, institutions, businesses, and systems within a city to survive, adapt, and grow no matter what kinds of chronic stresses and acute shocks they experience." There is no one answer to achieving community resilience, beyond the shared recognition that physical environment and social communities reach temporary thresholds between stability and change. Engineering resilience and ecological resilience illustrate different perspectives that, when bound together, form an improved basis for a robust *and* adaptive built environment. Louisiana's Master Plan proposes a range of top-down engineering and ecological solutions to encourage coastal sustainability. What is needed are equitable bottom-up approaches that support community-based resilience and adaptation planning and design at the architectural, neighborhood, and community scales. A range of frameworks, including IFC, are being implemented in Louisiana addressing architectural and urban challenges and providing community-based resilience benefits. While none of these approaches meet all of the needs of coastal and inland residents, they provide meaningful paths forward for bridging large-scale ecosystem restoration together with community-based design and adaptation.

## References

Abendroth, L. M., & Bell, B. (2015). *Public interest design practice guidebook: SEED methodology, case studies, and critical issues*. New York: Routledge.

Adger, W. N. (2000). Social and ecological resilience: Are they related? *Progress in Human Geography, 24*, 347–364. https://doi.org/10.1191/030913200701540465.

Anguelovski, I., Connolly, J., & Brand, A. (2018). From landscapes of utopia to the margins of the green urban life. *City, 22*(3), 417–436. https://doi.org/10.1080/13604813.2018.1473126.

Armitage, D. R., Plummer, R., Berkes, F., Arthur, R. I., Charles, A. T., Davidson-Hunt, I. J., et al. (2009). Adaptive co-management for social–ecological complexity. *Frontiers in Ecology, 7*(2), 95–102. https://doi.org/10.1890/070089.

Bahadur, A. V., Ibrahim, M., & Tanner, T. (2010). *The resilience renaissance? Unpacking of resilience for tackling climate change and disasters*. Strengthening Climate Resilience Discussion Paper 1, UK Department for International Development. Retrieved from https://assets.publishing.service.gov.uk/media/57a08b1ce5274a27b2000973/SCR-DiscussionPaper1-resilience-renaissance.pdf

Barrett, B. (1970). *Water measurements of coastal Louisiana*. Baton Rouge: Louisiana Wildlife & Fisheries Commission.

Beatley, T. (2000). Preserving biodiversity: Challenges for planners. *Journal of the American Planning Association, 66*(1), 5–20. https://doi.org/10.1080/01944360008976080.

Beatley, T. (2009). *Planning for coastal resilience: Best practices for calamitous times*. Washington DC: Island Press.

Berkes, F., & Folke, C. (1998). *Linking sociological and ecological systems: Management practices and social mechanisms for building resilience*. New York: Cambridge University Press.

Berkes, F., Gardner, J. S., & Sinclair, A. J. (2000). Comparative aspects of mountain land resources management and sustainability: Case studies from India and Canada. *International Journal of Sustainable Development & World Ecology, 7*(4), 375–390. https://doi.org/10.1080/13504500009470056.

Bethel, M. B., Brien, L. F., Esposito, M. M., Miller, C. T., Buras, H. S., Laska, S. B., et al. (2014). Sci-TEK: A GIS-based multidisciplinary method for incorporating traditional ecological knowledge into Louisiana's coastal restoration decision-making processes. *Journal of Coastal Research, 30*(5), 1081–1099. https://doi.org/10.2112/JCOASTRES-D-13-00214.1.

Bilskie, M. V., & Hagen, S. C. (2018). Defining flood zone transitions in low-gradient coastal regions. *Geophysical Research Letters, 45*(6), 2761–2770. https://doi.org/10.1002/2018GL077524.

Birch, T., & Brand, A. (2019, April). *Deep-seeded and uprooted: Colorblind resiliency planning as gentrification*. Paper presented at the American Association of Geographers National Conference, Washington DC.

Birch, T., & Carney, J. (2019). Delta urbanism: Aligning adaptation with the protection and restoration paradigm in Coastal Louisiana. *Technology|Architecture + Design, 3*(1), 102–114. https://doi.org/10.1080/24751448.2019.1571834.

Black, R., Bennett, S., Thomas, S., & Beddington, J. (2011). Migration as adaptation. *Nature, 478*(7370), 447–449. https://doi.org/10.1038/478477a.

Boesch, D. F. (2006). Scientific requirements for ecosystem-based management in the restoration of Chesapeake Bay and coastal Louisiana. *Ecological Engineering, 26*, 6–26. https://doi.org/10.1016/j.ecoleng.2005.09.004.

Brand, F., & Jax, K. (2007). Focusing the meaning(s) of resilience: Resilience as a descriptive concept and a boundary object. *Ecology and Society*, *12*(1), 23. Retrieved from http://www.ecologyandsociety.org/vol12/iss1/art23/

Brody, S. D. (2003). Implementing the principles of ecosystem management through local land use planning. *Population and the Environment*, *24*(6), 511–540. Retrieved from https://www.jstor.org/stable/40603370

Brody, S. D. (2008). *Ecosystem planning in Florida: Solving regional problems through local decision-making*. Hampshire, UK: Ashgate.

Burley, D. (2010). *Losing ground: Identity and land loss in coastal Louisiana*. Jackson: University Press of Mississippi.

Burley, D., Jenkins, P., Laska, S., & Davis, T. (2007). Place attachment and environmental change in coastal Louisiana. *Organization and Environment, 20*(3), 347–366. https://doi.org/10.1177/1086026607305739.

Burton, C. G. (2015). A validation of metrics for community resilience to natural hazards and disasters using the recovery from Hurricane Katrina as a case study. *Annals of the Association of American Geographers, 105*(1), 67–86. https://doi.org/10.1080/00045608.2014.960039.

Chabreck, R. (1972). *Vegetation, water, and soil characteristics of the Louisiana coastal region*. LSU Agricultural Experiment Station Reports. 147. Retrieved from http://digitalcommons.lsu.edu/agexp/147

Coastal Protection and Restoration Authority (CPRA). (2007). *Louisiana's comprehensive master plan for a sustainable coast*. Baton Rouge: Coastal Protection and Restoration Authority. Retrieved from http://coastal.la.gov/reports/louisianas-2007-comprehensive-master-plan-for-a-sustainable-coast/

Coastal Protection and Restoration Authority (CPRA). (2012). *Louisiana's comprehensive master plan for a sustainable coast*. Baton Rouge: Coastal Protection and Restoration Authority. Retrieved from http://coastal.la.gov/2012-coastal-master-plan/

Coastal Protection and Restoration Authority (CPRA). (2017a). *2017 Coastal Master Plan: Appendix E: Flood risk and resilience program framework*. Baton Rouge: Louisiana Coastal

Protection and Restoration Authority. Retrieved from http://coastal.la.gov/wp-content/uploads/2016/04/Appendix-E_FINAL_10.09.2017.pdf

Coastal Protection and Restoration Authority (CPRA). (2017b). *Louisiana's comprehensive master plan for a sustainable coast*. Baton Rouge: CPRA. Retrieved from http://coastal.la.gov/wp-content/uploads/2017/04/2017-Coastal-Master-Plan_Web-Book_CFinal-with-Effective-Date-06092017.pdf

Colten, C. E. (2014). Scenario building workshops. The Water Institute of the Gulf. Retrieved from https://thewaterinstitute.org/assets/docs/reports/9_02_2014_Scenario-Building-Workshops.pdf

Colten, C. E. (2017). Environmental management in coastal Louisiana: A historical review. *Journal of Coastal Research, 33*(3), 699–711. https://doi.org/10.2112/JCOASTRES-D-16-00008.1.

Costa, R. B. (2018). "A self-inflicted wound": The impact of coastal erosion and restoration on Louisiana's oyster industry. *Southern Cultures, 24*(1), 27–45. https://doi.org/10.1353/scu.2018.0003.

Cote, M., & Nightingale, A. (2012). Resilience thinking meets social theory: Situating social change into socio-ecological systems (SES) research. *Progress in Human Geography, 36*(4), 475–489. https://doi.org/10.1177/0309132511425708.

Couvillion, B., Beck, H., Schoolmaster, D., & Fischer, M. (2017). *Land area change in coastal Louisiana 1932 to 2016: U.S. Geological Survey Scientific Investigations Map 3381*. Reston: US Geological Survey.

Cretney, R. (2014). Resilience for whom? Emerging critical geographies of socio-ecological resilience. *Geography Compass, 8*(9), 627–640. https://doi.org/10.1111/gec3.12154.

Cutter, S. (2016a). The landscape of disaster resilience indicators in the USA. *Natural Hazards, 80*, 741–758. https://doi.org/10.1007/s11069-015-1993-2.

Cutter, S. (2016b). Resilience to what? Resilience for whom? *The Geographical Journal, 182*(2), 110–113. https://doi.org/10.1111/geoj.12174.

Cutter, S., Burton, C., & Emrich, C. (2010). Disaster resilience indicators for benchmarking baseline conditions. *Journal of Homeland Security and Emergency Management, 7*(1), 24. https://doi.org/10.2202/1547-7355.1732.

Davoudi, S. (2012). Resilience: A bridging concept or a dead end? *Planning Theory & Practice, 13*(2), 299–333. https://doi.org/10.1080/14649357.2012.677124.

Day, J. W., Boesch, D. F., Clairain, E. J., Kemp, G. P., Laska, S. B., Mitsch, W. J., et al. (2007). Restoration of the Mississippi Delta: Lessons from Hurricanes Katrina and Rita. *Science, 315*(5819), 1679–1684. https://doi.org/10.1126/science.1137030.

Doherty, T., & Clayton, S. (2011). The psychological impacts of global climate change. *American Psychologist, 66*(4), 265–276. https://doi.org/10.1037/a0023141.

Fainstein, S. (2015). Resilience and justice. *International Journal of Urban and Regional Research*, 157–168. https://doi.org/10.1111/1468-2427.12186.

Findlay, A. M. (2011). Migrant destinations in an era of environmental change. *Global Environmental Change-Human and Policy Dimensions, 21*(1), S50–S58. https://doi.org/10.1016/j.gloenvcha.2011.09.004.

Folke, C. (2006). Resilience: The emergence of a perspective for social-ecological systems analyses. *Global Environmental Change, 16*, 253–267. https://doi.org/10.1016/j.gloenvcha.2006.04.002.

Folke, C., Carpenter, S., Walker, B., Scheffer, M., Chapin, T., & Rockstrom, J. (2010). Resilience thinking: Integrating resilience, adaptability and transformability. *Ecology and Society, 15*(4), 20. Retrieved from http://www.ecologyandsociety.org/vol15/iss4/art20/

Gagliano, S. M. (1973). Canals, dredging, and land reclamation in the Louisiana coastal zone. In *L. S. U. Center for Wetland resources, hydrologic and geologic studies of Coastal Louisiana, Report No. 14*. Baton Rouge.

Gagliano, S. M., & Van Beek, J. L. (1975). An approach to multiuse management in the Mississippi delta system. In M. L. Broussard (Ed.), *Deltas, models for exploration*. Houston: Houston Geological Society.

Godschalk, D. R. (2003). Urban hazard mitigation: Creating resilient communities. *Natural*

*Hazards Review, 4*(3), 136–143. https://doi.org/10.1061/(ASCE)1527-6988(2003)4:3(136).
Groves, D. G., & Sharon, C. (2013). Planning tool to support planning the future of Coastal Louisiana. *Journal of Coastal Research, 67*(sp1), 147–161. https://doi.org/10.2112/SI_67_10.
Grumbine, R. E. (1994). What is ecosystem management? *Conservation Biology, 8*(1), 27–38. https://doi.org/10.1046/j.1523-1739.1994.08010027.x.
Grumbine, R. E. (1997). Reflections on "What is ecosystem management?". *Conservation Biology, 11*(1), 41–47. https://doi.org/10.1046/j.1523-1739.1997.95479.x.
Gunderson, L. (2000). Ecological resilience—In theory and application. *Annual Review of Ecology and Systematics, 31*, 425–439. https://doi.org/10.1146/annurev.ecolsys.31.1.425.
Holling, C. S. (1973). Resilience and stability of ecological systems. *Annual Review of Ecology and Systematics, 4*, 1–24. https://doi.org/10.1146/annurev.es.04.110173.000245.
Holling, C. S. (1996). Engineering resilience versus ecological resilience. In P. Schulze (Ed.), *Engineering within ecological constraints*. Washington DC: National Academies Press.
Holling, C. S. (2001). Understanding the complexity of economic, ecological, and social systems. *Ecosystems, 4*, 390–405. https://doi.org/10.1007/s10021-001-0101-5.
Ko, J.-Y., Day, J. W., Wilkins, J. G., Haywood, J., & Lane, R. R. (2017). Challenges in collaborative governance for coastal restoration: Lessons from the Caernarvon River diversion in Louisiana. *Coastal Management, 45*(2), 125–142. https://doi.org/10.1080/08920753.2017.1278145.
Kolker, A. S., Allison, M. A., & Hameed, S. (2011). An evaluation of subsidence rates and sea-level variability in the northern Gulf of Mexico. *Geophysical Research Letters, 38*(21). https://doi.org/10.1029/2011GL049458.
La. Rev. Stat. § 49:214.24 (2015).
Laska, S., Wooddell, G., Hagelman, R., Gramling, R., & Farris, M. T. (2005). At risk: The human, community and infrastructure resources of coastal Louisiana. *Journal of Coastal Research, 44*, 90–111. Retrieved from https://www.jstor.org/stable/25737051
Lazarus, R. (2009). Super wicked problems and climate change: Restraining the present to liberate the future. *Cornell Law Review, 91*(5). Retrieved from http://scholarship.law.cornell.edu/clr/vol94/iss5/8
Lopez, J. A. (2009). The multiple lines of defense strategy to sustain coastal Louisiana. *Journal of Coastal Research, 10054*, 186–197. https://doi.org/10.2112/SI54-020.1.
Louisiana Coastal Wetlands Conservation and Restoration Task Force (LCWCRTF). (1998). *Coast 2050: Toward a sustainable coastal Louisiana*. Baton Rouge: US ACE Louisiana Coastal Wetlands Conservation and Restoration Task Force.
Lyles, L. W., Berke, P., & Smith, G. (2014). Do planners matter? Examining factors driving incorporation of land use approaches into hazard mitigation plans. *Journal of Environmental Planning and Management, 57*(5), 792–811. https://doi.org/10.1080/09640568.2013.768973.
Manning-Broome, C., Dubinin, J., & Jenkins, P. (2015). *The view from the coast: Local perspectives and policy recommendations on flood-risk reduction in South Louisiana*. Baton Rouge: Center for Planning Excellence.
Markusen, A. (2003). Fuzzy concepts, scanty evidence, policy distance: The case for rigour and policy relevance in critical regional studies. *Regional Studies, 37*, 701–717. https://doi.org/10.1080/0034340032000108796.
Matyas, D., & Pelling, M. (2014). Positioning resilience for 2015: The role of resistance, incremental adjustment and transformation in disaster risk management policy. *Disasters, 39*(S1), S1–S18. https://doi.org/10.1111/disa.12107.
McCrea, R., Walton, A., & Leonard, R. (2015). A conceptual framework for investigating community wellbeing and resilience. *Rural Society, 23*(3), 270–282. https://doi.org/10.1080/10371656.2014.11082070.
Mehmood, A. (2016). Of resilient places: Planning for urban resilience. *European Planning Studies, 24*(2), 407–419. https://doi.org/10.1080/09654313.2015.1082980.
Montgomery, D. R., Grant, G. E., & Sullivan, K. (1995). Watershed analysis as a framework for implementing ecosystem management. *Water Resources Bulletin, 31*(3), 369–386. Retrieved

from https://andrewsforest.oregonstate.edu/sites/default/files/lter/pubs/pdf/pub2108.pdf
Nassauer, J. I., & Opdam, P. (2008). Design in science: Extending the landscape ecology paradigm. *Landscape Ecology, 23*(6), 633–644. https://doi.org/10.1007/s10980-008-9226-7.
National Research Council. (2012). *Disaster resilience: A national imperative*. Washington, DC: The National Academies Press.
NOAA NWS. (2016). Exceedance probability analysis for selected storm events. NOAA National Weather Service. Retrieved November 29 from http://www.nws.noaa.gov/oh/hdsc/aep_storm_analysis/
OCD-DRU. (2015). *National disaster resilience competition phase II application*. Baton Rouge: Louisiana Office of Community Development/Disaster Recovery Unit.
Pendall, R., Foster, K. A., & Cowell, M. (2009). Resilience and regions: Building understanding of the metaphor. *Journal of Regions, Economy and Society, 3*(1), 71–84. Retrieved from https://iurd.berkeley.edu/wp/2007-12.pdf
Perkes, D. (2009). A useful practice. *Journal of Architectural Education, 63*, 64–71. https://doi.org/10.1111/j.1531-314X.2009.1005.x.
Peyronnin, N., Green, M., Parsons Richards, C., Owens, A., Reed, D., Chamberlain, J., et al. (2013). Louisiana's 2012 Coastal Master Plan: Overview of a science-based and publicly informed decision-making process. *Journal of Coastal Research, 67*(sp1), 1–15. https://doi.org/10.2112/SI_67_1.1.
Picou, S., Marshall, B., & Gill, D. (2004). Disaster, litigation, and the corrosive community. *Social Forces, 82*(4), 1493–1522. Retrieved from https://pdfs.semanticscholar.org/26d4/ff9358f647686bca0b25deab85df11caec81.pdf
Prein, A. F., Rasmussen, R. M., Ikeda, K., Liu, C., Clark, M. P., & Holland, G. J. (2016). The future intensification of hourly precipitation extremes. *Nature Climate Change, 7*, 48–52. https://doi.org/10.1038/nclimate3168.
Rittel, H., & Webber, M. (1973). Dilemmas in a general theory of planning. *Policy Sciences, 4*(2), 155–169. Retrieved from https://link.springer.com/article/10.1007/BF01405730
Rockefeller Foundation. (2018). Resilient Cities. Retrieved from https://www.rockefellerfoundation.org/our-work/topics/resilientcities/
Taylor, P. (2005). Time: From hegemonic change to everyday life. In S. Holloway, S. Rice, & G. Valentine (Eds.), *Key concepts in geography* (pp. 146–152). London: Sage.
Tobin, G. A. (1995). The levee love affair: A stormy relationship? *Journal of the American Water Resources Association, 31*(3), 359. https://doi.org/10.1111/j.1752-1688.1995.tb04025.x.
Tompkins, E. L., Few, R., & Brown, K. (2008). Scenario-based stakeholder engagement: Incorporating stakeholders preferences into coastal planning for climate change. *Journal of Environmental Management, 88*(4), 1580–1592. https://doi.org/10.1016/j.jenvman.2007.07.025.
Trisolini, K. (2010). All hands on deck: Local governments and the potential for bi-directional climate change regulation. *Stanford Law Review, 62*(3), 669–746. Retrieved from https://www.stanfordlawreview.org
United States Department of Housing and Development (HUD). (2015). *National disaster resilience competition: Phase 2 fact sheet*. Washington DC: US Department of Housing and Urban Development.
US Army Corps of Engineers (USACE). (2004). *Louisiana Coastal Area: Ecosystem restoration study* (Vol. 1). New Orleans: U.S. Army Corps of Engineers. Retrieved from https://www.mvn.usace.army.mil/Portals/56/docs/LCA/Main%20Report.pdf?ver=2016-07-01-095948-907
Vale, L., & Campanella, T. (2005). *The resilient city: How modern cities recover from disaster*. Oxford, UK: Oxford University Press.
Vermeer, M., & Rahmstorf, S. (2009). Global sea level linked to global temperature. *Proceedings of the National Academy of Sciences of the USA, 106*(51), 21527–21532. https://doi.org/10.1073/pnas.0907765106.

Walker, B., & Salt, D. (2006). *Resilience thinking: Sustaining ecosystems and people in a changing world*. Washington DC: Island Press.

Wiseman, J., & Brasher, K. (2008). Community wellbeing in an unwell world: Trends, challenges, and possibilities. *Journal of Public Health Policy, 29*(3), 353–366. https://doi.org/10.1057/jphp.2008.16.

# Appendix 1: Twenty-First-Century Flooding in Louisiana: Sources and Methods

**Julie Torres**

All attempts were made to collect similar flood inundation data for the nine storms featured here. However, this was not possible as there is no comprehensive database of major flood footprints in Louisiana. We used multiple sources to create the flood maps in this chapter, each with their own method of collecting the flood data. These methods include the observation of high-water marks by persons on the ground, the use of satellite imagery to detect the land-water interface, and the analysis of detailed data collected at gaging stations run by governmental organizations. Here we provide detailed descriptions of the sources used and how we extracted the data.

## Flood Event Selection

The nine floods presented in the figure are the only ones for which we could find data and whose footprints weren't completely overlapped by a larger flood. Unsurprisingly, they also happen to be some of the largest events in terms of Louisiana flooding since 2000. While maps showing forecasted flooding, flood watches and warnings, local flooding, recorded rainfall, and FEMA flood insurance information were readily available for most or all 28 twenty-first-century flood events (Table 1.1), maps of or data pertaining to the actual footprint of flood inundation that occurred during those events was limited.

J. Torres (✉)
Independent Researcher, New Orleans, LA, USA



S. Laska (ed.), *Louisiana's Response to Extreme Weather*,
Extreme Weather and Society, https://doi.org/10.1007/978-3-030-27205-0

**Table 1.1** The 28 major LA flood events of the twenty-first century, including those we were able to map (in bold)

| | Event | Date | Landfall location | Parishes with flooding |
|---|---|---|---|---|
| 1 | Tropical Storm Allison | June 2001 | Morgan City, LA | Cameron, St. Tammany, St. Bernard, E. Baton Rouge |
| 2 | Tropical Storm Bertha | Aug 2002 | Mouth of MS River | St. Tammany, E. Feliciana |
| 3 | Hurricane Isidore | Sept 2002 | Grand Isle, LA | St. Tammany, upper Jefferson |
| 4 | Hurricane Lili (Cat 1) | Oct 2002 | Vermillion Bay, LA | Vermillion, Acadia |
| 5 | Tropical Storm Bill | June 2003 | Central LA coast | Terrebone, St. Tammany |
| 6 | Hurricane Ivan | Sept 2004 | Cameron, LA | Cameron, Calcasieu, Caddo |
| 7 | Tropical Storm Matthew | Oct 2004 | Central LA coast, Frenier, LA | Terrebone, Lafourche |
| 8 | Hurricane Cindy (Cat 1) | July 2005 | Grand Isle, LA | Orleans |
| 9 | Hurricane Katrina (Cat 3) | Aug 2005 | Mouth of MS River | Most coastal parishes |
| 10 | Hurricane Rita (Cat 3) | Sept 2005 | LA-TX border | Vermillion, Cameron, Calcasieu, Vernon, Beauregard, Jefferson Davis, Allen |
| 11 | Western/Northern River Flood | Oct 2006 | – | Vernon, St. Landry, Calcasieu |
| 12 | Hurricane Humberto (Cat 1) | Sept 2007 | Vinton, LA | Lafayette, Beauregard |
| 13 | Tropical Storm Edouard | Aug 2008 | Gilchrist, TX | Calcasieu |
| 14 | Hurricane Gustav (Cat 2) | Sept 2008 | Mouth of MS River | Most coastal parishes |
| 15 | Hurricane Ike (Cat 1) | Sept 2008 | Point Bolivar, TX | Cameron, Vermillion |
| 16 | Tropical Storm Bonnie | July 2010 | Mouth of MS River | Washington, W. Baton Rouge |
| 17 | Tropical Depression 5 | Aug 2010 | LA-MS border | Orleans, Avoyelles |
| 18 | Mississippi & Atchafalaya River Flood | Apr 2011 | – | Avoyelles, Iberville, W. Baton Rouge, Pointe Coupee, Iberia |
| 19 | Tropical Storm Lee | Sept 2011 | Vermillion Bay, LA | St. Tammany, Orleans, Plaquemines, St. Bernard, Jefferson |

| | | | | |
|---|---|---|---|---|
| 20 | Hurricane Isaac (Cat 1) | Aug 2012 | Port Fourchon, LA | Most coastal parishes |
| 21 | Red River Flood | June 2015 | – | Caddo, Bossier, Natchitoches, Rapides |
| 22 | Mississippi River Flood | Jan 2016 | – | W. Baton Rouge, W. Feliciana, Pointe Coupee |
| 23 | Northern LA Rivers Flood | Mar 2016 | – | Tangipahoa, St. Tammany, Washington, Livingston, Ascension, Calcasieu, Beauregard, Vernon, Ouachita, Caldwell, Richland, Madison, Bienville, Natchitoches |
| 24 | Eastern LA Rivers Flood | Aug 2016 | – | Ascension, Livingston, E. Baton Rouge, W. Baton Rouge, Tangipahoa, St. Helena, Rapides, Pointe Coupee, Vermillion, Acadia, Evangeline, St. Martin |
| 25 | Tropical Storm Cindy | June 2017 | LA-TX border | Cameron, Calcasieu, Vermillion, Iberia |
| 26 | Hurricane Harvey | Aug 2017 | Cameron, LA | Cameron, Jefferson Davis, Acadia, Vermillion, St. Martin, Acadia, Lafayette, Calcasieu, Iberia, Terrebone |
| 27 | Red River Flood | Mar 2018 | – | Natchitoches, Bossier, Caddo |
| 28 | Hurricane Nate (Cat 1) | 2017 | Mouth of MS River | St. Tammany |

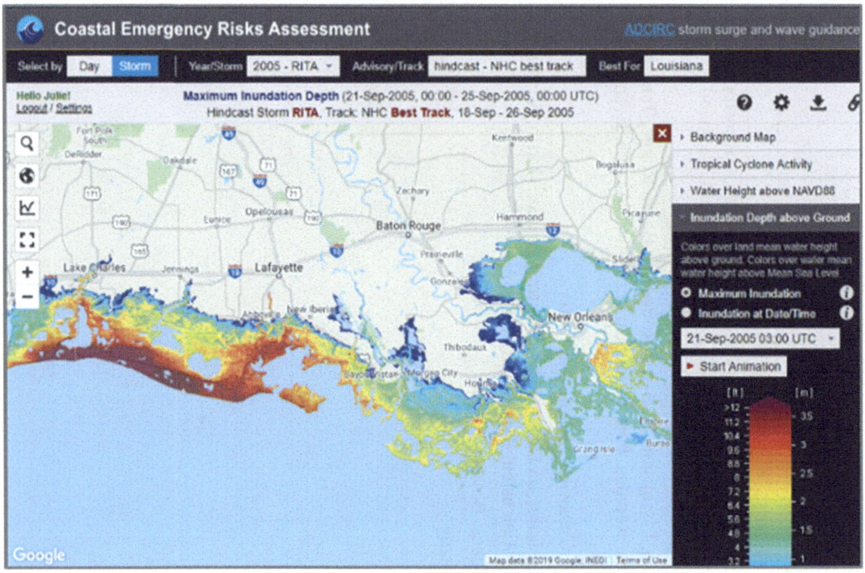


Fig. 1.1 An example of CERA's ADCIRC hindcast of inundation depth above ground for Hurricane Rita

## Data Sources

(a) The Coastal Emergency Risks Assessment (CERA) utilizes the Advanced Circulation (ADCIRC) model to conduct hindcast simulations of tropical storm surge with high levels of accuracy (Fig. 1.1). Researchers enter known or closely estimated inputs for past flood events into the model to see how well the output matches the known results. This tool has the "maximum inundation depth aboveground" shapefile polygon data for many named storms since 2004, including Hurricanes Katrina, Rita, Gustav, and Isaac.

(b) The Dartmouth Flood Observatory (DFO) provides to the public satellite-based measurement of surface water. For three of our nine mapped floods (Hurricane Ike, March 2016, Hurricane Harvey), we were able to access the DFO Rapid Response Inundation Map and associated shapefiles (Fig. 1.2). While DFO data was available for October 2006 flooding in extreme southwest LA, this flood footprint was completely overlapped by several other floods. Additionally, DFO has flood maps for Hurricane Rita and Hurricane Katrina, but only local, zoomed-in areas of flooding are available as opposed to statewide coverage, and the shapefiles are not available. Aspects of the DFO website are still in an experimental phase.

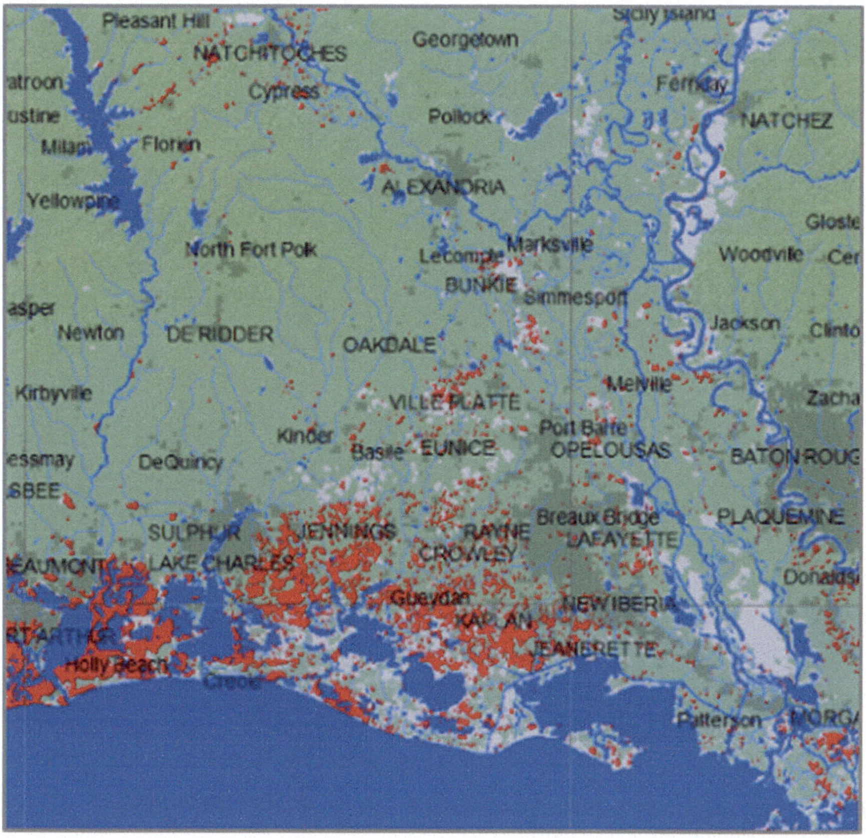


**Fig. 1.2** Screenshot of flooded area (red) as a result of Hurricane Harvey as identified by DFO

(c) The US Geological Survey (USGS) published two scientific investigations report characterizing peak streamflows and flood inundation of selected areas of Louisiana for the March and August 2016 flood events (Breaker et al. 2016; Watson et al. 2018; Fig. 1.3). For each event, the USGS documented hundreds of high-water marks in areas bordering the flooding rivers. The high-water marks were used to create inundation maps for selected Louisiana communities intended to estimate the aerial extent and depth of flooding.

(d) The Stephenson Disaster Institute (SDMI) flood map for the August 2016 flood event (Fig. 1.4) was created by taking reported addresses (when available) that have been flooded and general reported flood areas and creating a projected aggregate flood area. The newly compiled flood area is then refined to elevation data derived from Light Detection and Ranging, or LiDAR, imagery to create an elevation-appropriate flood area which was then mapped.

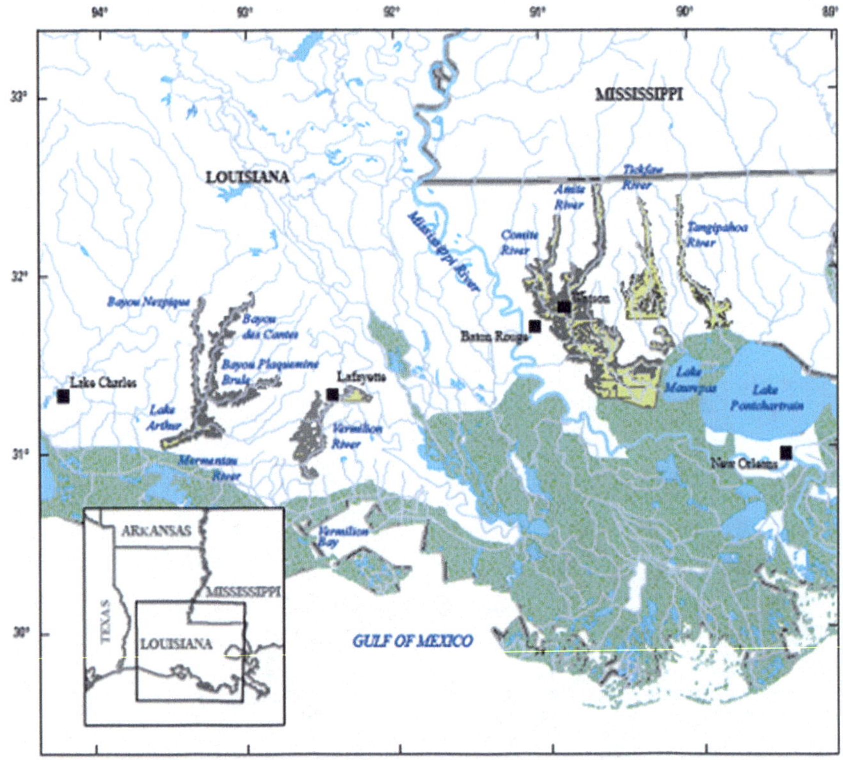


**Fig. 1.3** Map of selected river flooding (yellow) as a result of the August 2016 event from the USGS publication Breaker et al. (2016)

(e) The only flood map we were able to locate for the spring 2011 event was in a Louisiana Geological Survey (LGS) report of investigations by Carlson et al., (2011) assessing inundation and damage that occurred in the Atchafalaya Basin. The report includes a map of estimated areas impacted by the spring 2011 flooding of the Atchafalaya River (Fig. 1.5). Extensive flooding occurred along the lower Mississippi River as well, but there are no maps available of this flooding.

## Data Extraction and Application Techniques

Flood shapefiles were available and easily downloadable from the CERA tool for Hurricanes Katrina, Rita, Gustav, and Isaac. These were imported into ArcMap and layered on a shaded relief base map of Louisiana.

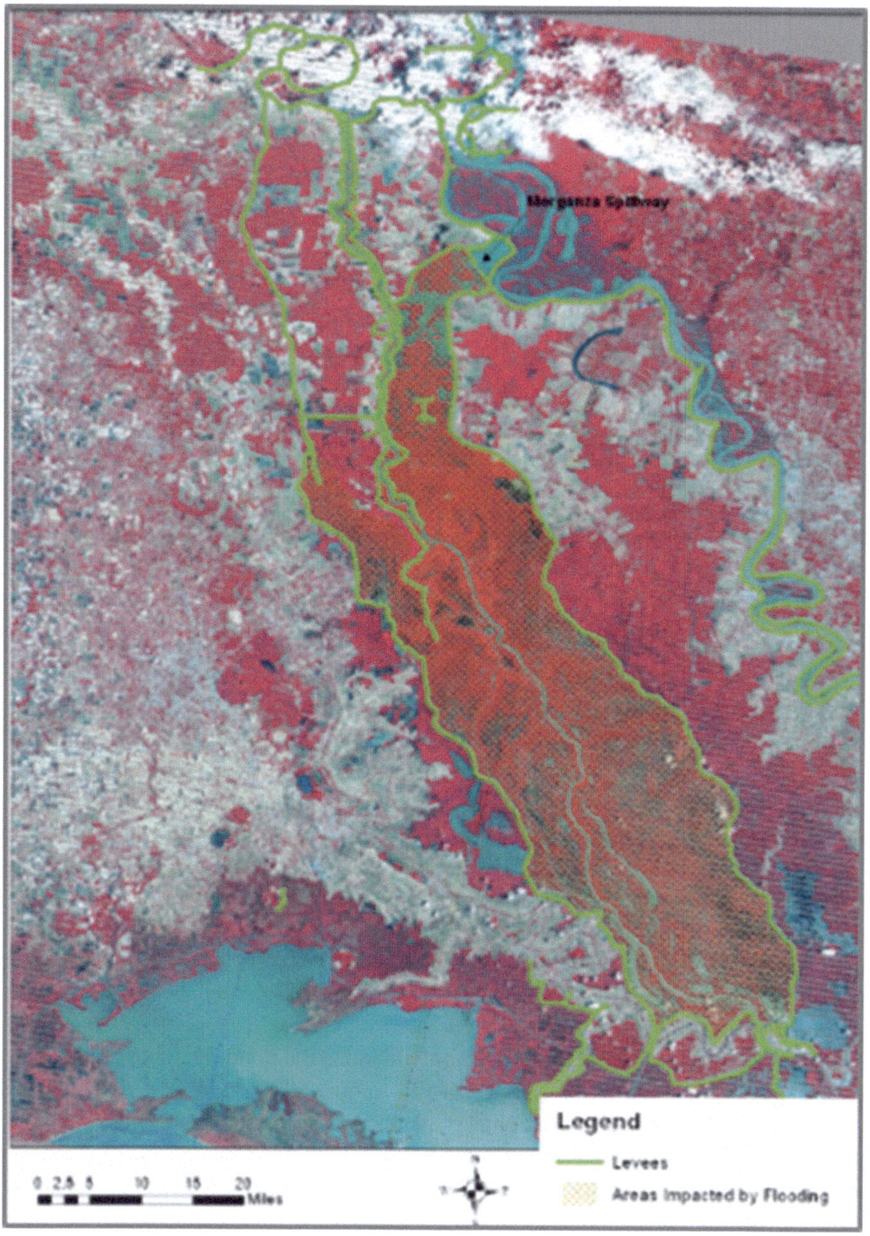


**Fig. 1.4** August 2016 flooding as mapped by SDMI

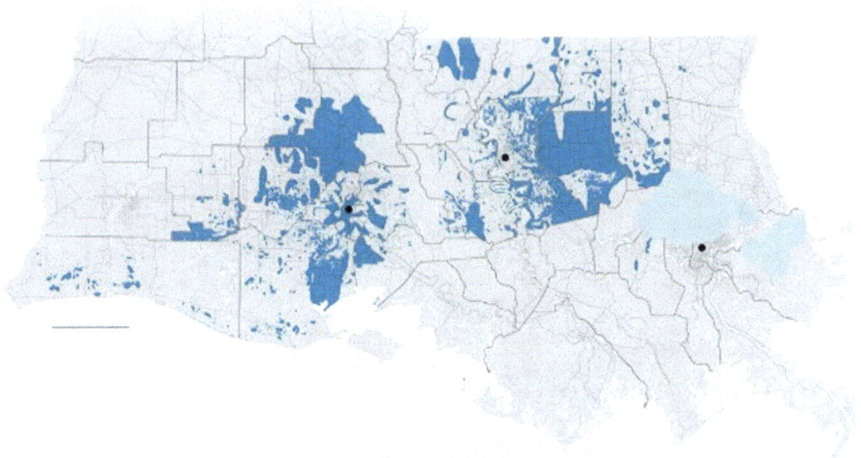

**Fig. 1.5** Atchafalaya flooding during the spring 2011 event from LGS publication Carlson et al. (2011)

For Hurricane Ike, March 2016, and Hurricane Harvey, flood imprint shapefiles were downloaded from DFO (http://floodobservatory.colorado.edu) and imported into GIS software for standard map generation. However, the shapefile for the March 2016 event seemed incomplete to those with firsthand knowledge of the flooding extent. The shapefile omitted significant river flooding located on Lake Pontchartrain's north shore. The flood map in the USGS publication Breaker et al. (2016) showed this river flooding based on high-water mark reporting. We therefore decided to combine the two sources to create a composite map for this event. To extract the flood imprint from the USGS map, polygons were traced with great care and thoroughness (Fig. 1.6), imported into ArcMap, and georeferenced. The same procedure was used to extract the flood footprint for the August 2016 flood from Watson et al. (2018). Again, this source only mapped the flooding around major rivers, so we used an additional source for the August 2016 flood. The SDMI flood map (Fig. 1.4) had enough color contrast that in a program, such as PowerPoint, one can remove the background, or unwanted portion, with good accuracy (Fig. 1.7); we were unable to catch only the smallest specks of flooding. With background removed, the remainder was georeferenced and combined with the USGS map footprint.

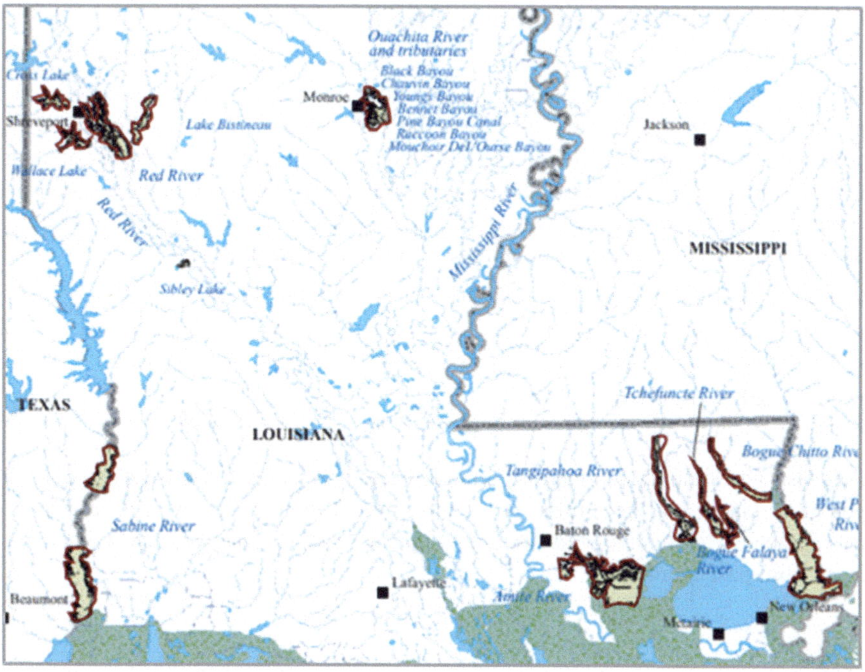


**Fig. 1.6** To supplement DFO flood footprint for the March 2016 event, the flood polygons (yellow) in Watson et al. (2018) were traced (red) and combined

**Fig. 1.7** To supplement the USGS flood footprint for the August 2016 event, the flooded areas in blue were separated from the gray background in the SDMI map

## Complications and Recommendations

We were surprised at how difficult it was to find statewide flood inundation maps for major flood events. Those we used were created using different methods of data collection. Additionally, different methods had to be used to extract the useful data from the maps. We understand that tracing polygons and georeferencing them is subject to human error, which is why a reliable database of flood inundation data for all major floods should exist. A database of this kind should:

- Be frequently updated
- Include all types of floods (storm surge, flash floods, and river floods)
- Use data collected from a standard source(s) (i.e., satellite imagery)
- Cover all areas, urban and rural
- Be accessible from a central location

Forecasts and simulated hindcasts certainly have their places as they warn people of potential danger and help improve weather model inputs. However, it is only by having access to standardized flood data that we can accurately detect trends and learn from past events.

## References

Breaker, B. K., Watson, K. M., Ensminger, P. A., Storm, J. B. & Rose, C. E. (2016). *Characterization of peak streamflows and flood inundation of selected areas in Louisiana, Texas, Arkansas, and Mississippi from flood of March 2016* (No. 2016–5162, pp. 1–33). US Geological Survey. https://doi.org/10.3133/sir20165162.

Carlson, D., Horn, M., Van Biersel, T., & Fruge, D. (2011). *2011 Atchafalaya Basin inundation data collection and damage assessment project*. Retrieved from https://www.lsu.edu/lgs/publications/products/report-of-investigations-series.php

Watson, K. M., Harwell, G. R., Wallace, D. S., Welborn, T. L., Stengel, V. G. & McDowell, J. S. (2018). *Characterization of peak streamflows and flood inundation of selected areas in southeastern Texas and southwestern Louisiana from the August and September 2017 flood resulting from Hurricane Harvey* (No. 2018–5070). US Geological Survey. https://doi.org/10.3133/sir20185070.

# Index





S. Laska (ed.), *Louisiana's Response to Extreme Weather*,
Extreme Weather and Society, https://doi.org/10.1007/978-3-030-27205-0

**D**

The manufacturer's authorised representative in the EU is Springer Nature Customer Service Centre GmbH, Europaplatz 3, 69115 Heidelberg, Germany. If you have any concerns regarding our products, please contact ProductSafety@springernature.com

Printed and bound by CPI Group (UK) Ltd, Croydon, CR0 4YY
05/10/2026
02228145-0011